Fixed Mobile Convergence Handbook

Fixed Mobile Convergence Handbook

Edited by

Syed A. Ahson
Mohammad Ilyas

CRC Press
Taylor & Francis Group
Boca Raton London New York

CRC Press is an imprint of the
Taylor & Francis Group, an **informa** business

Contents

Preface

Service providers are in the process of transforming from legacy packet and circuit-switched networks to converged Internet protocol (IP) networks and consolidating all network services and business units on a single IP infrastructure. Future users of communication systems will require the use of data rates of around 100 Mbps at their homes while all services and applications will require high bandwidths. The next generations of heterogeneous wireless networks are expected to interact with each other and be capable of interworking with IP-based infrastructures. As Metcalfe's law estimates, "the value of a telecommunications network is proportional to the square of the number of connected users of the system." Although this may be debatable from a mathematical point of view, it does happen. Scalability and service accessibility have been the main drivers for the interconnection of telecommunication networks. The telephony network and the Internet are two highly interconnected services that achieve their value by connecting any user to any other user, and by providing access to services and content worldwide. The wired–wireless integrated network (WWIN) can be categorized as fixed mobile convergence (FMC). FMC means the convergence of the existing wired and wireless network. Mobile nodes (MNs) are equipped with multimode radio interfaces so that they can perform roaming among these different access technologies.

The last few years have seen an exceptional growth in the wireless local area network industry, with substantial increase in the number of wireless users and applications. This growth has been mostly due to the availability of inexpensive and highly interoperable network solutions based on Wi-Fi standards and to the growing trend of providing built-in wireless network cards into mobile computing platforms. Advancement in wireless technologies and mobile computing enables mobile users to benefit from disparate wireless networks such as wireless personal area networks (WPANs), wireless local area networks (WLANs), wireless metropolitan area networks (WMANs), and wireless wide area networks (WWANs) that use mobile telecommunication cellular network technologies such as Worldwide Interoperability for Microwave Access (WiMAX), Universal Mobile Telecommunications System (UMTS), General Packet Radio Service (GPRS), code division multiple access 2000 (CDMA2000), Global System for Mobile Communications (GSM), Cellular Digital Packet Data (CDPD), Mobitex, High-Speed Downlink Packet Access (HSDPA), or third generation (3G) to transfer data.

The requirements for next generation networks (NGNs) lead to an architectural evolution that requires a converged infrastructure where users across multiple domains can be served through a single unified domain. Convergence is at the core of IP-based NGNs. The aim of IP convergence is to build a single network infrastructure that is cost effective, scalable, reliable, and secure. The aim of standardization has been to enable a mix and match of services bundled to offer innovative services to end users. Service enabler, in this context, is the approach to eliminate the vertical silo structure and to transform into a horizontal-layered architecture. The integration of different communication technologies is one of the key features in NGNs. In integrated network environments, it is expected that users can access the Internet on an "anytime, anywhere" basis and with better quality of service (QoS) by selecting the most appropriate interface according to their needs. Although network integration enhances user experience, it raises several challenging issues such as candidate network discovery, call admission control, secure context transfer, and power management for multimode terminals. There have been several standard group activities to handle those issues in integrated heterogeneous networks. For example, the integration of 3GPP and non-3GPP accesses (e.g., CDMA2000, WLAN, and WiMAX) has actively been studied by the 3GPP consortium.

Such types of interconnections would be beneficial to consumers. The interconnection would enable consumers access to a wider range of content, namely, content available in other fixed and/or mobile networks. Roaming and mobility capabilities supported by the interconnection would also enable consumers access to contents from a wider range of access points, namely, from access points belonging to other fixed and/or mobile networks. In addition, it would provide consumers with a consistent, personalized, rich content, and service-rich user experience from any place and at any time.

FMC not only transforms technologies for the delivery of digital television but also helps users from being passive consumers of unidirectional broadcasted media to being active consumers of interactive, mobile, and personalized bidirectional multimedia communication. Users expect to be enabled to access any content, anytime, anyhow, anywhere, and on any device that they wish to be entertained with. The NGN has been considered as a fully converged architecture that can provide a wide spectrum of multimedia services and applications to end users.

Several industrial standard organizations and forums have been taking the initiative on NGNs in recent years. For instance, the European Telecommunications Standards Institute (ETSI)—Telecoms and Internet converged Services and Protocols for Advanced Networks (TISPAN) focuses on an NGN for fixed access network, which has published Release 1 in 2005. Meanwhile, the International Telecommunication Union's Telecommunication Standardization Sector (ITU-T) started the NGN Global Standards Initiative (NGN-GSI) and has published its first release in 2006. On the other hand, a similar effort has been made in wireless network domain: UMTS (i.e., W-CDMA) and CDMA2000 defined in 3GPP and 3GPP2 are categorized as third-generation mobile network technologies and are now evolving to fourth-generation mobile network—Evolved Packet System (i.e., Long-Term Evolution/Evolved Packet Core (LTE/EPC)), which can be regarded as network generation mobile networks. The common notion of a variety of NGNs is to transport all information and services (voice, data, video, and all sorts of multimedia applications) by utilizing packet network IP technology, that is, an "all-IP" network.

This book provides technical information about all aspects of FMC. The areas covered range from basic concepts to research grade material, including future directions. It captures the current state of FMC and serves as a comprehensive reference material on this subject. It consists of 16 chapters authored by 44 experts from around the world. The targeted audience for the handbook include professionals who are designers and/or planners for FMC systems, researchers (faculty members and graduate students), and those who would like to learn about this field.

The book is expected to have the following specific salient features:

- To serve as a single comprehensive source of information and as reference material on FMC.
- To deal with an important and timely topic of emerging technology of today, tomorrow, and beyond.
- To present accurate, up-to-date information on a broad range of topics related to FMC.
- To present material authored by experts in the field.
- To present information in an organized and well-structured manner.

Syed Ahson
Seattle, Washington

Mohammad Ilyas
Boca Raton, Florida

Acknowledgments

Although the book is not technically a textbook, it can certainly be used as a textbook for graduate courses and research-oriented courses that deal with FMC. Any comments from the readers will be highly appreciated.

Many people have contributed to this handbook in their unique ways. First and foremost, we would like to express our immense gratitude to the group of highly talented and skilled researchers who have contributed 16 chapters to this handbook. All of them have been extremely cooperative and professional. Also, it has also been a pleasure to work with Nora Konopka and Jill Jurgensen of CRC Press; we are extremely grateful to them for their support and professionalism. Our families have extended their unconditional love and support throughout this project and they all deserve very special thanks.

Editors

Syed Ahson is a senior software design engineer with Microsoft. As part of the Mobile Voice and Partner Services group, he is currently engaged in research on new and exciting end-to-end mobile services and applications. Before joining Microsoft, Syed was a senior staff software engineer with Motorola, where he contributed significantly in leading roles toward the creation of several iDEN, CDMA, and GSM cellular phones. He has extensive experience with wireless data protocols, wireless data applications, and cellular telephony protocols. Before joining Motorola, Syed worked as a senior software design engineer with NetSpeak Corporation (now part of Net2Phone), a pioneer in VoIP telephony software.

Syed has published more than 10 books on emerging technologies such as WiMAX, RFID, mobile broadcasting, and IP multimedia subsystem. His recent books include *IP Multimedia Subsystem Handbook* and *Handbook of Mobile Broadcasting: DVB-H, DMB, ISDB-T and MediaFLO*. He has authored several research articles and teaches computer engineering courses as adjunct faculty at Florida Atlantic University, Florida, where he introduced a course on Smartphone technology and applications. Syed received his MS in computer engineering from Florida Atlantic University, Boca Raton, in July 1998, and his BSc in electrical engineering from Aligarh Muslim University, India, in 1995.

Dr. Mohammad Ilyas is associate dean for research and industry relations at the College of Engineering and Computer Science at Florida Atlantic University, Boca Raton, Florida. Previously, he has served as chair of the Department of Computer Science and Engineering and interim associate vice president for research and graduate studies. He received his PhD degree from Queen's University in Kingston, Canada. His doctoral research was about switching and flow control techniques in computer communication networks. He received his BSc degree in electrical engineering from the University of Engineering and Technology, Pakistan, and his MS degree in electrical and electronic engineering at Shiraz University, Iran.

Dr. Ilyas has conducted successful research in various areas, including traffic management and congestion control in broadband/high-speed communication networks, traffic characterization, wireless communication networks, performance modeling, and simulation. He has published over 25 books on emerging technologies, and over 150 research articles. His recent books include *Cloud Computing and Software Services: Theory and Techniques* (2010) and *Mobile Web 2.0: Developing and Delivering Services to Mobile Phones* (2010).

He has supervised11 PhD dissertations and more than 37 MS theses to completion. He has been a consultant to several national and international organizations. Dr. Ilyas is an active participant in several IEEE technical committees and activities. Dr. Ilyas is a senior member of IEEE and a member of ASEE.

Contributors

H. S. Al-Raweshidy
Wireless Networks and
    Communications Center
Electronic and Computer
    Engineering Division
School of Engineering and Design
Brunel University
Middlesex, United Kingdom

Sahin Albayrak
Department of Electrical
    Engineering and Computer
    Sciences
Technische Universität Berlin
Berlin, Germany

José I. Alonso
Department of Signals, Systems
    and Radiocommunications
Telecommunications
    Engineering School
Technical University of Madrid
Madrid, Spain

Hüseyin Arslan
Department of Electrical
    Engineering
University of South Florida
Tampa, Florida

Victoria Beltran
Department of Telematics
Technical University of Catalonia
Barcelona, Spain

S. R. Chaudhry
Wireless Networks and
    Communications Center
Electronic and Computer
    Engineering Division
School of Engineering and Design
Brunel University
Middlesex, United Kingdom

M. Oskar van Deventer
TNO, Netherlands Organisation
    for Applied Scientific Research
Delft, the Netherlands

Alfonso Fernández-Durán
Alcatel-Lucent Spain
Madrid, Spain

Dario Gallucci
Institute of Systems for Informatics
    and Networking
Department of Technology and
    Innovation SUPSI
University of Applied Sciences
    of Southern Switzerland
Manno, Switzerland

Silvia Giordano
University of Applied Sciences
    of Southern Switzerland
Manno, Switzerland

İsmail Güvenç
DOCOMO Communications
    Laboratories USA, Inc.
Palo Alto, California

Radovan Kadlic
Department of Telecommunication
Slovak University of Technology
Bratislava, Slovakia

Manzoor Ahmed Khan
Department of Electrical
    Engineering and Computer
    Sciences
Technische Universität Berlin
Berlin, Germany

Eun Cheol Kim
School of Electronics Engineering
Kwangwoon University
Seoul, South Korea

Jin Young Kim
School of Electronics Engineering
Kwangwoon University
Seoul, South Korea

Yong-Sung Kim
Systems R&D Team
Telecom Systems
Samsung Electronics
Suwon, South Korea

Dong-Hee Kwon
Department of Computer Science
    and Engineering
Pohang University of Science
    and Technology
Pohang, South Korea

Bu Sung Lee
School of Computer Engineering
Nanyang Technological University
Singapore, Singapore

Victor C. M. Leung
Department of Electrical and
    Computer Engineering
The University of British Columbia
Vancouver, British Columbia,
    Canada

Teck Meng Lim
Startlub Ltd.
Singapore, Singapore

Wan-Seon Lim
Department of Computer Science
    and Engineering
Pohang University of Science
    and Technology
Pohang, South Korea

Hisham A. Mahmoud
DOCOMO Communications
    Laboratories USA, Inc.
Palo Alto, California

Laurent Marchand
Ericsson Research Canada
Town of Mount Royal
Quebec, Canada

Eugen Mikoczy
Department of Telecommunication
Slovak University of Technology
Bratislava, Slovakia

Mariano Molina-García
Department of Signals, Systems
    and Radiocommunications
Telecommunications Engineering
    School
Technical University of Madrid
Madrid, Spain

Ramesh Nagarajan
Bell Laboratories, Alcatel-Lucent
Murray Hill, New Jersey

Pieter Nooren
TNO, Netherlands Organisation
    for Applied Scientific Research
Delft, the Netherlands

Dang Duc Nguyen
School of Computer Engineering
Nanyang Technological University
Singapore, Singapore

Josep Paradells
Department of Telematics
Technical University of Catalonia
Barcelona, Spain

Samuel Pierre
Mobile Computing and Networking
   Research Laboratory
Department of Computer
   Engineering
Ecole Polytechnique de Montreal
Montreal, Quebec, Canada

Pavol Podhradsky
Department of Telecommunication
Slovak University of Technology
Bratislava, Slovakia

Daniele Puccinelli
Institute of Systems for Informatics
   and Networking
Department of Technology
   and Innovation
University of Applied Sciences
   of Southern Switzerland
Manno, Switzerland

Mustafa E. Şahin
Department of Electrical
   Engineering
University of South Florida
Tampa, Florida

Fikret Sivrikaya
Department of Electrical
   Engineering and Computer
   Sciences
Technische Universität Berlin
Berlin, Germany

Young-Joo Suh
Department of Computer Science
   and Engineering
Pohang University of Science
   and Technology
Pohang, South Korea

Dong Sun
Bell Laboratories, Alcatel-Lucent
Murray Hill, New Jersey

Peyman TalebiFard
Department of Electrical and
   Computer Engineering
The University of British Columbia
Vancouver, British Columbia,
   Canada

N. Sai Krishna Tejawsi
Indian Institute of Technology
Department of Electronics and
   Electrical Communication
   Engineering
Kharagpur, India

Ahmet Cihat Toker
Department of Electrical
   Engineering and Computer
   Sciences
Technische Universität Berlin
Berlin, Germany

Salvatore Vanini
Institute of Systems for Informatics
   and Networking
Department of Technology
   and Innovation
University of Applied Sciences
   of Southern Switzerland
Manno, Switzerland

Yang Xia
School of Computer Engineering
Nanyang Technological University
Singapore, Singapore

Chai Kiat Yeo
School of Computer Engineering
Nanyang Technological University
Singapore, Singapore

Li Jun Zhang
Division Research and Development
Geninov Inc.
Montreal, Quebec, Canada

Liyan Zhang
School of Electronics and
    Information Engineering
Dalian Jiaotong University
Dalian, Liaoning, China

1

Fixed Mobile Convergence: The Quest for Seamless Mobility

Dario Gallucci, Silvia Giordano, Daniele Puccinelli,
N. Sai Krishna Tejawsi, and Salvatore Vanini

CONTENTS

1.1 Introduction

The last few years have seen an exceptional growth in the wireless local area network (WLAN) industry, with substantial increase in the number of wireless users and applications. This growth was mostly due to the availability of inexpensive and highly interoperable network solutions based on the Wi-Fi standard [26] and to the growing trend of providing built-in wireless network cards in mobile computing platforms. Today, public and private organizations are developing wireless mesh networks (WMNs), i.e., peer-to-peer multi-hop wireless networks based on the Wi-Fi technology whose nodes form a connectivity mesh. However, there are several business and technological challenges that need to be addressed to turn Wi-Fi-based WMNs into a global network infrastructure [4]. In particular, as the number of mobile Internet users increases and new emerging applications appear, it becomes very important to provide users with a level of quality of service (QoS) that compares favorably to the one enjoyed by wired Internet users (in terms of application reliability, throughput, end-to-end delay bounds, etc.). Existing solutions for WMNs suffer from reduced efficiency due to the lack of reliable self-configuration procedures that can dynamically adapt to varying network conditions, lack of efficient and scalable end-to-end QoS support, and lack of generalized and seamless mobility support. Further, as opposed to the case of the wired Internet, the convergence between data and multimedia networks is not really happening. Popular tools such as Skype for low-cost voice services or YouTube for low-cost video services still lack a counterpart for mobile networks. This is mainly due to the following reasons:

- Lack of ubiquitous coverage for mobile users. Users can enjoy continuous coverage when they are within range of any node within a given WMN, but as they move away they will eventually need to connect through a different network provider; unfortunately, this transition is not straightforward due to the lack of automatic switch support at the network layer.
- Even if some dedicated solution for an automatic switch is provided, there is no assurance that the applications will continue to work properly.
- A seamless handover, if at all possible, is hard to perform among heterogeneous networks; for example, it may not possible to exploit GPRS or UMTS coverage where the mesh network is not available.

In the literature, there are several architectural solutions for the management of global node mobility, but none of them is suitable to provide mobility support with QoS in WMNs. To fully exploit mesh network solutions with advanced services, it is therefore necessary to provide an innovative architectural solution that offers

- Seamless mobility over multiple networks with local and wide coverage
- Optimized global node mobility to exploit multiple heterogeneous networks in parallel
- Adequate QoS

In the framework of the FP7 project EU-MESH [1], we have developed this architectural solution for mobile users in WMNs. Our solution is based on a cross-layer approach and performs a seamless handover as well as application layer persistence. Moreover, our solution is capable of selecting the optimal connection among heterogeneous network technologies and is capable of handling specialized multimedia traffic. To support rapid and seamless handovers over heterogeneous networks and different operators, EU-MESH explores application layer solutions that exploit cross-layer monitoring and seeks to optimize them for WMNs using self-tuning and parameter adaptation, information collection, user interaction, and streamlined handover procedures. Cross-layer monitoring enables the effective adaptation of the delivered QoS to variable network conditions (for instance, due to operational anomalies). Furthermore, autonomic components based on proactive monitoring are used to self-optimize the internal parameters when the network context changes, and to self-configure mobile clients (e.g., autodetection features of the onboard hardware). All of this is obtained with a novel design approach (based on autonomic components and cross-layer monitoring and control) to optimize the performance of the WiOptiMo system [8,10,11], which provides seamless inter-network roaming by handling mobility at the application layer. This architecture is enriched with a heuristic approach to service classification and a traffic classification algorithm for the management of multimedia traffic at the network and application layers for an improved characterization of the traffic.

1.2 Architectural Mobility Support

Global node mobility requires the support of a seamless vertical and horizontal handover. A vertical handover occurs among different type of networks, while a horizontal handover occurs among networks of the same type. In a macro-mobility scenario, handovers are performed between different domains (inter-domain handover), while in a micro-mobility scenario, they are performed within the same domain (intra-domain handover). Handover management may or may not require a modification of the mobile client network stack. In [28], architecture and protocols are proposed for link-layer and network-layer intra-domain handoff for unmodified 802.11 devices with DHCP support. This approach operates in ad hoc mode, controls the handoff from the mesh infrastructure, and multicasts data using multiple paths to the mobile client at the cost of network overhead. A very similar approach for mobile clients in infrastructure mode is presented in [17], where a layer 3 handoff is performed using a Mobile IP [18]-like solution (that suffers from performance degradation due to encapsulation and the lack of QoS capabilities) or, alternatively, a flat routing protocol. In [3], a mechanism for multi-homed WMNs is presented. This mechanism optimizes routing and provides inter-domain handoff between Internet-connected access points that may be located on different networks with a significant network overhead for client management during the handoff and for network topology maintenance. In [7], micro- and macro-mobility is supported with a novel network layer mobility module in the mobile client network stack. This solution requires changes to the standard DHCP protocol and the integration of mobility management inside the architecture. DHCP introduces latency, and making a 3G network node mobile is challenging because the required architectural components lie in the PDSN (packet data serving node) of the network. To improve the timing of the handoff decision, a scheme is proposed in [22] that reduces the 802.11 handoff delay by passively monitoring other channels for the presence of nearby access points. The drawback of this scheme is a regular overhead caused by the continuous passive monitoring process. Support for applications with QoS constraints cannot be achieved with any of the handoff mechanisms adopted in the solutions described above, all of which are threshold-based (they initiate a handoff when service degrades below a given threshold). To overcome these limitations, WiOptiMo [9] employs a cross-layer monitoring of the communication status to predict the interruption of the current connection and select a new connection according to the network context. In Section 1.2.1, we describe WiOptiMo's architecture and its characteristics. For the sake of clarity, we focus on the special case of a client–server architecture. In Section 1.2.2, we present an extension of WiOptiMo, specifically optimized for WMNs.

1.2.1 WiOptiMo Description

WiOptiMo is an application layer solution for seamless mobility across heterogeneous networks and domains. It provides persistent connectivity to users moving across different wired and wireless networks. WiOptiMo's design is entirely based on the use of currently deployed network protocols and drivers. It does not require any ad hoc modification or adaptation in the current 802.x standards, but it can be easily adapted to accommodate and exploit future improvements in these standards. Moreover, it has been designed to have minimal impact on the CPU load and, consequently, on the energy consumption. All these design characteristics make WiOptiMo immediately deployable on a large scale over a wide range of mobile devices.

WiOptiMo enables a handover initiated by the mobile device. While it is designed for applications based on the client–server paradigm, its extension to peer-to-peer models is straightforward.

Mobility management and seamless handover are handled by two main components: the Client Network Address and Port Translator (CNAPT) and the Server Network Address and Port Translator (SNAPT). These two components form an interface between the client and the server, and give them the illusion that there is no mobility in the system. Both the CNAPT and the SNAPT can be either installed locally on the client and the server, or on a connected machine. The CNAPT and the SNAPT collectively act as a middleware (Figure 1.1) so that the client believes to be running either on the same machine as the server or on a machine with a stable direct connection to the server (depending on the configuration adopted). The CNAPT is an application that can be installed on the same device as the client application or on a different device in the same mobile network. For instance, in the case of a team of consultants or auditors that require mobility while working together, the CNAPT can be installed on the mobile device of one team member and the whole team can share the seamless handover provided by that one device. Similarly, the SNAPT is an application that can be installed on the same device as the server application, on a different device of the same network, or on any Internet server (e.g., on a corporate front-end server, on a home PC connected to Internet, on a mesh router, or on an 802.11 access point). This flexibility of the SNAPT installation is particularly important because it avoids scalability issues. This is a completely new approach: The mobility of multiple users can be handled either using a star topology with a central server with large computational capabilities and large bandwidth, or with a distributed topology where the SNAPT is installed on the accessible nodes (i.e., mesh routers), saving transmission costs.

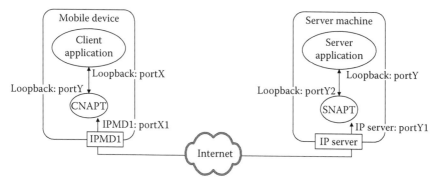

FIGURE 1.1
The CNAPT and the SNAPT collectively act as a middleware.

1.2.1.1 The CNAPT Module

The CNAPT module is an application that emulates the server's behavior on the client side and, at the same time, the client's behavior on the Internet side.

The CNAPT can be installed either on the same machine of the client, or on an additional mobile device that acts as a connectivity box that manages many different network access technologies and has a broadband IP connection to the original mobile device. Figure 1.2 refers to the case where the CNAPT and the client application are installed on the same device. In this case, they use the loopback address to communicate with each other. The most commonly used IP address on the loopback network is 127.0.0.1 for IPv4 and ::1 for IPv6.

The CNAPT emulates the client and the server applications behavior by providing the following sockets:

- A server socket on the client side for each service the client can request from the Internet. This server socket listens on the real server service port. It is named Server Service Emulator Server Socket (SSESS).

- A client request emulation socket on the Internet side for each service request sent via the Internet. This socket is bound to the current IP address of the node and relays packets to the right SSESS provided by the SNAPT.

- A server socket on the Internet side for each client service (services that can be used by the server for publish/subscribe communication models). This socket listens on the client service emulator port, which is different from the client service real port to avoid binding errors. This socket is termed Client Service Emulator Server Socket (CSESS).

- A server request emulation socket on the client side for each client service request. This socket relays packets to the real client service server socket.

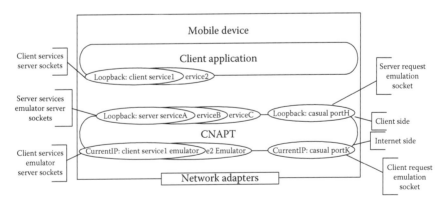

FIGURE 1.2
The client network address and port translator (CNAPT).

1.2.1.2 The SNAPT Module

The SNAPT is a software application that emulates the client's behavior on the server side along with the server's behavior on the Internet side (Figure 1.3). It provides

- A server socket on the Internet side for each server service. This server socket listens on the server service emulator port (the SSESS). This port is different from the server service real port in order to avoid a binding error if the SNAPT is installed on the same machine as the server.

- A client request emulation socket on the server side for each server service request. This socket relays packets to the real service server socket.

- A server socket on the server side for each client service. This server socket listens on the real client service port (the CSESS). The CSESSs are grouped by a CNAPT ID. If they use the same port, they are bound to different virtual IP addresses to avoid binding errors.

- A server request emulation socket on the Internet side for each client service request. This socket relays packets to the right CSESS provided by the corresponding CNAPT ID.

- A server socket on the Internet side for each dynamic request to instantiate a server service coming from the CNAPT.

- A control socket on the Internet side used for the CNAPT–SNAPT communication. This socket is used for transmitting handshake packets during the handover.

The SNAPT can emulate several client requests simultaneously and in parallel and it can be used by more than one server.

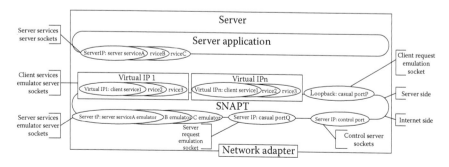

FIGURE 1.3
The server network address and port translator (SNAPT).

1.2.1.3 Flow Control

During normal communication, the CNAPT relays the client requests to the SNAPT that manages the server. Upon receiving client requests, the SNAPT processes them and in turn relays them to the corresponding servers. The server response path mirrors the client request path.

During a handover phase, the CNAPT and the SNAPT enable a transparent switch to the new selected connection. The application's data flow is interrupted at the CNAPT, which stops forwarding the outgoing IP packets generated by the client. The SNAPT also stops forwarding all the outgoing packets generated by the server. The packets already stored in the transmission buffers of both the CNAPT and the SNAPT sockets will be forwarded, respectively, to the SNAPT and the CNAPT after the completion of the handover.

The TCP window mechanism for flow control is exploited to pause the application, avoiding the need for a possibly large amount of extra buffer space for the outgoing packets during the handover.

1.2.2 WiOptiMo and WMNs

The WiOptiMo system can be used in multi-radio multi-operator WMNs with an extra component, the *Controller*.

1.2.2.1 Definitions and Challenges

Seamless mobility support requires providing mechanisms for a user to maintain the same identity irrespective of the terminal used and its network point of attachment, without interrupting any active network sessions and avoiding or minimizing user intervention. This implies supporting a handover between different network providers and technologies. The handover process can be practically broken down into three functional blocks.

Handover initiation: The proactive monitoring of the current connection and/or possible alternative connections in order to (1) effectively anticipate or explicitly deal with a loss of connectivity, (2) trigger alternative handovers in order to optimize costs and performance. A decision task manages the handover initiation process: it monitors the reliability and the performance of the current connection, and possibly searches for new network providers and connectivity.

Network selection: Selection of a new connection point according to decision metrics such as signal quality, cost, and bandwidth. Information about these metrics can be gathered proactively and/or reactively.

Handover execution: A set of procedures to be carried out for the authentication and re-association of the mobile terminal (switching procedure).

The handover of the mobile terminal is said to be network executed if it is totally under the control of the network (as is the case between UMTS/GSM/GPRS cells). In a mobile executed handover, the handover decision is autonomously taken by the mobile terminal. The handover can be either soft (or alternative) when it is performed for the sole purpose of optimization of the connection cost or QoS, or hard (also called imperative) when it is performed due to an imminent loss of connectivity.

Most of the proposed schemes for mobility management follow a common pattern: motion detection at layer 3 or below, selection of next attachment point, discovery and configuration of a new IP address, and signaling for the redirection of incoming data packets. An efficient solution is one that provides transparency and low handover latencies, is robust with respect to a wide range of situations, scales well, minimizes user costs and resource usage, guarantees security, and can be easily deployed on top of existing protocols and technologies.

1.2.2.2 *WiOptiMo in a Heterogeneous and Multi-Operator Mesh Network*

1.2.2.2.1 *Motivation*

In order to support seamless and fast handoffs in a heterogeneous and multi-operator mesh network environment, only schemes that have a minimum impact on the network layer should be adopted, so that complex rerouting mechanisms can be avoided. Application layer solutions comply with this requirement and can act as middleware, taking into account the QoS requirements of applications as well as user preferences. This is not feasible with approaches that operate at the lower layers and have no way of matching the needs and expectations of mobile users. An approach to mobility management should provide the end user with the freedom to choose among different carriers to create spontaneous service establishment and provide a distributed/bottom up seamless handover. From this point of view, a system that can be easily and quickly adapted to new network providers and to changes in wireless networks standards would be intrinsically advantageous and would pave the way to a more dynamic and competitive operators/providers market.

WiOptiMo contains all the features described above. In addition to that, WiOptiMo does not require changes to the traditional OSI protocol stack and can be easily adapted to different operating systems. It also provides backward compatibility (for instance, in case non-IP-based protocols are adopted in the future). Finally, it does not introduce any additional network overhead (i.e., no IP encapsulation or control information is added to the user payload).

WiOptiMo is also very flexible and scalable. With WiOptiMo, mobility management can be distributed and centralized. Its SNAPT server module can be installed on any access node, allowing to reroute network traffic to a set of alternative hubs in case the SNAPT in use is overloaded.

In addition to the architectural motivations listed above, WiOptiMo has other key advantages that are related to the performance in the presence of mobility. In fact, to improve the handoff latency and increase the quality of a client's connectivity, application layer mechanisms can be implemented that employ cross-layer information. The goal of such mechanisms is to provide reactive and proactive handoff procedures that allow the reduction of the overhead of a handoff operation. For example, regular patterns in the user's geographic movements (measured, for instance, with an onboard GPS equipment or with a Wi-Fi-enabled mobile device) can be exploited to predict the onset of the handover, and network traffic can be classified to select optimized paths within the mesh network.

1.2.2.2.2 Adaptation

As previously stated, WiOptiMo's client component (CNAPT) runs on the user's device (or on a device directly connected to the user's device), while the server component (SNAPT) can be installed on any access node. In order to improve scalability in a WMN, multiple SNAPTs can be positioned on different mesh routers across the network (Figure 1.4). WiOptiMo, originally written in Java, has been ported to C so that it can run on mesh routers, which are typically embedded platforms that provide no Java support.

In the original design of the WiOptiMo system, the CNAPT can connect simultaneously with multiple SNAPTs, but this behavior can only be statically set a priori. In the highly dynamic environment that is typical of WMNs, this design is not suitable. An extra component, the Controller, is needed for the selection of the appropriate SNAPT at run-time. The task

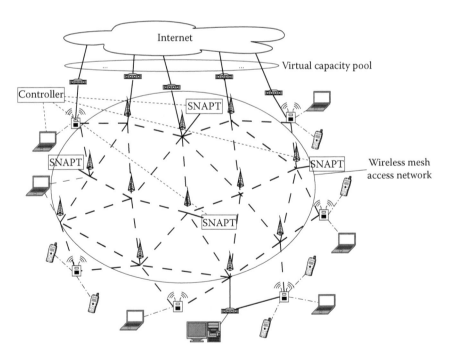

FIGURE 1.4
WiOptiMo configuration for the EU-MESH project.

of the Controller is to select the SNAPT to which the CNAPT can connect. The Controller is not needed if the system has a star topology, because the CNAPT always connects to the same SNAPT.

At start-up, the CNAPT contacts the Controller to get the SNAPT's address. At run-time, active SNAPTs inform the Controller about their connected CNAPTs and their network load. If during the handover a SNAPT should no longer be available, or a better SNAPT is available, the Controller notifies the CNAPT. The metric to quantify the quality of a SNAPT is a function of

- *Available bandwidth*: If the current SNAPT is overloaded, a SNAPT with a smaller network load is preferred.
- *Network latency*: If we can classify the user traffic (see Section 1.4.1), we can improve the user's experience by connecting to the SNAPT that is more appropriate for the features of the current traffic (this is important for applications with low latency requirements, such as voice).
- *Packet delay*: The priority goes to the SNAPT that ensures the smallest packet delay.
- *Packet loss*: TCP-like protocols employ the packet loss ratio as a measure of congestion and therefore reduce the throughput over a noisy wireless channel even in the case of high bandwidth availability.

1.3 Proactive Mobility Support for the End User

All the phases in the handover process (refer to Section 1.2.2) can have a major impact on the overall handover latency, and, consequently, on the QoS perceived by the user. The phases of handover initiation and network selection, however, are particularly critical and deserve special attention. For example, in order to effectively trigger a handover and perform an optimal network selection, it is important to collect fresh and reliable information about the ongoing communication patterns and the features of the various candidate networks through some form of cross-layer monitoring. Layers 1 and 2 can provide information about the connection quality (for instance, in terms of signal strength and packet loss); layer 3 can report about the existence and the quality of the routing path toward the destination; layer 4 and the upper layers can provide important information concerning local and end-to-end traffic load.

Furthermore, proactive and reactive mechanisms can be applied on the monitoring of the current connection and/or of possible alternative connections in order to (1) effectively predict handovers, or (2) trigger alternative handovers to optimize costs and performance. Network selection can be done according to the requirements of the user and of the application.

Finally, if information between access points and the mobile terminal are available, the handover could be optimized. This is the case of the recent 802.11r standard that provides secure and fast handover but only intra-domain.

WiOptiMo has already been designed according to this combination of cross-layering and autonomic approach, but it is not optimized for a mesh context. To achieve this goal, WiOptiMo has been empowered with reactive and proactive handoff mechanisms based on the awareness of **user behavior** and **content type**.

1.3.1 Awareness of User Behavior and Content Type

The periodic patterns of user mobility are typically dictated by routines and schedules (i.e., home-to-work travel). User mobility patterns can be exploited to predict future scenarios and optimize the performance based on past and current events. Information on user encounters, geographic patterns, user location, and motion detection may lead to motion pattern recognition and possibly allow a mobile device to engage in a proactive handoff. In particular, user location data can allow the automatic configuration of wireless network settings based on the position of the device and the estimated user's speed, as well as the automatic selection of the optimal data transfer technology.

In Section 1.3.2, we describe a possible technique for the localization of the user, a piece of information that we employ to streamline a network-executed handover. In general, a localization system should be available ubiquitously (it should not be limited to specific environments) and should provide good accuracy.

Users offer very different amount and type of traffic to the network. Widely used applications have variegate requirements in terms of either end-to-end or handoff latency. For example, to be perceived as acceptable, the one-way delay of VoIP should not exceed 150 ms, service interruption for handoff is annoying if it is longer than 50 ms, while e-mail or web browsing do not have to phase out users' attention, which is in the order of 10 s. In Section 1.4.1, we show how network traffic can be categorized to improve the user experience. Categories are generated from some distinctive traffic characteristics. As said before, the requirements in terms of latency is one of them, while the packet loss rate and the available bandwidth are also effective metrics for classification.

1.3.2 Location Prediction

In WiOptiMo's basic strategy, the client triggers a handoff. The CNAPT continuously verifies the reliability and performance of the current Internet connection: if the reliability/performance metrics drop below a critical threshold or if the current Internet connection has been interrupted, the CNAPT triggers a handover to a new network provider.

In order to predict an imminent loss of connectivity or to effectively deal with a lack of connectivity, WiOptiMo can be complemented by a mechanism for a network-executed handover, which consists in the prediction of future user locations. We propose a mechanism to track user movements based on the received signal strength indication (RSSI) without the need for extra hardware such as GPS or accelerometers.

There exists a significant body of work about the tracking of user movements based on RSSI signatures. Most of the proposed strategies are empirical approaches that employ probabilistic methods requiring location-dependent RSSI calibration [14]. The performance of such algorithms depends on the training data. Popular techniques for the estimation of future user locations include hidden Markov models and particle filtering because of their self-adaptivity in nonstationary environment (noisy environments due to multipath fading) [6,23,25]. For the mobility management issues in cellular networks in [23], authors have proposed various techniques such as the LZ-parser algorithm and $O(k)$ Markov predictors to estimate the next base station or access point and predict the handover. These techniques fail in tracking the intracellular movements of the user.

In the remainder of this section, we describe the role of SNAPT and CNAPT in building the location graph/location history and predicting the next location, a strategy for a more general solution and to deal with shortcomings of using low threshold value, and, finally, the demonstration of the proposed algorithms and our results.

1.3.2.1 Location Mapping

We define the RSSI vector $R = \{R_0 \dots R_{N-1}\}$ as the list of the RSSI values measured at the mobile device with respect to N access points. Each access point i is uniquely identified by its MAC address.

Let $\mathcal{A}$ and $\mathcal{B}$ represent the set of access points employed to populate, respectively, RSSI vectors A and B. If $\mathcal{A} \cap \mathcal{B} \neq \varnothing$, then the Euclidean error $E(A, B)$ between two RSSI vectors A and B is defined as

$$E(A,B) = \sqrt{\frac{\sum_{i \in \mathcal{A} \cap \mathcal{B}} (A_i - B_i)^2}{\mathcal{A} \cap \mathcal{B}}}, \quad \text{while } E(A,B) \triangleq \infty \quad \text{if } \mathcal{A} \cap \mathcal{B} = \varnothing \quad (1.1)$$

As raw RSSI is notoriously unstable, in [6], a number of techniques for the conditioning of raw RSSI have been proposed for purposes of location mapping and location prediction. Since our objective is to predict the handovers, accurate location tracking is not really needed, and we simply bind the RSSI data to a large area (namely, a circle of a 10 m radius).

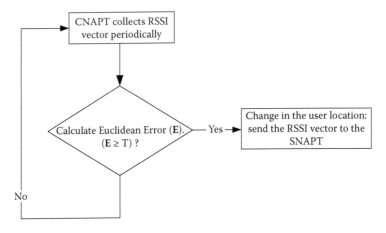

FIGURE 1.5
The role of the CNAPT in location mapping.

1.3.2.2 The Role of the CNAPT

The CNAPT on the mobile device creates an RSSI vector with the RSSI from all the access points found in the scanned network list. The RSSI vector is updated at fixed time intervals and the latest two instances, P(Previous) and C(Current), are stored. If the $E(P,C) \geq \tau$, where τ is a fixed threshold, the CNAPT detects that the user has moved to a different location and sends C to the SNAPT to which it is connected. This mechanism is shown in Figure 1.5.

1.3.2.3 The Role of the SNAPT

The SNAPT keeps track of the user's route, maintaining information regarding the current and previous user locations. From the SNAPT's viewpoint, a newly received RSSI vector may correspond to a new user location or to a previously visited location (it is therefore an unestablished location, since its nature is undetermined at this point).

The SNAPT maintains a *location graph* $\mathcal{L}(R_m) = \lambda_m$ for $m = 0 \ldots l{-}1$, where l denotes the number of established locations contained in the location graph, λ_m is the mth established location, and R_m is the RSSI vector measured at the mth established location. The SNAPT also maintains the location history $H = \{h_0 \ldots h_{v-1}\}$, which is the sequence of all the v visited locations (v is incremented each time the user changes her location). When the CNAPT signals that the user has moved to a different location by sending C, i.e., the RSSI vector pertaining to the user's latest location, the SNAPT computes the Euclidean error $E(C, R_m)$ for $m = 0 \ldots l{-}1$. If $E(C, R_m) > \theta$ for all $m = 0 \ldots l{-}1$, l and v are incremented, a new location λ_{l-1} is established, $\mathcal{L}(R_{l-1} \triangleq C) = \lambda_{l-1}$, and $h_{v-1} \triangleq \lambda_{l-1}$. Otherwise, C is assumed to correspond to the location λ_m with

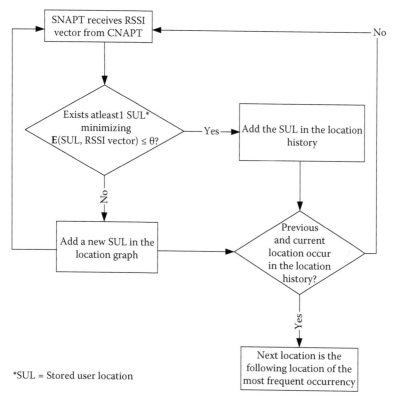

FIGURE 1.6
The role of the SNAPT.

$m = \mathbf{argmin}_{m=0...l-1}E(C, R_m)$, v is incremented, and $h_{v-1} \triangleq \lambda_m$. This mechanism is shown in Figure 1.6. The number of visits ω_i to a given location λ_i is equal to

$$\omega_i = \sum_{j=0...v-3} 1_{h_j=h_{v-2}} 1_{h_{j+1}=h_{v-1}} 1_{h_{j+2}=\lambda_i}. \qquad (1.2)$$

If $v \geq 3$, the next location that the user will visit is predicted to coincide with one location λ_m with $m = \mathrm{argmax}_{i=0...l-1}\omega_i$ (ties are broken randomly); if $v < 3$, no prediction is possible. If the next location predicted by the SNAPT is not covered by the current network, the SNAPT can proactively prepare for a handover, thereby reducing the handover latency and providing the user with seamless mobility.

1.3.3 Clustering of Stored User Locations

The performance of our location prediction scheme is highly sensitive to the calibration of the threshold θ used by the SNAPT to map RSSI vectors received from the CNAPT to its stored user locations. Stored user locations λ_k and λ_j are said to be equivalent if $\lambda_{k-1} = \lambda_{j-1}$ and if $\lambda_{k+1} = \lambda_{j+1}$ with $k \in [0 \ldots l-1]$ and $j \in [0 \ldots l-1]$. We propose the adaptation of the threshold θ based on the number of equivalent stored locations (an overly low value of θ would boost the number of equivalent stored locations). Namely, for the mth stored user location, we employ

$$\theta_m = \theta_0 + W_m \theta_W, \tag{1.3}$$

where

θ_0 and θ_W are the fixed thresholds
W_m is the number of equivalent stored locations for the mth stored user location

The use of an adaptive threshold streamlines the clustering of stored user locations that are estimated to be sufficiently close together.

1.3.4 Empirical Results

1.3.4.1 Calibration of τ

We collected a dataset over the course of a week using $N = 8$ access points and a total of $l = 4$ locations using a Dell laptop equipped with a Wireless 1395 WLAN Mini-Card (and running Microsoft's Packet Scheduler). The RSSI appeared to be relatively stable because the acquisitions were purposefully run at times when the level of human activity in the deployment area was insignificant. This was done to avoid induced fading (induced by human activity [19,20]), which is a primary cause of RSSI instability. Similarly to [27], we measured the Euclidean error in the dB domain to assign a stronger weight to small power values that would not influence the value of the error.

Based on Figure 1.7, which shows the empirical distribution of the Euclidean error between all the combinations of RSSI vectors measured in the same location as well as different locations, we can set $\tau \in [5.5 \ldots 7.7]$ dB.

Figure 1.8 shows the number of stored user locations for different values of the threshold (τ) and the number of RSSI vectors that are mapped incorrectly.

Based on these results, we empirically set $\tau = 7.4$ dB.

1.3.4.2 Calibration of θ_0 and θ_W

We obtain a dataset from the real-life motion of a user along a predetermined path on two different floors, whose layout is represented in Figures 1.9 (ground floor) and 1.10 (first floor).

For simplicity of representation, we indicate each stored user location with a number i indicating the stored user location λ_{i-1} and a letter representing the associated RSSI vector.

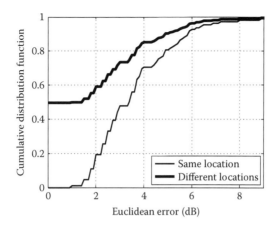

FIGURE 1.7
Empirical cumulative distribution function of the Euclidean error between the RSSI vectors pertaining to the same location as well as to different locations.

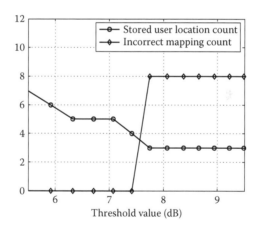

FIGURE 1.8
Number of nodes formed per different threshold values.

We first examine a path that covers three different locations shown in Figure 1.9. The location graphs corresponding to different 2-tuple (θ_0, θ_W) are shown in Table 1.1. For the 2-tuple (5.5, 3.2), a larger number of user locations are stored due to significant fluctuation of the RSSI at location c. Due to the overly low θ_W, clustering fails to kick in. For the 2-tuples (5.5, 4.5) and (6.3, 3.2), however, the location mapping algorithm works correctly. We now examine a second path covering four different locations shown in Figure 1.10. The location graphs corresponding to different 2-tuple (θ_0, θ_W) are shown in Table 1.2. For the 2-tuple (7.7, 3.2), the location graphics are severely flawed: RSSI vectors a and c are mapped incorrectly owing to the overly high value of θ_0. For the 2-tuple (5.5, 3.2), we also have a similar problem. For the 2-tuples

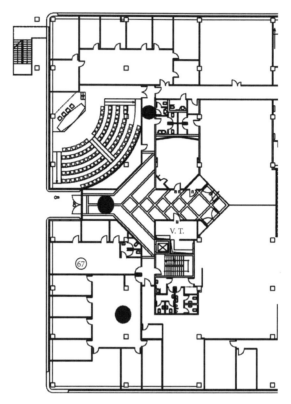

FIGURE 1.9
Locations on the ground floor.

(6.3, 3.2) and (5.5, 4.5), however, the location mapping algorithm works correctly. Our empirical data therefore suggest the calibration $\theta_0 \in [5.5, 6.3]$ dB and $\theta_W \in [3.2, 4.5]$ dB.

1.4 QoE and Resource Optimization

Current networks must ensure the efficient delivery of heterogeneous traffic because telecommunication operators are constantly increasing end users' bandwidth requirements.

1.4.1 Heuristic Service Classification

The efficient delivery of heterogeneous traffic is essential to the Quality of Experience (QoE) of the end user. A careful analysis of the network usage patterns of each application in the network can provide the best solution

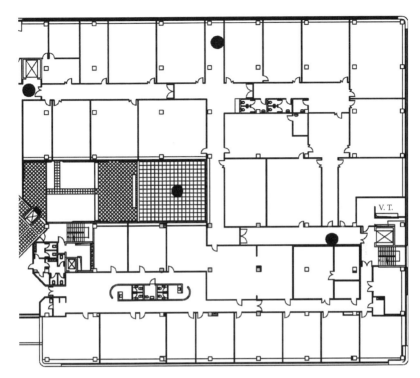

FIGURE 1.10
Locations on the first floor.

for the networking requirements of any user. While many common applications can be profiled directly by analyzing a significantly long traffic trace, a technique known as network fingerprinting [16], an exhaustive traffic classification is a daunting task. To make matters worse, widespread rich web applications generate traffic that can seldom be distinguished from traffic due to generic web browsing.

Since a complete payload analysis only offers marginal benefits at a much higher computational cost than header analysis, a header-based approach to traffic classification is typically preferred. One key element is the transport protocol (TCP, UDP, or other), which tells us whether the traffic under test tolerates unreliable delivery. However, there are well-known applications that use an unreliable protocol even if unreliable delivery may cause a significant drop in the QoE level. DNS is a case in point: a dropped name lookup packet may cause an extended delay for the application (compensated by protocol-level caching). Another key element is the packet length, which correlates well with the delay tolerance of the application data. For instance, a transmission flow composed by small packets is certainly delay sensitive, while a flow whose packets typically match the maximum transmission unit (MTU) size contains delay tolerant data because the generating application

TABLE 1.1

Location Graphs Formed for Different 2-Tuples (θ_O, θ_W) Based on a Path Covering the Locations Shown in Figure 1.9

θ_O, θ_W	Location Graph	Comments
(7, 3.2)		Three different locations are established
(6.3, 3.2)		Multiple locations are assigned to *c* and are all clustered together
(5.5, 3.2)		Four locations are assigned to *c* and two sets of clusters are formed
(5.5, 4.5)		Multiple locations are formed for *a* and *c*; clustering is performed correctly

can wait for the whole data before starting transmission. We propose a traffic classification based on the above metrics, as shown in Table 1.3.

1.4.2 Network Modeling and Network Utility

Network resources are limited, and dedicated strategies are needed to fulfill the conflicting requirements of multiple users of different applications.

1.4.2.1 Network Utility

Network utility (NU) offers a formal method for the allocation of a limited distributed resource to connect multiple traffic source–destination pairs. The goal is to maximize the usage of a network resource while meeting at least the minimum requirements of each pair. The original concept of NU [13]

TABLE 1.2

Location Graphs Formed for Different 2-Tuples (θ_0, θ_W) Based on a Path Covering the Locations Shown in Figure 1.10

θ_0, θ_W	Location Graph	Comments
(7.7, 3.2)		The setting $\theta_0 = 7.7$ is too high and causes an incorrect mapping
(7, 3.2)		Correct operation
(6.3, 3.2)		Correct clustering
(5.5, 4.5)		Correct clustering
(5.5, 3.2)		Failed clustering

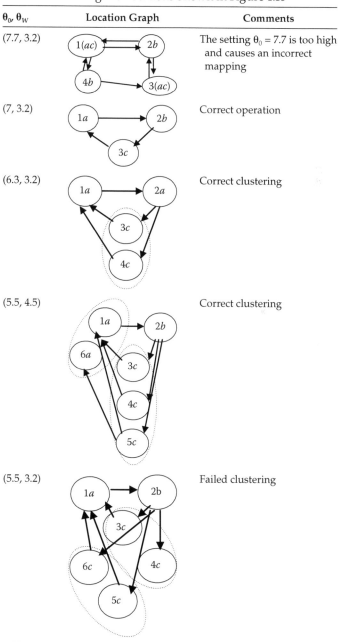

TABLE 1.3

Traffic Categories Determined by Heuristic Analysis

	Delay Tolerant	Delay Sensitive
Packet loss tolerant	Large packets, without or with limited protection	Small packets, without acknowledgment
Packet loss sensitive	Large packets, over reliable transport protocol	Small packets, with protection (repetition or FEC) or reliable transport protocol

does not support multiple metrics and does not consider requirements as latency or packet loss tolerance. It also does not take into account the additional constraints provided by the wireless medium, like transmission rate reduction or packet loss due to noise or interference. To be used in WMNs with QoE constraints, the definition of NU must be adapted to handle multiple metrics and additional constraints.

We can define the QoE ratio (quality ratio) as a monotonically decreasing function of a given performance metric (mapping the values of the metric to satisfaction level). Specifically, we focus on packet loss and delay, which we choose to represent, along with bandwidth, the network state. The packet loss has a strong impact on the QoE, and its effects on a wide range of known traffic types and sources are the objects of several studies, such as [12,21]. The behavior of two popular applications, HTTP and VoIP, with opposite reactions to packet loss, are shown in Figure 1.11.

Delay also has a strong impact on the QoE. The effect of delay on known traffic flows is the object of [2,15]. The behavior of HTTP and VoIP in the presence of delay is shown in Figure 1.12.

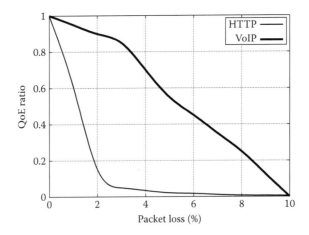

FIGURE 1.11

Quality ratio as a function of packet loss for known types of traffic.

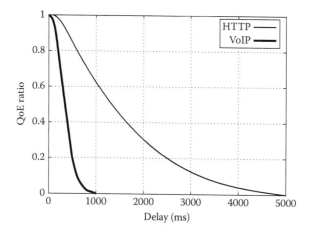

FIGURE 1.12
Quality ratio as a function of delay for known types of traffic.

The bandwidth itself has a major impact on the QoE due to its direct effect on packet loss and delay. In fact, even in the absence of a significant amount of noise and interference, packet loss may still occur due to buffer overflows. Similarly, the delay of a packet is due to the time it takes the packet to cross the network plus the router dispatch time, which varies depending on the available bandwidth.

1.4.2.2 Alternatives to Network Utility

In the networking literature, there exist two main classes of approaches to optimizing the QoE while allocating a limited resource to network users. One, which we refer to as the *penalty approach* [13], is to just reject flows that fail to meet quality requirements after the allocation phase, while the other is to reformulate the allocation problem as a routing problem.

The penalty approach has been proposed in [13] along with the NU optimization algorithm: a penalty that accounts for the delay and packet loss requirements of the users is added to the cost of the flow (computed according to a given metric), and flows that exceed a given overall cost are removed.

Examples of the second class are usually found in quality-aware routing schemes. In [5], the allocation problem is solved as a routing problem using a routing cost metric that encompasses quality requirements. The optimal solution is obtained by running Dijkstra's algorithm over all possible permutations of the resource allocation to the flows in the network.

1.4.2.3 Modifications to Network Utility

We seek to tailor the NU method to the fundamental features of the network architecture resulting from the use of the WiOptiMo technology. Since

WiOptiMo redirects each outgoing data flow from the client to the server through the CNAPT and the SNAPT, every single connection from the user's device to the intended destination can be managed separately. Therefore, the NU method can be modified to take advantage of the available information on the established connections transforming the fundamental unit from user activity to application activity and possibly operates on each independent data flow.

We start from the original NU optimization algorithm and replace the user utility function in the maximization problem, (1.4), with an individual utility function for each user application flow:

$$\text{USER}_r(U_r; \lambda_r) = \text{maximize } U_r(x_r) - w_r, \tag{1.4}$$

where $U_r(x_r)$ is the utility of the flow x_r, with $x_r = w_r / \lambda_r$, where w_r is the amount that the user elects to pay for per unit time and λ_r is the flow cost per unit of traffic. Therefore, x_r can be viewed as (the share of) the flow returned to the user.

In the returned flow x_r, we insert a multiplicative coefficient (α_r) that represents the product of the values of the quality ratio for the required values of packet loss and delay; we obtain

$$\text{USERAPP}_r(U_r; \lambda_r) = \text{maximize } U_r(\alpha_r w_r \lambda_r) - w_r \tag{1.5}$$

The user application utility is thus decreased depending on the condition of the network links to which the traffic is allocated.

The selection of the network links for the specific kind of traffic is therefore affected by the QoE the user expects from the network, and unsuitable routes are less likely to be selected by the optimization algorithm because they will imply a less favorable utilization/cost ratio.

Nonetheless, while the penalty approach only makes use of the penalty a posteriori (to reject the paths that exceed a given cost), our solution actively seeks out the paths that maximize the value-to-cost ratio to maximize the QoE. In economic terms, this approach can be described as getting the most out of your money.

1.4.3 Experiments

To evaluate the effects of the proposed modifications to the NU method, a wireless hybrid ad hoc network has been simulated with ns2 [24].

The simulated network consists of a total of 32 nodes: 12 nodes operate as both backbone nodes and access points, and 20 more nodes are associated as clients to the access points. The client nodes generate traffic over the network starting at different times and with a distinct traffic type. The backbone nodes have reliable links to three or four other nodes, while the client nodes are mobile and move across the network and connect to different access points during the simulation. The simulated network is represented in Figure 1.13.

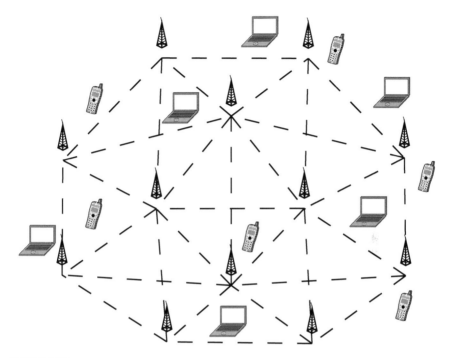

FIGURE 1.13
Network layout. Client nodes are moving.

In this network, the user applications are simulated using the presented optimization algorithms. To increase the network load to a specific value, UDP background traffic is introduced between adjacent nodes after the completion of a run of the resource allocation algorithm. The results are reported in Table 1.4.

Table 1.4 presents the comparison of our scheme with the traditional NU method with the penalty approach and QoE-aware Dijkstra. The latter serves as a benchmark for the optimal behavior (it is not possible to implement it in practice because of its computational complexity). We show the average

TABLE 1.4

Simulation Results, with Network Load at 95%

	Average Available Bandwidth [%]	Worst Case Available Bandwidth [%]	Rejected Applications [%]
QoE-aware Dijkstra	>100	98.0	2.8
NU with penalty approach	>100	>100	12.7
Proposed NU modifications	>100	97.9	2.9

Note: Over 100% of bandwidth means that the obtained bandwidth is more than the requested.

results of 100 simulation runs with same network conditions but different initial client locations. Traffic delay and packet loss are not reported because they are always within the constraints. Our results show that the performance of the proposed modifications to the standard NU method compares favorably with QoE-aware Dijkstra, without requiring to iterate over all permutations of user service requests (which would mean a complexity of $O(n!)$, as it happens with the penalty approach). This is reflected in the number of rejected flows, which is smaller with the proposed solution that promotes a flow based on QoE satisfaction.

1.5 Conclusion

In this chapter, we presented an architectural solution for mobility support within WMNs. A research implementation of this solution has been developed within the scope of the FP7 project EU-MESH [1] and is currently under experimental evaluation. In this chapter, we also proposed two enhancements to this solution to manage (1) practical issues due to mobility (moving resources among consecutive point of attachments during handover), and (2) the user satisfaction. We presented a solution for mobility support that minimizes the handover time by predicting the next location of a mobile device. For QoE support, we defined a modified version of NU to handle traffic flows based on the QoE requirements. Both solutions have been validated empirically with very promising results.

References

1. EU-Mesh. Available from: http://www.eu-mesh.eu/
2. ITU-T Recommendation g.114. Technical report, International Telecommunication Union, Geneva, Switzerland, 1993.
3. Y. Amir, C. Danilov, R. Musaloiu-Elefteri, and N. Rivera. An inter-domain routing protocol for multi-homed wireless mesh networks. In *IEEE International Symposium on a World of Wireless, Mobile and Multimedia Networks, 2007 (WoWMoM 2007)*, Helsinki, Finland, pp. 1–10, June 2007.
4. A. Balachandran, G. Voelker, and P. Bahl. Wireless hotspots: Current challenges and future directions. In *Proceedings of First ACM Workshop on Wireless Mobile Applications and Services on WLAN Hotspots (WMASH'03)*, San Diego, CA, September 2003.
5. P. Baldi, L. De Nardis, and M.-G. Di Benedetto. Modeling and optimization of UWB communication networks through a flexible cost function. *IEEE Journal on Selected Areas in Communications*, 20(9): 1733–1744, December 2002.

6. P. Bellavista, A. Corradi, and C. Giannelli. Evaluating filtering strategies for decentralized handover prediction in the wireless internet. In *Proceedings of the Computers and Communications, 2006 (ISCC '06)*, Cagliari, Italy, pp. 167–174, 2006. IEEE.

7. M. Buddhikot, A. Hari, K. Singh, and S. Miller. Mobilenat: A new technique for mobility across heterogeneous address spaces. *Mobile Networks and Applications*, 10(3): 289–302, June 2005.

8. G. A. Di Caro, S. Giordano, M. Kulig, D. Lenzarini, A. Puiatti, F. Schwitter, and S. Vanini. Deployable application layer solution for seamless mobility across heterogenous networks, Ad Hoc & sensor wireless networks, 4(1–2): 1–42, May 2007.

9. S. Giordano, M. Kulig, D. Lenzarini, A. Puiatti, F. Schwitter, and S. Vanini. WiOptiMo: Optimised seamless handover. In *Proceedings of IEEE WPMC*, Aalborg, Denmark, September 2005.

10. D. Lenzarini. Method and system for seamless handover of mobile devices in heterogenous networks. US patent 7,620,015.

11. S. Giordano, D. Lenzarini, A. Puiatti, and S. Vanini. Wiswitch: Seamless handover between multi-provider networks. In *Proceedings of the 2nd Annual Conference on Wireless On demand Network Systems and Services (WONS)*, St. Moritz, Switzerland, January 2005.

12. D. Guo and X. Wang. Bayesian inference of network loss and delay characteristics with applications to TCP performance prediction. *IEEE Transactions on Signal Processing*, 51(8):2205–2218, August 2003.

13. F. P. Kelly, A. K. Maulloo, and D. K. H. Tan. Rate control for communication networks: Shadow prices, proportional fairness and stability. *SIGMOBILE Mobile Computing and Communications Review*, 49(3): 237–252, March 1998.

14. P. Kontkanen, P. Myllymaki, T. Roos, H. Tirri, K. Valtonen, and H. Wettig. Probabilistic methods for location estimation in wireless networks. In *Emerging Location Aware Broadband Wireless Ad hoc Networks*, R. Ganesh, S. Kota, K. Pahlavan, and R. Agusti. Eds., Springer Science + Business Media, Inc., Chapter 11, New York, 2004.

15. R. B. Miller. Response time in man-computer conversational transactions. In *Proceedings of the AFIPS Fall Joint Computer Conference*, vol. 33, San Francisco, CA, pp. 267–277, 1968.

16. A. W. Moore and K. Papagiannaki. Toward the accurate identification of network applications. In *Passive and Active Network Measurement*, Boston, MA, pp. 41–54, 2005.

17. V. Navda, A. Kashyap, and S. R. Das. Design and evaluation of iMesh: An infrastructure-mode wireless mesh network. In *6th IEEE International Symposium on a World of Wireless Mobile and Multimedia Networks, 2005 (WoWMoM 2005)*, Taormina, Italy, pp. 164–170, June 2005.

18. C. Perkins. IP mobility support for IPv4. RFC 3344, Internet Engineering Task Force, August 2002. Available from: http://www.ietf.org/rfc/rfc3344.txt

19. D. Puccinelli and S. Giordano. Induced fading for opportunistic communication in static sensor networks. In *10th IEEE International Symposium on a World of Wireless, Mobile and Multimedia Networks (WoWMoM'09)*, Kos, Greece, June 2009.

20. D. Puccinelli and M. Haenggi. Spatial diversity benefits by means of induced fading. In *3rd IEEE International Conference on Sensor and Ad Hoc Communications and Networks (SECON'06)*, Reston, VA, September 2006.

21. A. Raake. Short- and long-term packet loss behavior: Towards speech quality prediction for arbitrary loss distributions. *IEEE Transactions on Audio, Speech, and Language Processing*, 14(6): 1957–1968, November 2006.

22. I. Ramani and S. Savage. Syncscan: Practical fast handoff for 802.11 infrastructure networks. In *Proceedings of the 24th Annual Joint Conference of the IEEE Computer and Communications Societies (INFOCOM 2005)*, vol. 1, Miami, FL, pp. 675–684, March 2005, IEEE: Washington, DC.

23. L. Song, D. Kotz, R. Jain, and H. Xiaoning. Evaluating location predictors with extensive Wi-Fi mobility data. *SIGMOBILE Mobile Computing and Communications Review*, 2(1): 19–26, 2004.

24. UCB/LBNL/VINT. Network simulator—ns version 2.31. Available from: http://www.isi.edu/nsnam/ns

25. Widyawan, M. Klepal, and S. Beauregard. A novel back tracking particle filter for pattern matching indoor localization. In *Proceedings of the 1st ACM International Workshop on Mobile Entity Localization and Tracking in GPS-Less Environments (MELT '08)*, San Francisco, CA, 2008.

26. IEEE 802.11 working group. Part 11: Wireless LAN medium access control (MAC) and physical layer (PHY) specification/amendment 2: Higherspeed physical layer (PHY) in the 2.4 GHz band. IEEE Standard, IEEE, November 2001.

27. K. Woyach, D. Puccinelli, and M. Haenggi. Sensorless sensing in wireless networks: Implementation and measurements. In *2nd International Workshop on Wireless Network Measurement (WinMee'06)*, Boston, MA, 2006.

28. A. Yair, C. Danilov, M. Hilsdale, R. Musǎloiu-Elefteri, and N. Rivera. Fast handoff for seamless wireless mesh networks. In *Proceedings of the 4th International Conference on Mobile Systems, Applications and Services (MobiSys '06)*, Uppsala, Sweden, pp. 83–952006. ACM: New York.

2

User-Centric Convergence
in Telecom Networks

Sahin Albayrak, Fikret Sivrikaya, Ahmet Cihat Toker,
and Manzoor Ahmed Khan

CONTENTS

2.1 Introduction

The business models of telecommunication operators have traditionally been based on the concept of the so-called closed garden: they operate strictly in closed infrastructures and base their revenue-generating models on their capacity to retain a set of customers and effectively establish technological and economical barriers to prevent or discourage users from being able to utilize services and resources offered by other operators. After the initial monopoly-like era, an increasing number of (real and virtual) network operators have been observed on the market in most countries. Users benefit from the resulting competition by having a much wider spectrum of choices for more competitive prices.

On the other hand, current practices in the telecommunication business still tie the users to a single operator even though the number of players in the market has long been growing. The users tend to manually combine their subscriptions to multiple operators in order to take simultaneous advantage of their different offers that are suited for a variety of services, as illustrated in Figure 2.1. For example, a user might hold two SIM cards/phones from two distinct operators, one of which provides a flat-rate national calling plan while the other provides low cost, high-quality international calling with pay-as-you-go option. Extending this example to a case where there are a large number of operators with a multitude of service options and offers in future all-IP telecommunication networks, manual handling of such multi-operator service combinations is clearly tedious and impractical for the user.

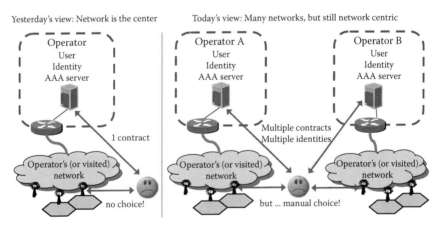

FIGURE 2.1
Past and current networking paradigm in the telecom world.

2.2 User-Centric Networking Paradigm

In its most generic sense, the *user-centric* view in telecommunications considers that the users are free from subscription to any one network operator and can instead dynamically choose the most suitable transport infrastructure from the available network providers for their terminal and application requirements [1]. In this approach, the decision of interface selection is delegated to the mobile terminal enabling end users to exploit the best available characteristics of different network technologies and network providers, with the objective of increased satisfaction. The generic term *satisfaction* can be interpreted in different ways, where a natural interpretation would be obtaining a high quality of service (QoS) for the lowest price. In order to more accurately express the user experience in telecommunications, the term QoS has been extended to include more subjective and also application-specific measures beyond traditional technical parameters, giving rise to the quality of experience (QoE) concept. We elaborate this in detail in Sections 2.2.1 and 2.2.2.

The PERIMETER project [2], funded by the European Union under the Framework Program 7 (FP7), has been investigating such user-centric networking paradigm for future telecommunication networks, where the users not only make network-selection decisions based on their local QoE evaluation but also share their *QoE evaluations* with each other for increased efficiency and accuracy in network selection, as depicted in Figure 2.2. This section provides a high-level view of a distributed QoE framework, as introduced by the PERIMETER project, for user-centric network selection and seamless mobility in future telecom networks. The focus is kept on the exploitation of QoE at a conceptual level, while keeping the technical details and implementation issues, e.g., the distributed storage of QoE reports, out of the scope of this section.

2.2.1 Quality of Experience

QoE reflects the collective effect of service performances that determines the degree of satisfaction of the end user, e.g., what the user really perceives in terms of usability, accessibility, retainability, and integrity of the service. Until now, seamless communications is mostly based on technical network QoS parameters, but a true end-user view of QoS is needed to link between QoS and QoE. While existing 3GPP or IETF specifications describe procedures for QoS negotiation, signaling, and resource reservation for multimedia applications, such as audio/video communication and multimedia messaging, support for more advanced services, involving interactive applications with diverse and interdependent media components, is not specifically addressed. Such innovative applications, likely to be offered by third-party application providers and not the operators, include collaborative virtual environments, smart home applications, and networked games. Additionally, although the QoS parameters required by multimedia applications are well known, no

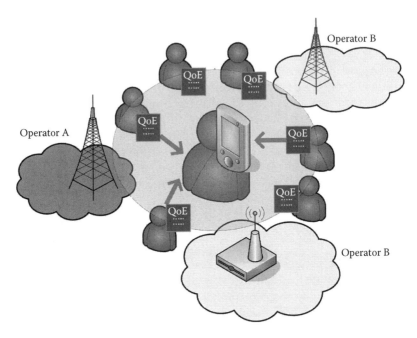

FIGURE 2.2
Future user-centric networking paradigm based on a QoE framework.

standard QoS specification is enabled to deploy the underlying mechanisms in accordance with the application QoS needs.

For future Internet to succeed and to gain wide acceptance of innovative applications and service, not only QoS objectives but also QoEs have to be met. Perceived quality problems might lead to acceptance problems, especially if money is involved. For this reason, the subjective quality perceived by the user has to be linked to the objective, measurable quality, which is expressed in application and network performance parameters resulting in QoE. Feedback between these entities is a prerequisite for covering the user's perception of quality.

There is no standard yet on evaluating and expressing QoE in a general context. However, there have been recommendation documents or publications that suggest mainly application-specific QoE metrics, objectives, and considerations. Among those, the Technical Report 126 of the DSL Forum (Digital Subscriber Line Forum) is a good source of information on QoE for three basic services composing the so-called triple play services. Regardless of the specific service context, there are some common factors that have a major influence on the user QoE:

- *End-user devices* such as an iPhone, Android G1/G2 phone, Blackberry handset, or laptop with a 3G Modem. Various device characteristics, e.g., CPU, memory, screen size, may have a significant influence on the user QoE. It is also useful for service providers to know those aspects in order to maximize QoE.

- *The application* running on the terminal is of paramount importance, determining the actual network requirements for a satisfactory QoE level.

- *Radio network* of the operator is usually the bottleneck in terms of capacity, coverage, and mobility aspects, and hence can greatly influence QoE.

- *Operator's application servers* can also have an effect on QoE. Content servers, various gateways, MMSC (Multimedia Messaging Service Center), and streaming servers are typical examples of serving entities. The connection of these servers and their amount on the network might impact QoE as well.

- *Price and billing* is one of the major factors in determining the user satisfaction level for most user groups, and therefore could be regarded as part of the QoE specification. High prices for services or billing errors can negatively influence a subscriber's QoE.

- *Network security* has also a big influence on QoE, with the major issues of data hacking attempts or malicious software. QoE can greatly drop when subscribers do not feel that the network is secure.

- *Privacy* is an increasingly common concern in today's digital society. Users would like to ensure that their identity, communications, and digital actions are well preserved from being exposed or misused by unauthorized parties. Therefore, privacy is an important aspect of QoE specification for most services.

- *Core network* components, though not visible directly to subscribers, also have a strong effect on the end-to-end service quality experienced by the user. The core network can affect subscribers' QoE by affecting connection aspects, such as latency, security, and privacy.

2.2.2 QoE Aggregation and Exploitation

This section presents a partial view on our QoE framework proposed within the PERIMETER project. The aim is not to present a complete picture on the assessment and utilization of QoE, but to set a basis for the second part of the chapter where cooperation and resource sharing among operators is investigated, with this QoE framework acting as an enabler for inter-operator mediation. Figure 2.3 depicts a local-level view of the PERIMETER middleware running on the user terminal, which is responsible for acquiring, processing, and exploiting the QoE-related information.

2.2.2.1 Data Network Processor

In order to make user-centric decisions and share user experiences based on QoE, a software entity must first evaluate and quantify QoE for a given set of inputs including the network interface and the application running

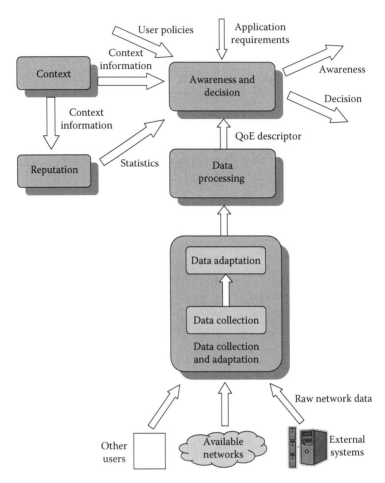

FIGURE 2.3
Evaluation and exploitation of the quality of experience data in PERIMETER.

on the user terminal. Named as the *data network processor* (DNP), this entity is responsible for calculating, from network performance measurements, user's context information, and user's feedback, a *QoE descriptor* (QoED). This QoED will be used to take a handover action based on user's policies.

The main responsibility of the DNP is generating QoED reports. Each QoED item is an aggregate and synthetic description of the quality of the user's experience. It consists of a set of key parameters that summarize the quality of service from a user's point of view:

- Mean opinion score (MOS) for different types of applications
- Cost rating
- Security rating
- Energy-saving issues

Once the QoED is calculated, it is uploaded onto a distributed *knowledge base* (KB), which is a peer-to-peer storage module running on user terminals and on the so-called *support nodes* specifically deployed by the operators with the incentive of obtaining user QoE reports more efficiently. The distributed KB of QoE reports can then be probed with a QoED query (QoEDq) in order to obtain past QoE reports of other users for decision making, which is described later in more detail. A QoEDq consists of a set of optional parameters that are used to filter network performance and user's context information stored both locally and globally. These filters apply to

- Network connection: to get performance information and QoED items associated to it.
- Application information: to get QoED items calculated for applications of the same class.
- Geographical location: to get QoED items calculated at the same area.
- User's ID: to get QoED items calculated by a certain user.

A QoEDq item may contain all or just a reduced set of parameters, allowing a wide variety of queries, for example, QoEDs associated to a certain *provider* or a certain *technology*. The calculated QoED items are mainly utilized by the decision maker (DM), which is described in Section 2.2.2.2.

The DNP may generate QoED reports in two different ways: (1) Subscription-based reports, where a certain component, which acts as a client from the DNP's point of view, subscribes to the reception of QoED reports according to a specific QoEDq. (2) Unsolicited reports, where the DNP takes the initiative and sends a QoED report to all the components that offer a receiving interface for this type of events. The unsolicited reports are triggered by events that are related to an imminent handover action due to a significant change of network conditions, for example, signal loss. In this case, the QoED specifies the network that triggered the event and the actual user's context description (location, application under use, etc.).

2.2.2.2 Decision Maker

The DM is the entity that makes use of the knowledge gathered by the DNP, user context information, and the preferences entered by the users to take allocation decisions for all the applications running on the terminal. It resides in the awareness and decision component of the PERIMETER middleware, as shown in Figure 2.3. The decisions that the DM is responsible for taking are what we call *allocation decisions*, where different applications running on the terminal are allocated to different access networks operated by different network providers. From this perspective, the atomic decision is the movement of an application from a certain point of attachment (PoA) to another. This decision is made based on local and remote QoE reports, abstracting the network and subjective user satisfaction, context reports, and user preferences.

The main purposes of the DM can be listed as follows:

- Take allocation decisions on which operator will be chosen for the applications
- Utilize local and remote QoE reports for the decisions
- Utilize context reports for the decisions
- Utilize user preferences for the decisions
- Infer the failure mode that has led to degradation in the QoE

The novel PERIMETER approach, in which users share their experiences, allows novel decision algorithms to be developed. Within this scope, the DM differentiates itself from the state-of-the-art decision mechanism in the following aspects:

- *Failure mode inference*: The DM is able to discern the cause of the problem that has led to the degradation in QoE. The degradation can be due to a problem at the application service provider side, core network side, access network side, or at the air interface. This novelty has two advantages. First, it minimizes the number of allocations that require handovers, which puts burden on network components, and degrades the QoE even more for their durations. Second, the users are not concerned with the actual cause of degradation in the QoE. They have a holistic view of the application and the service agreement. If an application is not running on an operator network properly, they will most likely blame the network operator and give a bad MOS input. Thus, there is an incentive for the operators to select decision mechanisms that are able to discern the causes of the connection problems. This information can also be used for network optimization purposes.
- *Reasoning*: The fact that users will be exchanging information about subjective measures on their applications requires a common understanding and agreement on the concepts that make up these subjective measures. This necessitates semantic information to be embedded in the stored information. Reasoning algorithms will be used for taking failure mode inference (FMI) and taking the appropriate decisions based on the inferred failure mode.
- *Distributed probing*: Thanks to the PERIMETER middleware, a distributed database of network performance data as experienced from different locations is available. This allows a practical implementation of the distributed probing of the network. This approach is used for FMI at the first stage, but it will be investigated for further utilization purposes that may benefit the network operators as well.

The DM requests a set of remote QoE reports, which are used to calculate a description map, a mathematical representation of the received reports. These description maps are compared with previously calculated description maps, called *failure profiles*, which are stored in the KB. Each failure profile reflects a specific failure in some part or multiple parts of the network. The comparison of user reports (also called samples) with the failure profiles is based on the assumption that users connected via the same access point share the same or at least parts of the route to a certain service, and thus experience similar problems accessing their service or using a specific application (Figure 2.4).

In order to deduce which part of the network is affected by impairments (e.g. congestion), those specific QoE reports must be selected that complement the view on the network. The process of selecting the most useful QoE reports and deducing possible network problems is facilitated by ontological reasoning and rule-based reasoning. The outcome of the reasoning process in the FMI component is either that a failure in a specific part of the network could be deduced (in the access network (AN), the core network (CN) or in the service domain (SD)) or the cause for impairments might remain unsolved. This is done by naïve Bayesian network type of inference. Specifically, summary statistics obtained using the QoE reports generated by different subgroups of users are compared with failure profiles in order to find the source of network failure. Following this procedure, a second inference process called *allocation decision* is started to deduce how to react to the deduced network failure. Again, remote QoE reports may be requested to provide the inference algorithm with information on network performance, this time focusing on the best allocation of applications considering the result of the previous process.

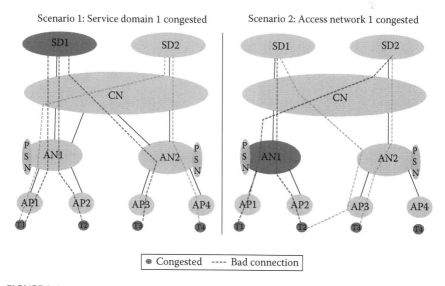

FIGURE 2.4
Different modes of failure in a multi-operator, multi-access-technology environment.

In user-centric networks, the users' freedom to switch operators in real time and the availability of a distributed KB that stores individual QoE reports will naturally have significant implications on the network operators as well. In the remainder of this chapter, we focus on this perspective and study the interaction among operators in such a setting.

2.3 Convergence from the Network Operator Perspective

Telecommunication network management practices are strongly rooted in the monopolistic telecom operators. The liberalization of the operators has only changed the landscape in a way that there were multiple closed operators rather than one closed operator. As a result, they are usually centrally managed, poorly integrated with outside components, and strictly isolated from external access. On the other hand, the IP world has been about internetworking from its conception (hence the name IP, Internetworking Protocol). Furthermore, the exposure of users to the prolific Internet services means that similar service models will have to be provided by the next generation telecom networks. The clash between these two opposite approaches poses important challenges for network operators. This is due to the fundamental risk associated with their networks turning into mere bit pipes. In order for future telecom networks to be economically viable, they should provide similar user experience with Internet services, albeit in a more managed and reliable manner.

There lies the grand challenge of the so-called Telco 2.0 operators. The operators have to offer even more data-intensive applications on their networks to make their operations profitable. This comes in a time when the increasing data traffic is starting to hurt user experience and poses itself as the biggest risk facing the operators [8].

2.3.1 Motivation

The increase in the demand for more networking resources is evident from the discussions above. There are three strategies that the network operators and broadband service providers can follow under these circumstances.

2.3.1.1 Capacity Expansion

The most direct method of combating missing capacity is investing directly into infrastructure. This has been the case for most of the operators who flagshipped the adoption of Apple's iPhone, such as AT&T in the United States. In a press release in March 2009 [9], the company announced that its investment for the state of Illinois alone was $3.3 billion. Industry analysts put the projected capital expenses of the company in the range of $18 billion and discern it as an industry-wide trend. Clearly, this is a brute-force solution to the problem and can only be extended to the point when the investment costs drive access prices beyond market prices. Even if one assumes that the market would adjust all prices accordingly, the emergence of "Greenfield operators" employing new technologies such as WiMAX, or a possible decrease in revenues due to the falling data traffic mean that this strategy is not sustainable.

2.3.1.2 Employing Untapped Networking Resources

The concept of community communication networks goes back to the mid 1990s [10]. The goal of community networks is to reduce the investment costs for the most expensive part of the end-to-end path in communication networks, the access part. The main idea is to combine access points of end users into a single access network, which is then offered to other foreign users in exchange of a fee, or to new members in exchange of access point. Early incarnations of this idea used wired connections such as cable, fiber, and twisted copper networks [11]. With the ubiquity of wireless access networks, realized by the popularity of 802.11-based wireless LANs, the idea has experienced a revival. Companies such as FON [12] are already offering commercial community networks, and free communities are burgeoning in European (Berlin [13], Rome [14], Athens [15]) and U.S. (San Francisco [16]) cities employing the 802.11 technology. The 802.16-based solutions for lower population density rural environments are also being proposed in the literature [17], which is yet to become a reality.

The essential role of the community networks from the perspective of mobile fixed convergence is the opening up of last mile wired connectivity to the wireless domain. This new untapped wireless capacity can be used by the network operators to extend their networking resource pool. In fact, the concept of operator-assisted community networks has been developed in the literature for the coexistence of community networks with wireless network operators. It has been shown [7] recently that the coexistence of a community network and a licensed operator is viable, under the condition that community network fees are below a threshold value. Such a scenario can be seen as cooperation between the wired ISP that provides the backhaul connectivity to the wireless operator via the proxy of community network.

2.3.1.3 Mutual Resource Sharing

The final strategy that the operators can follow is to establish strategic partnerships with other operators in order to (1) reduce down the investment costs or (2) make use of trunking gains in the case of asymmetric service demand profiles. The first option involves sharing varying portions of the end-to-end communication network, which we investigate further in Section 2.3.3.

However, it is worth noting that the agreements of this sort are off-line in nature that can only be reached after long legal and financial investigations by the negotiating parties.

In the second scenario, an operator gives access to the users of a cooperating operator. This scenario is only viable when the operators are not competing for the same users. For example, one operator might concentrate on rural users, who are rarely in the metropolitan area, and the other on urban users. This scenario can also be extended to the case that these operators are virtual operators who depend on the services of a third operator that provides the infrastructure. This scenario does not require long legal and financial investigations, and is more dynamic in nature. But there are technological and trust-related obstacles that need to be addressed before this can be realized.

We believe that the dynamic resource sharing between two licensed or virtual operators and cooperation between a licensed operator and a wireless community network are of similar nature, and face similar obstacles that we want to address. These main challenges are

- Lack of analytical solutions to model load balancing
- Information asymmetry and lack of transparency between different operators

The dynamic nature of the problem requires analytical solutions available to the operator networks to take "cooperate"/"do not cooperate" decisions. An analytical solution has been provided by Tonguz and Yanmaz [18] for the case of load balancing between two access networks of the same operator. However, this formulation necessitates the availability of access network internal information to both of the cooperating parties. This information transparency and symmetry is not applicable to the multi-operator environment. We formulate the problem of modeling resource sharing between access networks with multi-operator assumptions. Furthermore, we utilize the user-centric networking principle presented earlier in this chapter in order to alleviate the information asymmetry and transparency problem. Finally, we propose a game-theoretic framework to be employed in user-centric networking scenario to model the interaction between network operators.

Before presenting the developed framework, we first compare our dynamic resource-sharing proposal to the other resource-sharing approaches in the literature, namely, the network sharing and spectrum sharing. We then provide a formal problem formulation, and finally present our framework as a possible solution approach.

2.3.2 Relation to the State of the Art

The need for more effective usage of networking resources is self-evident. Achieving this by sharing resources has been approached in great detail in the scientific literature and is an industry practice.

2.3.2.1 Spectrum Sharing

We discern different levels of where the network resource sharing can be realized. In the lowest layer, spectrum sharing and cognitive radio techniques aim at intelligently sharing unused spectrum between users and operators. Ref. [23] is an excellent survey on these topics. The main difference to our proposal of sharing resources at the network layer is the fact that both cognitive radio and spectrum sharing require new radio technologies, and are not applicable in the short term.

2.3.2.2 Network Sharing

Network sharing [5,6,22] is a fairly new industry trend, where operators share varying portions of the access networks to leverage the initial investment and reduce the operation costs of the most expensive part of their networks. Depending on the level of network sharing, the resources shared between operators may involve radio spectrum, backhaul links, and even some network layer links. The main difference to our approach lies in the dynamicity of the sharing agreements.

2.3.2.3 CRRM

Current wireless telecommunications involve many different radio access technologies, which are specialized for different environments and user contexts. The development as well as the business cycles of these technologies can assure us that they will be available simultaneously for the years to come. *Common radio resource management* (CRRM) is the concept that such multiple radio access technologies (RAT) can be combined in an operator network to diversify the service offer, as well as for making use of trunking gains [3,4]. Our proposal may be seen as an extension of CRRM methods to multi-operator scenarios.

2.3.3 Problem Formulation

The problem we are addressing is the minimization or avoidance of possible degradation in user-perceived QoE in an access network as the number of users increases in an open user-centric network environment. In this section, we adhere to the ITU recommendation [19] that relates QoS values to QoE in an exponential manner in order to abstract the QoE assessment level. This relation was defined for voice services, and is being extended for more general data services in the literature [20]. The QoS value we choose is the user experiences in an access network. We therefore make the implicit assumption that the delay in the transport/core network is negligibly small in relation to the access network delay.

The method with which the avoidance or minimization is achieved is by borrowing network layer resources from an access network that belong to another operator (community, virtual or real operator). In a user-centric environment, the operators have to find additional resources, not to degrade the QoE, otherwise the users will move away to alternative operators. Therefore, the borrowing operator has the incentive to look for additional resources. By making the choice of resource sharing at the network layer, we are making our solution agnostic of the actual mechanism with which resources are shared, which can be realized by allocating explicit spectrum, serving users from the borrowing operator or by giving backhaul bandwidth.

What would be the incentives for the donor operator to lend some of its resources to the borrower? A quick answer would be that if the donor operator is underutilized at that particular point of time, then it could increase its utilization to a point where it still can serve its current users, thereby maximizing its revenues. However, the challenge of user centricity comes from the fact that users can instantaneously decide on the operators they choose. Therefore, the donor operator may choose to ignore the borrowing operator, in an attempt to drive the QoE in the borrowing network down, and gain more users. Therefore the dynamics of the resource sharing between two operators become strategy dependent, and not trivial. We aim to bring all the players, the users, the operators, and their resource allocation schemes and strategies under a single framework that makes use of queuing networks, game theory, and mechanism design.

The user-centric networking approach makes the aforementioned problem very challenging. However, it is due to this user-centric networking paradigm that this problem is manageable. A key component of the user-centric networking is the sharing of user experience through a distributed database, as explained in the first part of this chapter. We assume that this will be an open database, which the users as well as operators will be able to query. We also propose inference methods that can be used by the users and the operators to overcome the lack of inherent information transparency in the resource sharing problem we described above.

In addition to providing information transparency to the players of this complex resource sharing problem, the distributed user experience database also allows mechanism design principles to be applied to the interaction between the donor and borrowing operators. The key intuition here is the following. If an operator knows that its internal state can be inferred to a certain degree of certainty by the other operator, there is an incentive for both operators to tell the truth about the amount of resources they commit or request. This property is desirable, as with it we are able to formulate and solve the problem without requiring a neutral third party for which we have provided a solution in [21].

2.4 User-Centric Networking as an Enabler for Network Operator Cooperation

In this section, we present the queuing and game theory framework we propose to model the inter-operator resource sharing problem in a user-centric network environment.

2.4.1 Stakeholders

The players we consider for modeling the resource-sharing interaction are the end users and two operators that serve the users at that particular location. Each of these players has different concerns and incentives for participating in the resource-sharing interaction. We explore these individually.

For the initial analysis, we assume that there is a single application class that the users use. After an application session is created, the users or the agents on their mobile devices choose the operator that maximizes their QoE. We follow a network-controlled approach to mobility management, which means that users do not change their network after their session has been allocated to a certain operator. It is the responsibility of the operators to hand over the user sessions between each other in a seamless manner. The users also publish their QoE reports of the operator to an open accessible database. By publishing their data, in an anonymous form, they also get access to the data of the other users, which they utilize to take better decisions. Therefore, it can be deduced that the users have a strong incentive to publish their data, as long as anonymity is guaranteed. As noted earlier, this database can be implemented in a distributed manner via a peer-to-peer network composed of other end users. In our earlier investigation [24], we have found that the performance of such a mobile peer-to-peer network can be greatly increased by the introduction of fixed, high-capacity *support nodes*. We further assume that the network operators will invest in such nodes, in exchange of accessing the anonymous QoE reports. On the other hand, the main concerns of the users are the maximization of their QoE and service continuity.

Our intuition is that the resource sharing is viable, when there is an asymmetry in the utilization of the operators. Therefore, we discern a *donor* operator and a *borrowing* operator. The borrower needs additional networking resources in order to keep the QoE of the users it is currently serving. Therefore, it has a strong incentive to borrow resources from the donor operator. Its main concern is the continuation of the QoE of the users that would be served by the donor operator. In other words, if the donor offers to share a certain amount of resources, the borrower should trust that these resources will be available throughout the sessions of the transferred users, without degradation in their QoE levels. On the other hand, the donor operator has the incentive to lend resources in order to increase its utility. However, it is able to do this only until the additional traffic coming from the borrower starts to reduce the QoE of the users that the donor operator is currently serving. Thus, the main concern of the donor operator is the QoE of the users it is serving. Furthermore, it has to make sure that there is a utilization asymmetry between the operators; otherwise the resource sharing is counterproductive, given that the users can choose both operators.

2.4.2 Queuing Model

Figure 2.5 depicts the queuing network model used to model the interaction among operators. Queuing networks [25] are generalization of the classical single-node queues. In order to define a queuing network, one has to define the node types, the arrival process, and the internode traffic matrix, which is composed of probabilities p_{ij} representing the probability that a job leaves node i and enters node j.

In our model, we chose to use model access networks owned by different operators by a processor sharing (PS) node model. PS model was first proposed by Kleinrock in his seminal paper [27], as an idealization of round-robin style feedback queue. PS is equivalent to a round-robin service discipline, where the time that each job gets during a round is infinitesimally small. The result of this limit is the load-dependent behavior of the queue, such that it is as if each user is seeing a queue of capacity C/k when there are k jobs in a queue of capacity C. This generalization is very suitable for wireless networks, where the performance experienced by individual users degrade with the increasing number of users in the wireless network, as long as the technology is interference-limited. Telatar and Gallager [26] were the pioneers of application of PS to wireless multi-access systems, which has been used numerous times since then. Finally, we model the user decision-making process with an infinite server queue.

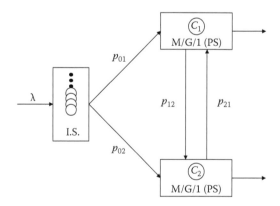

FIGURE 2.5
Queuing model for operator interaction.

The users choose the operator i with probability p_{0i}, which reflects the users decisions. Note that these probabilities are functions of the number of users in different networks, since this number affects the QoE, which is the decision criterion. The probabilities p_{12} and p_{21} are the transfer traffic probabilities. With these definitions, we are able to write down the traffic equations:

$$\lambda_1 = \lambda \cdot \frac{p_{01} + p_{02}p_{21}}{1 - p_{12}p_{21}}$$

$$\lambda_2 = \lambda \cdot \frac{p_{02} + p_{01}p_{12}}{1 - p_{12}p_{21}}$$

These represent the effective throughput that each different operator sees, which reduces to

$$\lambda_1 = \lambda \cdot p_{01}$$

$$\lambda_2 = \lambda \cdot (p_{02} + p_{01}p_{12})$$

in our scenario, when 1 is the borrowing operator and the 2 is the donor operator. This reflects the fact that the donor operator is able to increase its utility by allowing more traffic, and the borrower is able to keep its input traffic at a level where it can support the QoE demands of its current users. If we call p_{12} the *rate of borrowing* agreed between the operators and denote it by p_B, we are able to represent the additional utility the donor operator gains in terms of p_B, p_{01} and p_{02}. Since there are only two operators in this scenario, the condition $p_{01} + p_{02} = 1$ holds, and hence the donor operator can use p_B as a decision variable in the negotiation with the borrowing operator.

We have to note that we have made a simplifying assumption to come up with these basic traffic equations. The modeling logic behind queuing networks assumes that jobs leave a node after a service is completed. However, this is not the case for transfer jobs. We can deal with this by assuming that the general distribution for service times includes not only regular jobs, but also shorter length jobs that represent transfer jobs, which leave the borrowing operator after a short stint. We elaborate how we deal with this assumption as a future work in Section 2.5.

The main reason behind the choice of queuing networks and PS discipline is the product-form solutions that these type of models have. Generally, a three-node network such as ours can be described in an infinite three-dimensional state space, whose solution would require extensive numeric algorithms to run for long time. However, this is not possible in the dynamic scenario the user-centric networking necessitates. Baskett et al. have shown that the solution for the state probabilities can be expressed as the product of individual state probabilities [28]. For the simple model, we get the following solution for state probabilities $\Pi(k_1, k_2)$, which correspond to the probability that there are k_1 users in operator 1 and k_2 users in operator 2 networks:

$$\Pi(k_1, k_2) = (1 - \sigma_1) \cdot \sigma_1^{k_1} \times (1 - \sigma_2) \cdot \sigma_2^{k_2}$$

where $\sigma_i = \lambda_i(p_B)/C_i$. This result is very important, since both of the operators can calculate performance parameters such as blocking probability, throughput and delay, making use of the state probabilities, which is a function of p_B, the transfer probability. This transfer probability can be interpreted as the ratio of requests which enter the borrower operator, but leave from the donor operator. This is the negotiation variable between the operators. As we demonstrate in Section 2.4.3, the existence of a QoE database that the users and operators can access makes possible a strategy-proof negotiation mechanism possible. In such a mechanism, fooling of the negotiating partner is not beneficial. We are working on the development of such a mechanism, and therefore present not the mechanism itself, but the procedure to find it.

2.4.3 Trust Establishment in Inter-Operator Resource Sharing

We model the interaction between mobile user and network operator as a noncooperative game. This interaction consists of the users choosing one of the operators, and the operators publishing their spare capacities to the user database. The question here is whether or not it is beneficial to the operator to publish their actual spare capacities, rather than lying. We present here the game-theoretic formulation of this interaction.

Players of the game are network operators and users. Let Σ be the set of operators with elements ω_1 and ω_2. The set of strategies available to the users is to choose ω_1 or ω_2. Payoff of user depending on its strategy is tied to his perceived QoE in the chosen network. This QoE is a function of the number of users in different operator networks. Therefore, the user needs this information to maximize his utility. With the aid of QoED database, the user can infer this value with a certain confidence level.

The operators have two choices in their strategy set, i.e., to give the correct or false spare capacity information. Intuitively, when the fact that the users can infer these values is known to the operators, the truth-telling strategy dominates.

One can show this dominance by modeling the utility functions of the players appropriately. We model the payoff u_a of users with

$$u_a = \varphi e^{-\beta d} + \gamma,$$

which is the IPQX model for the QoE value associated with average delay d [20], which could be obtained utilizing the queuing model. For the operators, the payoff function is clearly the revenue maximization, which can be translated in resource utilization and is given by

$$u_w = \begin{cases} \mu\alpha & \text{if } a = \text{associated} \\ 0 & \text{otherwise} \end{cases}$$

where
  μ represents price per unit bandwidth
  α the allocated bandwidth

Since the problem is formulated as a noncooperative game, one would have to find the Nash equilibrium strategy profile, and demonstrate that this profile corresponds to a truth-telling strategy for both operators.

After proving our intuition about the truth-revealing capability of the user-centric networking, we proceed with modeling the interaction between the donor and borrowing operators.

Let the two network operators be enabled to borrow and donate their resources when needed, thus each operator at a particular time can behave as either resource borrower or resource lender. We also consider that each operator has multiple indivisible items termed as network resource, which may correspond to spectrum, throughput, or set of users. Let $\omega_b \in \Sigma$ represent the borrower operator and $\omega_d \in \Sigma$ represent the lender operator. ω_b knows the amount of resources to be borrowed in order to keep the QoE levels of its users in an acceptable range. It also has a private valuation v_b of this resource. ω_d is interested in designing a lending mechanism such that it gets the maximum additional utility. In order to achieve this, it requires private information of the borrowing operator, such as the amount of spare resources. We have already argued based on intuition that this information would be published by the borrowing operator to the users, which means that this information is not private anymore, but has become public. In a similar fashion, the amount of spare bandwidth in the donor operator is also public. Based on this public information, it is possible to design a mechanism for finding out the amount of resources to be shared and the payment for these resources. The mechanism would have to be designed to maximize a social choice function, which balances the gain of the borrower and the cost of the donor.

2.5 Summary and Future Work

The increasingly dynamic nature of the telecommunications scene is expected to go beyond the technical domain and also cover business models and socioeconomic aspects of telecommunications, eventually giving rise to the user-centric network vision presented in this chapter. There are many challenges, both technical and socioeconomic, that needs to be addressed for this vision to come true, such as the need for a standardized view of QoE among all stakeholders that should act as a common performance and valuation criterion. This chapter has focused on the exploitation of an open QoE KB for resource sharing among network operators. We have presented a queuing network model that was simplified to introduce the main ideas. Specifically, the transfer traffic has not been modeled. It remains as a future work to introduce a separate handover traffic class, and associated traffic class switching probabilities, which become the actual negotiation variable to make our model more realistic. Furthermore, we plan to introduce load dependence of the transition probabilities, which is very important to link the user decisions to the operators' sharing decisions. The idea is that the initial network selection probabilities will favor the operator that has fewer users normalized to the overall capacity. Finally, we plan to extend our queuing model to support multiple application classes.

Apart from solving the operator user game and formulating the mechanism, we also investigate the range of user distributions over the operators, for which resource sharing makes sense from instantaneous and mean utility maximization. Building up on the intuition that network sharing will be strategically viable in the case of load asymmetry, we will investigate the limits of the level of symmetry. The methodology we will follow for this purpose is the following. Depending on the user distribution between operators, each operator has two choices in their strategy profiles. They can either cooperate, or not cooperate. In the instantaneous utility maximization assumption, the operators compare the instantaneous utility gains from the two strategies, that is, they do not consider the future. In the mean utility maximization assumption, the operators consider the benefits of altruistic behavior by taking into account that the other operator might help him in the future, if they happen to be in congestion. This is an application of the well-known iterated game concept.

Acknowledgments

The authors would like to thank Sebastian Peters for his valuable contributions and the European Commission for their support through the PERIMETER project.

References

1. T.G. Kanter, Going wireless: Enabling an adaptive and extensible environment, *ACM Journal on Mobile Networks and Applications*, 8, 37–50, 2003.
2. Perimeter, European Union ICT Project #224024, www.ict-perimeter.eu

3. J. Perez-Romero, O. Sallent, R. Agusti, P. Karlsson, A. Barbaresi, L. Wang, F. Casadevall, M. Dohler, H. Gonzalez, and F. Cabral-Pinto, Common radio resource management: Functional models and implementation requirements, in *Proceedings of the 16th IEEE International Symposium on Personal, Indoor and Mobile Radio Communications*, vol. 3, Berlin, Germany, pp. 2067–2071, 2005.
4. F. Gabor, F. Anders, and L. Johan, On access selection techniques in always best connected networks, in *Proceedings of the ITC Specialist Seminar on Performance Evaluation of Wireless and Mobile Systems*, Antwerp, Belgium, August 2004.
5. C. Beckman and G. Smith, Shared networks: Making wireless communication affordable, *IEEE Wireless Communications*, 12(2), 78–85, 2005.
6. J. Hultell, K. Johansson, and J. Markendahl, Business models and resource management for shared wireless networks, in *Proceedings of the 60th IEEE Vehicular Technology Conference*, vol. 5, Los Angeles, CA, pp. 3393–3397, 2004.
7. M.H. Manshaei, J. Freudiger, M. Felegyhazi, P. Marbach, and J.-P. Hubaux, Wireless social community networks: A game-theoretic analysis, *in 2008 IEEE International Zurich Seminar on Communications*, Zurich, Switzerland, pp. 22–25, March 12–14, 2008.
8. R. Marvedis, Mobile operators threatened more by capacity shortfalls than growth of WiMAX, September 30, 2009 [Online]. Available: http://maravedis-bwa.com/Issues/5.6/Syputa_readmore.html [Accessed: October 01, 2009].
9. AT&T, AT&T investment in 2009 will add more than 40 new cell sites throughout Illinois, March 23, 2009 [Online]. Available: http://www.att.com/gen/press-room?pid=4800&cdvn=news&newsarticleid=26690 [Accessed: October 01, 2009].
10. T. Miki, Community networks as next generation local network planning concept, in *Proceedings of the 1998 International Conference on Communication Technology*, 1998 (ICCT '98), vol. 1, Beijing, China, pp. 8–12, October 22–24, 1998.
11. G. Casapulla, F. De Cindio, and O. Gentile, The Milan Civic Network experience and its roots in the town, in *Proceedings of the Second International Workshop on Community Networking, 1995 'Integrated Multimedia Services to the Home,'* Princeton, NJ, pp. 283–289, June 20–22, 1995.
12. FON Corporate Homepage [Online]. Available: http://www.fon.com/ [Accessed: October 01, 2009].
13. Berlin Freifunk Homepage [Online]. Available: http://berlin.freifunk.net/ [Accessed: October 01, 2009].
14. Ninux Rome Homepage [Online]. Available: http://www.ninux.org/ [Accessed: October 01, 2009].
15. Athens Wireless Metropolitan Network Homepage [Online]. Available: http://www.awmn.net/ [Accessed: October 01,2009].
16. Free the Net San Francisco Homepage [Online]. Available: http://sf.meraki.com/ [Accessed: October 01, 2009].
17. K. Sibanda, H.N. Muyingi, and N. Mabanza, Building wireless community networks with 802.16 standard, in *Third International Conference on Broadband Communications, Information Technology & Biomedical Applications, 2008*, Pretoria, South Africa, pp. 384–388, November 23–26, 2008.
18. O. Tonguz and E. Yanmaz, The mathematical theory of dynamic load balancing in cellular networks, *IEEE Transactions on Mobile Computing*, 7(12), 1504–1518, 2008.
19. ITU-T, Recommendation P.862: Perceptual evaluation of speech quality (PESQ), an objective method for end-to-end speech quality assessment of narrowband telephone networks and speech codecs, International Telecommunication Union, Geneva, Switzerland, 2001.

20. T. Hoßfeld, P. Tran-Gia, and M. Fiedler, Quantification of quality of experience for edge-based applications, in *20th International Teletraffic Conference on Managing Traffic Performance in Converged Networks*, Ottawa, Canada, pp. 361–373, June 17–21, 2007.
21. M.A. Khan, A.C Toker, C. Troung, F. Sivrikaya, and S. Albayrak, Cooperative game theoretic approach to integrated bandwidth sharing and allocation, in *International Conference on Game Theory for Networks, 2009 (GameNets '09)*, Istanbul, Turkey, pp. 1–9, May 13–15, 2009.
22. T.Frisanco, P. Tafertshofer, P. Lurin, and R. Ang, Infrastructure sharing for mobile network operators from a deployment and operations view, in *International Conference on Information Networking, 2008 (ICOIN 2008)*, Busan, Korea, pp. 1–5, 2008.
23. I.F. Akyildiz, W.Y. Lee, M.C. Vuran, and S. Mohanty, Next generation/dynamic spectrum access/cognitive radio wireless networks: A survey, *Computer Networking*, 50(13), 2127–2159, September 2006.
24. F. Cleary, M. Fiedler, L. Ridel, A.C. Toker, and B. Yavuz, PERIMETER: Privacy-preserving contract-less, user centric, seamless roaming for always best connected future internet, in *22nd Wireless World Research Forum Meeting*, Paris, France, May 5–7, 2009.
25. G. Bolch, S. Greiner, H. de Meer, and S. Trivedi, *Queueing Networks and Markov Chains: Modeling and Performance Evaluation*, 2nd edn. Wiley, Hoboken, NJ, 2006.
26. I. E. Telatar and R.G. Gallager, Combining queueing theory with information theory for multiaccess, *IEEE Journal on Selected Areas in Communications*, 13(6), 963–969, 1995.
27. L. Kleinrock, Time-shared systems: A theoretical treatment, *Journal of the ACM*, 14(2), 242–261, 1967.
28. F. Baskett, K. Chandy, R. Muntz, and F. Palacios, Open, closed, and mixed networks of queues with different classes of customers, *Journal of the ACM*, 22(2), 248–260, April 1975.

3

Femtocell Networks: Technologies and Applications

Eun Cheol Kim and Jin Young Kim

CONTENTS

3.1 Introduction

Recently, *femtocell* system has attracted much attention in the community of wireless communications. The femtocell (called as "access point base station") is a subminiature base station that is used in indoor environments such as home or office. It can connect mobile phone with Internet and provide wireless and wireline convergence services with inexpensive cost. Therefore, a femtocell allows service providers to extend service coverage indoors where wireless link would be limited or unavailable. In this section, we briefly describe the concept and historical background of femtocell system. Through comparison with other similar systems, its key benefits and current standardization issues are discussed.

3.1.1 History of Femtocell System

A concept of *compact self-optimizing home cell site* has been reported since 1999. Alcatel announced in March 1999 that it would bring a GSM (Global System for Mobile telecommunication) home base station to communication market. The base station would be compatible with existing standard GSM phones. The system design reused and modified cordless telephony standards (as used by digital DECT (Digital Enhanced Cordless Telecommunications) cordless phones), which was a forerunner of UMA (Unlicensed Mobile Access) standard used in dual-mode Wi-Fi/cellular solutions today. At that time, dual-mode DECT/GSM (rather than GSM/Wi-Fi) was the mainstream approach. Although demonstration units were built and proven to work through a standard POTS (Plain Old Telephone Service) phone line, the cost of equipment was too high to make it commercially viable in the market.

And then, the focus was shifted to the UMA standard where the operators such as France Telecom/Orange and T-Mobile launched in several countries. The growing range of UMA-capable handsets was expanded to include 3G (third-generation) phones in September 2008. Rather than requiring any special base station, the UMA system uses standard Wi-Fi access points with a

UMA-capable handset and a UMA network controller (UNC) to connect into the mobile operator's core network.

The first complete 3G home base station was launched in 2002. Several research groups were striving to down in cost and size of the equipment. These were termed as *picocell* (because they were installed and maintained by operators, and were aimed at large business customers) and were simply smaller and lower cost versions of larger equipment. In mid-2004, two UK-based startup companies were using 3G chipsets to develop their own 3G cellular home base stations (*Ubiquisys* and *3Way Networks*).

Since around 2005, the term *femtocell* was adopted for a standalone, self-configuring home base station. Since 2006, the femtocell products were demonstrated by several vendors. The femtocell forum was formed in 2007 and grew to represent industry players and advocate this approach. Commercial femtocell service was first launched by Sprint with their Airave CDMA (code division multiple access) offering in 2008. Softbank Japan also launched their 3G femtocell system in 2009.

3.1.2 Concept of Femtocell System

With increasing demand for higher data rates, higher speed, and higher accuracy in wireless communication systems, mobile communication standards adopting new schemes have been designed and developed so far. They include WiMAX (Worldwide Interoperability for Microwave Access) (802.16e) [1–5], 3GPP's HSDPA/HSUPA (High Speed Downlink Packet Access/High Speed Uplink Packet Access) and LTE (Long Term Evolution) standards [6–9], and 3GPP2's EV-DO (EVolution-Data Optimized (or Only)) and UMB (Ultra Mobile Broadband) standards [10–14]. In addition, the Wi-Fi mesh networks have been developed to guarantee nomadic high data rate services in a more distributed fashion [15]. In order to be competitive for home and office utilization, cellular systems will require offering services roughly comparable to those provided by Wi-Fi networks even if the Wi-Fi networks will not be able to support the identical level of mobility and coverage as the cellular standards.

A great many efforts to enhance the capacity of wireless networks have been made by reducing cell sizes and transmit distance, reusing spectrum, and enhancing spectral efficiency [16]. When we try to make this effort in micro networks, high cost is required for establishing the network infrastructure. One promising solution for this problem can be adopting a *femtocell*. The term *femto* is generally a unit of a very small quantity of 10^{-15}. In telecommunication field, the femtocell represents a small cellular access point base station or a home base station. It is a kind of short-range, low-cost, and low-power base stations which are installed by the consumer for better indoor voice and data reception. Although the term *picocell* is also utilized with the identical sense, the term *femtocell* contains a more progressive and evolutionary meaning.

It is typically developed to be used in residential or small business environments, and is installed by its user and connects to the service provider's cellular network over a broadband connection such as DSL (digital subscriber line), cable modem, or a separate RF (radio frequency) backhaul channel. It can provide voice and data services of 2G/3G/3G-beyond mobile communications (and possibly in the future, 4G and its beyond services) to the users. Without expensive cellular equipment, it allows service providers to extend service coverage inside user's home, especially where access would otherwise be limited or unavailable. Moreover, it can decrease backhaul costs since it can route user's mobile phone traffic through IP (Internet Protocol) network. A conceptual femtocell network structure is described in Figure 3.1.

While conventional approaches require dual-mode handsets to deliver both in-home and mobile services, an in-home femtocell deployment can support fixed mobile convergence with existing handsets. Compared with other techniques for increasing system capacity, such as distributed antenna systems [17] and microcell systems [18], the key advantage of femtocells is that there is very little upfront cost to the service provider. Table 3.1 provides a detailed comparison of the key traits of these three approaches [19].

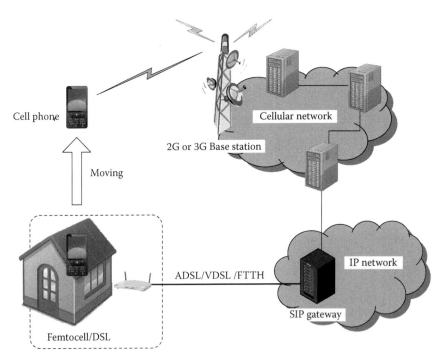

FIGURE 3.1
Femtocell network structure.

TABLE 3.1

Characteristics of Femtocells, Distributed Antennas, and Microcells

Infrastructure	Expenses	Features
Femtocell system: Consumer installed wireless data access point inside home, which backhauls data through a broadband gateway (DSL/Cabel/Ethernet/WiMAX) over the Internet to the cellular operator network.	**CAPEX:** Subsidized femtocell hardware. (CAPEX: CAPital EXpenditure) **OPEX:** (a) Providing a scalable architecture to transport data over IP; (b) upgrading femtocells to newer standards. (OPEX: OPerating EXpenditure)	**Benefits:** (a) Lower cost, better coverage, and prolonged handset battery life from shrinking cell size; (b) capacity gain from higher SINR and dedicated base station (BS) to home subscribers; (c) reduced subscriber churn. **Shortcomings:** (a) Interference from nearby macrocell and femtocell transmissions limits capacity; (b) increased strain on backhaul from data traffic may affect throughput.

(continued)

TABLE 3.1 (continued)

Characteristics of Femtocells, Distributed Antennas, and Microcells

Infrastructure	Expenses	Features
Distributed antennas: Operator installed spatially separated AE (Antenna Element) connected to a macro BS via a dedicated fiber/microwave backhaul link.	**CAPEX**: AE and backhaul installation. **OPEX**: AE maintenance and backhaul connection.	**Benefits**: (a) Better overage since user talks to nearby AE; (b) capacity gain by exploiting both macro- and micro-diversity (using multiple AEs per macrocell user). **Shortcomings**: (a) Does not solve the indoor coverage problem; (b) RF interference in the same bandwidth from nearby AEs will diminish capacity; (c) backhaul deployment costs may be considerable.

Microcells: Operator installed cell towers, which improve coverage in urban areas with poor reception.

CAPEX: Installing new cell towers.
OPEX: Electricity, site lease, and backhaul.

Benefits: (a) System capacity gain from smaller cell size; (b) complete operator control.
Shortcomings: (a) Installation and maintenance of cell towers is prohibitively expensive; (b) does not completely solve indoor coverage problem.

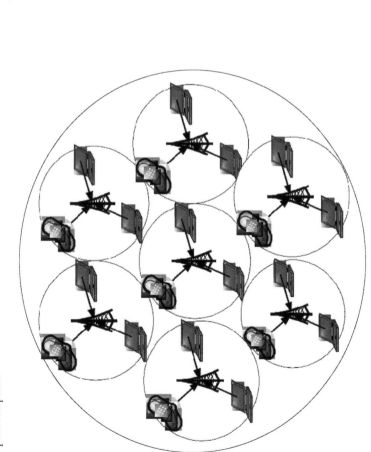

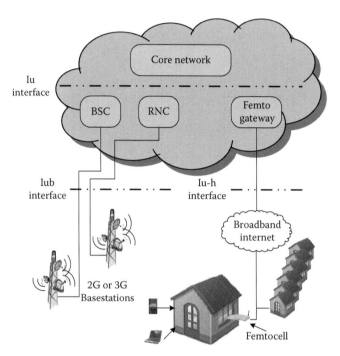

FIGURE 3.2
System architecture and context for femtocell operation.

Figure 3.2 illustrates system architecture and context for femtocell operation [20]. MSC (Mobile Switching Center) and SGSN (Serving GPRS Support Node) also communicate to the femtocell gateway in the same way as other mobile calls. All the services, including phone numbers, call diversion, voicemail, etc., operate in exactly the same way and appear the same to the end user. The connection between the femtocell and the femtocell controller uses a secure IP encryption (IPsec), which avoids interception and there is also authentication of the femtocell itself to ensure it is a valid access point.

Additional functions are also included in RNC (Radio Network Controller) processing, which would normally reside at the mobile switching center. Some femtocells also include core network element so that data sessions can be managed locally without need to flow back through the operator's switching centers. One of the essential capabilities of femtocell system is "self-configuration" (or "self-installation"). This requires considerable extra software that scans the environment to determine the available frequencies, power level, and/or scrambling codes to be used. This is a continuous process to adapt to changing radio conditions, for example, if the windows are opened in a room containing the femtocell. Within the operator's network, femtocell gateways aggregate large numbers of femtocell connections which are first securely held through high-capacity IP security firewalls.

3.1.3 Key Benefits of Femtocell System

The key benefits of femtocell systems are listed and briefly explained in terms of subscribers and operators. Through femtocell approach, the subscriber can achieve higher data rates and reliability while the operator can reduce the amount on traffic on his expensive macrocell network [19,21].

3.1.3.1 Enhanced Indoor Coverage

Femtocells can automatically give a good signal in the home setting with little impact on the wider network. Typical radiating power in the milliwatts range ensures that signals will not interfere with the wider cellular network.

3.1.3.2 Capacity Gain

Due to short transmit-receive distance, femtocells can greatly lower transmit power, prolong handset battery life, and achieve a higher signal-to-interference-plus-noise ratio (SINR). These are translated into improved reception and higher capacity. Because of the reduced interference, more users can be supported in a given area with the same region of spectrum, thus increasing the area spectral efficiency [16], or equivalently, the total number of active users per Hz per unit area.

3.1.3.3 Reduced Backhaul Traffic

Backhaul costs are one of the major operating expenses for cellular operators, along with power facilities and real estate. Femtocells allow backhaul traffic to be offloaded to the core IP network through customer-funded DSL subscriptions.

3.1.3.4 Termination Fees

Calls that are currently terminated in the home generate revenue for fixed-line operators who are competitors of mobile carriers. Femtocells will reduce this revenue stream and will reroute it to the cellular carriers.

3.1.3.5 Simplistic Handset Approach

Femtocells need no changes in handsets and no cost increments to carriers. This is a significant advantage compared with other solutions where limited handset availability is considered as a major gating factor.

3.1.3.6 Home Footprint and Quadruple Play

Many cellular operators do not have a fixed-line offering and therefore do not have any presence in the average home. Femtocells can provide a presence in the home where carriers can layer additional services such as Wi-Fi access and IPTV (Internet Protocol TeleVision).

3.1.3.7 Maximizing Spectral Investments

Carriers have already invested large amounts in 3G licenses and developing 3G networks but have yet to see financial benefits. Femtocells can offer the opportunity to both stimulate revenue and migration.

3.1.3.8 Churn Reduction

Femtocells can bring entire families or groups into a single agreement. These agreements tend to reduce churn to other operators and return higher revenue per user. The enhanced home coverage provided by femtocells will reduce motivation for home users to switch operators.

3.1.3.9 Promotion by Subsidization

Operators may substitute subsidization on handsets for subsidization on femtocells that can be easily recouped by additional subscription revenues of femtocell users.

3.1.3.10 Value-Added Services

Operators can layer value-added services on top of intelligent femtocells that can derive revenue growth.

3.1.3.11 Improved Macrocell Reliability

When the traffic originating from indoors can be absorbed into the femtocell networks over the IP backbone, the macrocell base station can redirect its resources toward providing better reception.

3.1.3.12 Cost Benefits

Femtocell deployments will reduce the OPEX (OPerating EXpenditure) and CAPEX (CAPital EXpenditure) for operators. The deployment of femtocells will reduce the need for adding macro base station towers [22,23].

3.1.4 Current Status of Femtocell Standards

Until 2006, most of the development with respect to femtocells was proprietary and there was a general lack of standardization and harmony in the femtocell market. In July 2007, the femtocell forum was formed in order to support and promote wide-scale deployments of femtocell globally. The forum could produce opportunities for driving standardization and economies of scale to lower development costs across the industry. The industry body comprised principally of equipment manufacturers, including Airvana, ip.access, NETGEAR, picoChip, RadioFrame, Tatara, Ubiquisys, and others.

In April 2009, the world's first femtocell standard [24] was officially published by 3GPP, paving the way for standardized femtocells to be produced in large volumes and enabling interoperability between different vendors' access points and femto gateways. The femtocell standard covers four main areas: (1) network architecture, (2) radio and interference aspects, (3) femtocell management, and (4) provisioning and security. The standard also uses a combination of security measures, including IKEv2 (Internet Key Exchange v2) and IPSec (IP Security) protocols, to authenticate the operator and subscriber and then guarantee the privacy of the data exchanged.

3.2 Femtocell Network Architectures

Unlike cellular or macro networks, femtocells only support several to a few tens (maximally) concurrent subscribers with an effective range, but adequate for covering a residence or small office. To provide connectivity with macro networks, operators can integrate a femtocell network into their macro cellular network with different approaches, primarily known as five different network architectures [25]. Each of architectures has its pros and cons, and most vendors typically adopt at least two different approaches for flexibility in future connections. The five approaches are described and compared in terms of many kinds of aspects.

3.2.1 Macro RNC Approach

In Figure 3.3, traffic from the femtocells is aggregated at one central location before connecting to the core network through the operator's macro RNC. Depending on the strategy of the operator and the wireless protocol, the interface between the femtocell concentrator and the RNC uses IP (Internet Protocol) or TDM (Time Division Multiplexing).

An advantage is that it can be quickly introduced with a relatively low cost while a significant disadvantage is that the existing RNC can quickly run out of capacity due to the high throughput capabilities of the femtocells. The data traffic passes through the operator's core network; therefore transport costs should be taken into account as well as scaling considerations for other network elements.

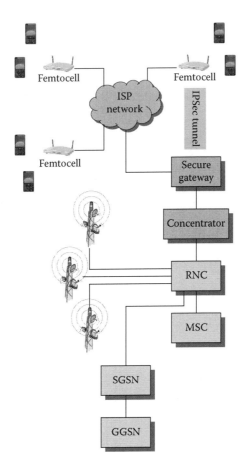

FIGURE 3.3
Network architecture of macro RNC.

3.2.2 Proprietary RNC Approach

Operators with this approach in Figure 3.4 must deploy a proprietary RNC in their networks. The proprietary RNC connects to the core network through the SGSN supporting packet switched data traffic and the MSC supporting circuit switched voice traffic. The approach is very comparable to the macro RNC approach with the difference that this RNC is exclusively used for femtocells. The main disadvantage is that it requires a new femtocell RNC from the same vendor that supplies the femtocells due to the proprietary nature of the interface between the femtocells and the RNC. Traffic from the femtocells also travels through the operator's core network, which means that the operator must scale its other core network elements to handle the additional traffic.

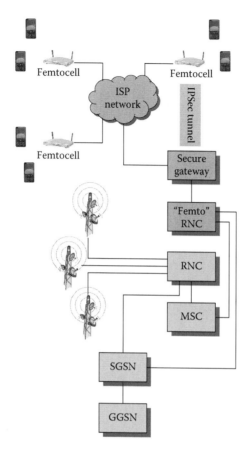

FIGURE 3.4
Network architecture of proprietary RNC.

3.2.3 Femtocell RNC Approach

This approach has a collapsed (or flat) network architecture where RNC, SGSN, GGSN (Gateway GPRS Support Node), and even MSC functionality resides within the femtocell. This architecture has a number of distinct advantages which include reduced latency, lower OPEX due to the increased dependence on IP transport, and future proofing for next-generation networks. The femtocell can be configured to support new features, such as wirelessly interconnecting multiple devices assigned to the femtocell without having to backhaul the data traffic to/from the operator's core network.

Figure 3.5 illustrates the collapsed network architecture of 3GPP2 standards. A softswitch is introduced and handles all of the circuit switched traffic from the femtocell network while a PCF (Packet Control Function)

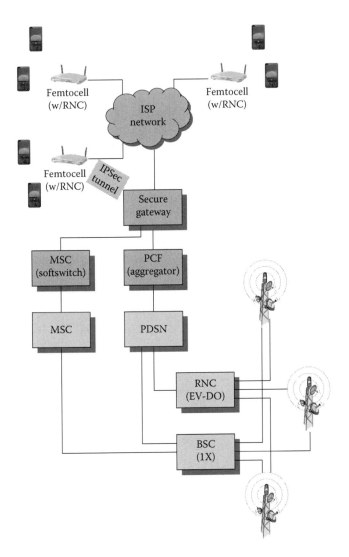

FIGURE 3.5
Network architecture of femtocell with an integrated RNC.

aggregates the 1X and EV-DO data traffic before sending it directly to the operator's PDSN (packet switched data network).

3.2.4 UNC Approach

This approach of Figure 3.6 is very comparable to UMA. The only discernable difference is that air interface between the terminal and the femtocell AP (Access Point) uses a cellular technology instead of Wi-Fi. A slightly modified UMA protocol stack resides at the femtocell while the UNC provides the connectivity between the network of femtocells and the cellular

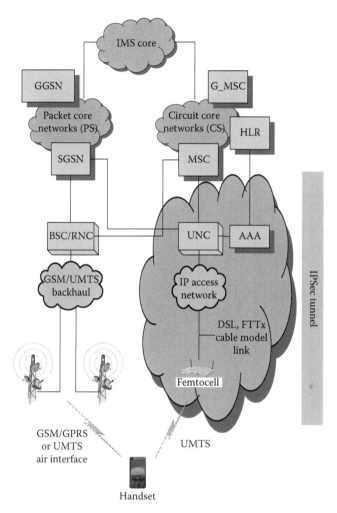

FIGURE 3.6
Network architecture of UNC.

core network. The circuit switched voice traffic is routed to the MSC while the packet switched data traffic is routed to the SGSN. The advantage is that it does not require the use of dual-mode devices, yet it is able to offer all of the benefits associated with UMA. The disadvantage is that all IP traffic must still pass through the operator's core network, deriving an impact on transport costs as well as potentially introducing congestion and higher latency.

3.2.5 IMS/VCC Approach

Among suggested approaches above, IMS (IP Multimedia Subsystems)/ VCC (Voice Call Continuity) approach illustrated in Figure 3.7 is the most

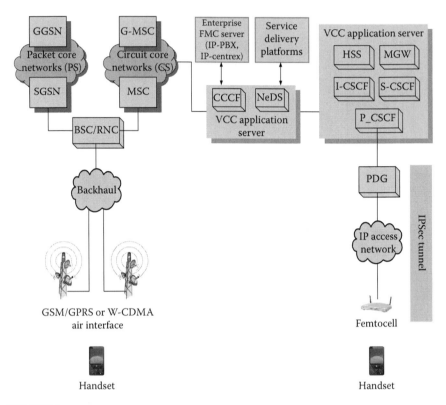

FIGURE 3.7
Network architecture of IMS/VCC.

futuristic and still being developed within 3GPP and 3GPP2. The PDG (Packet Data Gateway) is comparable to the UNC that is used in the UMA architecture. The PDG is connected to IMS architecture. The advantage of the VCC approach is that it allows IP traffic to remain outside of the operator's core network. Moreover, the use of IMS allows the operator to track, monitor, and manage the traffic.

3.3 Technical Challenges for Femtocell Systems

Despite recent great advances in wireless communication technologies, a number of potential technical challenges remain with femtocells and their deployment into carrier networks in order for femtocells to be viable option in real-world communication market [21,26]. The challenges are related to the integration of femtocells into the core network, scalability of the core network to support frequency reuse and interference issues, providing secure access to femtocell coverage, and so forth.

3.3.1 Seamless Network Integration

The primary issue is to ensure that a femtocell is interoperable with a carrier's core network. This challenge is worsened by the fact that core networks themselves are in a state of flux—some are undergoing an upgrade to packet-based platforms, while others remain circuit switched. Some carriers may have both circuit- and packet-based elements in their network as well.

The technical requirements of 2G femtocell and 3G femtocell may differ, and the elements of integration may require transitioning circuit-switched data to packet-switched one and vice versa. For example, Samsung's Ubicell is a CDMA 1x device and may only work with Samsung's core network technology, potentially limiting the place where it can be deployed. One promising solution to this problem is that some vendors can relieve this issue by partnership agreements. For example, Airvana's agreement with Siemens is to ensure the interoperability of Airvana's UMTS (Universal Mobile Telecommunications System) femtocells with Siemens' core networks.

3.3.2 Scalability

A large-scale femtocell deployment is expected to result in thousands of mini base stations that will be added to the network through a single RNC. The main issue is that the RNC is scalable to support the additional base stations. One of the solutions is migration to soft switches which may help RNCs handle the additional load. Its usefulness will ultimately depend on usage and adoption. An alternative way may be the adoption of concentrators, which will aggregate signals from individual femtocells before transmitting them to the RNC.

3.3.3 Management of Dynamic Network

A dynamic deployment of femtocells could create a dynamic network with constantly changing access points. As the femtocells are designed to be "plug and play," these access points can be connected at different locations, resulting in unplanned additions and movement of cell sites within a network. Network planning and provisioning can be a challenging issue for carriers to deploy femtocells on a broad scale. The carriers will remain the primary distribution hub of femtocell CPE (Customer Premises Equipment) in order to retain some control over their initial placement. Some carriers may distribute femtocells at a retail point of sale. Others may rent the device with monthly charge in order to retain control of network access point.

The promising solution is to employ intelligence into the femtocell to support auto-configuration functionality, given the inherent portability of the devices. For efficient network management of the femtocell, the following questions listed below should be resolved:

1. Prevention of the femtocell from interfering with the macro network
2. Making hand-in and hand-off
3. Integration of the femtocell solution with the core network
4. Requirements of control and management installations
5. Interface of the femtocell offering with CRM (Customer Relationship Management) and billing systems
6. Fault detection and network optimization

3.3.4 Interference from Frequency Reuse

The use of femtocells may cause interference with the macro cell site due to frequency reuse. In addition, there may also be interference from more than one femtocell being placed within the vicinity of each other. One of the possible solutions to overcome this issue would be for carriers to use cheaper sub-prime frequency bands for femtocells known as *guard bands*. The other source of interference comes from other electronic devices within buildings.

3.3.5 Seamless Handover and Secure Access with Macro Cells

From the very beginning of the femtocell suggestion, the primary question has been whether femtocells can operate and coexist with the macro cell network. This involves both challenges at the radio access levels and issues at higher levels such as security, authentication, and billing. It can be noted that trials conducted by the Femto Forum suggest that deployment of femtocells may not affect the existing macro system by providing measured data on indoor interference, external leakage, impact on macro network, and dropped call figures at base stations and in the macro network.

3.3.6 Cost

Carriers are placing the most significant importance on the price aspect. They are actively looking for compromise on the feature sets for femtocell products in order to meet the price points. A few years ago, the introductory price line for a femtocell AP was a few hundred dollars while today it is getting down to around $100 as the market nears initial rollout. These aggressive price demands associated with the femtocell may place significant strain on the supply manufacturers. This stringent cost pressure is mainly dependent on the semiconductor (chip) industry.

It is likely that chip vendors may not gain a profit from the femtocell products until some level of economy of scale is attained. Without widespread investment and innovative production of chip vendors, the femtocell AP around $100 may not be found in the near future.

3.3.7 Lack of Precedent

Until the first launching of the femtocell products, there was little evidence that fixed mobile convergence (FMC)-based services such as the femtocell would be worthy of the development efforts. There has not been much in the way of successful precedents regarding comparable solutions. This fact makes the femtocell market a high-risk, high-return situation to wireless vendors. Some large vendors may hesitate to invest to take technological lead in the femtocell industry. For example, it may cost a few tens of millions of dollars to develop a femtocell reference design.

3.3.8 Frequency Reallocation

Frequency reallocation can be considered as one of the crucial issues in deploying the femtocell systems. This issue seems to be somewhat peripheral, but sometimes critical to the operators. For example, especially in European countries, 900 MHz spectrum is used for UMTS and has far better in-building coverage. This fact may shrink the operators from adopting the femtocell systems.

3.3.9 Cellular Modality

One of the most keenly debated arguments is the choice of air interface technology. Many kinds of air interfaces are currently under consideration and they have significant impacts on the femtocell technology and market generation. The comparative features of many air interfaces are discussed below.

3.3.9.1 2G vs. 3G Mobile Communication Systems

As summarized in Table 3.2, the air interfaces of GSM (2G) and WCDMA (Wideband Code Division Multiple Access) (3G) femtocell systems have merits and demerits in cost and implementation features supporting service generation. In a current stage, it is widely accepted that one of them would lead the air interface standard for the femtocell market. As timeline comes

TABLE 3.2

Comparison Between GSM and WCDMA Systems

	Advantages	Disadvantages
GSM (2G)	Handset silicon reuse (cost)	Pure voice service
	Frequency planning	Spectral efficiency
	Installed subscriber base station	Future proofing
WCDMA (3G)	Multimedia data support	Cost
	Innovative service flexibility	Interference
	Maximizing spectrum investment	

near the market, 2G GSM-based femtocells or 3G WCDMA-based femtocells will encompass the majority of the market. There is a possibility that the market will be segmented at an early stage of femtocell deployment. In other words, the operators may employ GSM solutions for the cost-effective parts of their subscriber base station and adopt WCDMA for the premium parts accordingly.

3.3.9.2 HSDPA, HSUPA, and HSPA+ Systems

The operators are already taking migration to more advanced cellular systems such as HSDPA and HSUPA into account. For upgradation process, the operator will use future-proof devices by allowing for over-the-air (or IP network) software upgrade. Currently, HSPA+ upgrade is unclear because it should encompass MIMO (Multiple Input Multiple Output) and spatial diversity which are inherent in HSPA+ system. The motivation for migration to HSPA series comes from the fact that greater bandwidth (especially symmetrical bandwidth) is a critical part in supporting newly generated services to handsets.

3.3.9.3 Mobile WiMAX, UMB, and 3GPP-LTE Femtocells

The mobile WiMAX, UMB, and 3GPP-LTE systems are widely identified as the strong candidates for 4G technologies. They are all based on OFDMA (Orthogonal Frequency Division Multiple Access) and encompass MIMO solutions. In case of UMB and 3GPP-LTE femtocell systems, there is not any finalized specification yet. In the future femtocell market, they will be discussed as an upgrade version of current networks. However, in case of mobile WiMAX femtocell system, some operators are already in the process of rolling out or in the final planning stages for wireless networks. In a current situation, it is more likely that there would be a time lag between the mobile WiMAX and UMB/3GPP-LTE systems to enter the market.

3.3.9.4 CDMA

Some operators have indicated interest in CDMA-based femtocell systems. They include Verizon Wireless, KDDI in Japan, Sprint, etc. Also, some manufacturers emerged to produce cost-optimized CDMA femtocell chip. Limited trials for the CDMA femtocell system have been done, and they have been reported to successfully operate in the field trial.

3.3.9.5 Multimodality

One of the most critical points for the operators is how and when to achieve economy of scale in the femtocell market by addressing the widest possible base stations of potential subscribers. Correspondingly, a possible and promising solution would be a flexible one with multimodality functionality. The multimodality should support the incorporation of new services into their networks in order to cater to their current subscriber base station, and also allow them the chance to migrate to advanced services through technology upgrades. A recent example of multimodality includes a system supporting both WiMAX and CDMA systems or dual-mode CDMA and WCDMA systems. Currently, the primary interest is directed toward multimodal products between GSM and WCDMA systems. This multimodal approach offers many kinds of benefits while it has a demerit that it takes a lot of costly products compared with single-mode devices.

3.3.9.6 Feature Sets

The requirements of feature sets are still changing to match feature requirements with service goal within a constraint on cost. The product features of each set are discussed below.

3.3.9.6.1 Standalone System

The standalone system can be viewed as the most common approach where it simply incorporates the cellular radio aspects of the femtocell system, especially in the WCDMA case. The major motivation is to satisfy the price line challenged by operators. However, it retains the following possibilities: (1) potential for service disruption from outside of the femtocell, (2) reduction in plug-and-play capability, (3) support overheads, (4) service limitations, and (5) home clutter. In this approach, balancing between cost benefits and service requirements should be considered for smooth operation of the femtocell system.

3.3.9.6.2 ADSL Gateway

In order to enable feature set, ADSL (asymmetric digital subscriber line) gateway can be added so that the signal can be effectively routed through the equipment while preventing the set from supporting services generated by the third party. This set can facilitate much smoother plug-and-play offering, but still arouse cost issue in the development phase.

3.3.9.6.3 Wi-Fi Access Point

The installation of Wi-Fi access point is typically considered as less important thing rather than ADSL gateway or router, especially when the cost constraints are crucial. However, it can reduce the number of boxes needed in the home and can make additional services to be layered on nonhandset devices such as content distribution to devices outside the handset.

3.3.9.6.4 TV Set-Top Box

The inclusion of TV set-top box in the femtocell seems to be more innovative and forward looking. This feature can support quadruple play and attractive IPTV services. The success of this feature highly depends on the successful employment of IPTV as well as successful deployment of the femtocell products. The realization of this feature may be many years away from now.

3.3.9.6.5 *Video Distribution Mechanism*

Video distribution mechanism can bridge the gap between the femtocell set-top box and the point of viewing. This feature can control program menu and interface with IPTV. The 802.11n or UWB (UltraWideBand) technologies may play an important role in supporting this feature.

3.3.9.6.6 *Segmentation*

The widely recognized feature of the femtocell market is that it will be segmented in an early stage to tailor service requirements to meet cost constraints. It is expected that a single operator will provide different types of femtocells where the price is varying in two to five steps. Soon after the introduction of the femtocells, the segmentation is expected to rapidly occur no later than 2010. The effective segmentation of the femtocell products will be enabled by modular design for functionality implementation from the femtocell vendors.

3.4 Conclusions and Future Opportunities

Although some technical challenges exist in the way of the success of femtocell deployments, a great many opportunities also remain. Through femtocell systems, better communication quality can be maintained both in the indoor and outdoor environments. In the indoor environment, a personal base station offers higher signal power in the downlink, which incurs higher throughputs. In the uplink, transmission power can be reduced. Hence, battery power can be saved so that the lifetime of handsets is expanded, which is still a critical challenging issue for 3G (or 3G beyond) handsets with greedy multimedia processing. In addition, since all the users connect their links to their personal femtocells when they come home, the macrocell load can be lightened. This possibility has an implication that the macrocells can offer their remaining capacity to outdoor users with much better communication conditions.

In the indoor environment, the femtocell has advantageous features over Wi-Fi systems where both quality of voice service and offered data throughput significantly decrease with the number of users, although the IEEE 802.11 series standard is applied [27]. However, the femtocell system can be viewed as a natural technology for voice as it depends on cellular network standards designed for voice communications.

Viewed from an economic point, it is no longer necessary to exchange the conventional handset with a dual mode one in order to continue communications in both indoor and outdoor environments. This fact indicates a substantial cost saving as it spares the user the purchase of a new device with additional expense. Besides, some incentive measures to foster the technology by the operator could be given, for example, totally free calls when initiated from the femtocell. We need to keep in mind that the user will probably buy the femtocell. Then, it is necessary to find good reasons to encourage him or her to do so. Thus, the user will probably enjoy a slimmed bill.

From the cellular operator's standpoint, the benefits through femtocell systems become clearer. From the femtocell services, the users can maintain their communication services via the cellular network even at home. Moreover, the macrocell load can be lightened by adopting the femtocell systems. This implies a great amount of savings in an expensive infrastructure deployed in the cellular networks. If the femtocells are properly utilized, the operator will not need to deploy more macrocells in order to service more users or to provide higher throughput to its existing users. The operation and maintenance costs will also probably decrease as even the density of deployment could be reduced.

In conclusion, the femtocell system suggests a great many challenges and opportunities in the future direction of wireless communication services. For the future communication service environments, ubiquitous nature of services would be preferred by the users. The femtocell will surely play an important role both in the indoor and outdoor communication services.

Acknowledgment

This work has been, in part, supported by the MKE (Ministry of Knowledge Economy), Korea, under the ITRC (Information Technology Research Center) support program supervised by the IITA (Institute for Information Technology Advancement), and in part by Kwangwoon University in 2009 (IITA-2009-C1090-0902-0005).

References

1. S. J. Vaughan-Nichols, Achieving wireless broadband with WiMAX, *IEEE Comput.*, 37(6), 10–13, June 2004.
2. L. Kejie, Q. Yi, and C. Hsiao-Haw, A secure and service-oriented network control framework for WiMAX Networks, *IEEE Commun. Mag.*, 45(5), 124–130, May 2007.
3. A. Ansari, S. Dutta, and M. Tseytlin, S-WiMAX: Adaptation of IEEE 802.16e for mobile satellite services, *IEEE Commun. Mag.*, 47(6), 150–155, June 2009.

4. W. Fan, A. Ghosh, C. Sankaran, P. Fleming, F. Hsieh, and S. Benes, Mobile WiMAX systems: Performance and evolution, *IEEE Commun. Mag.*, 46(10), 41–49, Oct. 2008.
5. L. Kejie, Q. Yi, C. Hsiao-Hwa, and F. Shengli, WiMAX networks: From access to service platform, *IEEE Netw.*, 22(3), 38–45, May 2008.
6. C. Kappler, P. Poyhonen, N. Johnsson, and S. Schmid, Dynamic network composition for beyond 3G networks: A 3GPP viewpoint, *IEEE Netw.*, 21(1), 47–52, Feb. 2007.
7. H. N. Man, L. Shen-De, J. Li, and S. Tatesh, Coexistence studies for 3GPP LTE with other mobile systems, *IEEE Commun. Mag.*, 47(4), 60–65, Apr. 2009.
8. T. Ylingming, Z. Guodong, D. Grieco, and F. Ozluturk, Cell search in 3GPP long term evolution systems, *IEEE Veh. Technol. Mag.*, 2(2), 23–29, June 2007.
9. A. Larmo, M. Lindstrom, M. Meyer, G. Pelletier, J. Torsner, and H. Wiemann, The LTE link-layer design, *IEEE Commun. Mag.*, 47(4), 52–59, Apr. 2009.
10. G. Patel and S. Dennett, The 3GPP and 3GPP2 movements toward and an all-IP mobile network, *IEEE Personal Wireless Commun.*, 7(4), 62–64, Aug. 2000.
11. P. Agrawal, Y. Jui-Hung, C. Jyh-Cheng, and Z. Tao, IP multimedia subsystems in 3GPP and 3GPP2: Overview and scalability issues, *IEEE Commun. Mag.*, 46(1), 138–145, Jan. 2008.
12. J. Gozalvez, Ultra mobile broadband, *IEEE Mag. Veh. Technol.*, 2(1), 51–55, Mar. 2007.
13. R. Rezahfar, P. Agashe, and P. Bender, Macro-mobility management in EVDO, *IEEE Commun. Mag.*, 44(2), 65–72, Feb. 2008.
14. P. Feder, R. Isukapalli, and S. Mizikovsky, WiMAX-EVDO interworking using mobile IP, *IEEE Commun. Mag.*, 47(6), 122–131, June 2009.
15. Tropos Networks, Picocell mesh: Bringing low-cost coverage, capacity and symmetry to mobile WiMAX, White Paper, Tropos Networks, Sunnyvale, CA, March 2007.
16. M. S. Alouini and A. J. Goldsmith, Area spectral efficiency of cellular mobile radio systems, *IEEE Trans. Veh. Technol.*, 48(4), 1047–1066, July 1999.
17. A. Saleh, A. Rustako, and R. Roman, Distributed antennas for indoor radio communications, *IEEE Trans. Commun.*, 35(12), 1245–1251, Dec. 1987.
18. I. Chih-Lin, L. J. Greenstein, and R. D. Gitlin, A microcell/macrocell cellular architecture for low- and high- mobility wireless users, *IEEE J. Select. Areas Commun.*, 11(6), pp. 885–891, Aug. 1993.
19. V. Chandrasekhar, J. Andrews, and A. Gatherer, Femtocell networks: A survey, *IEEE Commun. Mag.*, 46(9), 59–67, Sep. 2008.
20. ThinkFemtocell Home Page, available at http://www.thinkfemtocell.com
21. ABI Research Report, Femtocell market challenges and opportunities: Cellular-based fixed mobile convergence for consumers, SMEs, and enterprises, July 2007.
22. Second International Conference on Home Access Points and Femtocells, Dallas, TX. Available at http://www.avrenevents.com/dallasfemto2007/purchase_presentations.htm
23. Analysys Mason Home Page, available at http://research.analysys.com
24. 3GPP Home Page, available at http://www.3gpp.org

25. M. W. Thelander, Femtocells-who says size doesn't matter? *Signals Ahead*, 4(9), 1–20, May 2007.

26. J. Kvaal, A. Rozwadowski, S. Jeffrey, T. O. Seitz, N. Swatland, A. M. Gardiner, and A. Ahuja, Femtocells, *Lehman Brothers, Global Equity Res.*, pp. 1–30, Oct. 2007.

27. Y. Haddad and G. L. Grand, Throughput analysis of the IEEE 802.11e EDCA on a noisy channel in unsaturated mode, in *Proceedings of the Third ACM Workshop on Wireless Multimedia Networking and Performance Modeling*, Crete Island, Greece, pp. 78–71, Oct. 2007.

4

Fixed Mobile Convergence Based on 3G Femtocell Deployments

Alfonso Fernández-Durán, Mariano Molina-García, and José I. Alonso

CONTENTS

4.1 Femtocells in Fixed Mobile Convergence Scenarios

The concept of fixed mobile convergence (FMC) consists of extending all or a part of the services provided by the wireless telecom service provider's core network to domestic and small and medium enterprise subscribers through the public Internet Protocol (IP) network, taking advantage of the proliferation of wireless local and personal area networks deployments in these scenarios. This convergence will require subscribers to be able to switch an active voice or data session between fixed wireless and mobile networks, ensuring a seamless network transition.

In this FMC context, the deployment of third-generation (3G) femtocells in residential and small enterprise scenarios has been adopted as one of the competing technologies to provide this convergence. A base station router (BSR) femto node is a low-cost and low-power Universal Mobile Telecommunications System (UMTS) Node B, mainly conceived for domestic use, that admits a limited number of simultaneous communications, and which is connected with the mobile core network through the user's digital subscriber line (DSL).

3G femtocells will provide solutions to different problems that are facing 3G networks. First of all, indoor coverage will be easily increased. UMTS indoor coverage is much more difficult to achieve than second-generation (2G) coverage, because UMTS typically operates at higher frequencies, making it more difficult to penetrate building structures. In addition to this, UMTS networks are expected to provide services with high throughput requirements, which will need higher signal levels, and therefore indoor solutions will be compulsory. Secondly, femtocells will have associated a better performance in terms of data rate. Unlike macrocells that support hundreds of users, femtocells will support fewer simultaneously active users, and therefore High Speed Packet Access (HSPA) connections will be able to deliver higher data rates per user than in the macro cellular environment. With higher data rates and fewer users, the quality of service requirements demanded in 3G networks within home and office scenarios are expected to be easily fulfilled, enhancing user satisfaction rates. This fact will lead to small femtocells that will be able to deliver a better voice and multimedia quality of experience (QoE). In addition to the obvious voice quality gains attributed to better coverage, femtocells will enable support for a new generation of higher rate voice codecs that leverage fewer users per access point and the proximity of the handset to the femtocell. Finally, operators will get other important business benefits due to the fact that femtocells will relieve the macro network from the indoor traffic that uses a substantial part of the mobile network resources, increasing the overall network capacity and reducing the cost of backhauling traffic to the operator's core network.

The femtocell deployment scenarios can be classified in different ways, depending on an architecture-centered or a user-centered vision. Using an architecture-centered approach, femtocell deployment scenarios can be classified taking into account the relationship between femtocells and macrocells within the architecture of the 3G network. Initially, femtocells were considered as a means to offload traffic from the existing congested mobile network and convey it through the data network. There are two types of deployments in such scenarios, depending on spectrum availability. If sufficient spectrum is available, the preferred solution is to devote a separate channel to femtocell deployments to minimize the macrocell–femtocell interaction. Since spectrum is a scarce resource, in many cases femtocell deployments have to share spectrum with the macrocell network. These conditions have been previously analyzed, as multitier and hotspot networks in [2,14,16,20] as an intermediate step toward the femtocell concept. The femtocell concept is studied in [4,5,11]. Performance improvements for co-channel deployments are also shown in [3,6,13].

The other type of deployment is focused on the extension of the cellular network in areas where user data networks are available and where macrocell base station deployment presents limitations. The dimensioning process of deployments in typical scenarios has not been analyzed in depth, since under these conditions the dominant effect is probably the femtocell–femtocell interaction. The interaction of femtocells is different from that of micro base stations or macro base stations, since it is assumed there is little coordination among cells at the radio interface. This chapter presents a simple dimensioning model for these types of deployments which can be used to estimate the number of simultaneous users that are acceptable for a given grade of service (GoS) while taking into account the interaction among independent femtocells. The model is validated by means of simulation, and additional results are obtained to assess the effect of femtocell positions in a massive deployment and the use of different frequency bands.

On the other hand, when selecting a user-centered approach, several types of deployments can be found, and some are selected to investigate the performance. When a BSR femtocell is deployed in a small business or office to provide better coverage, the model is called *open access*, and a controlled network is created with respect to radio interference and handover between BSRs. Coexistence between this type of network and the macrocell network seems consistent, as shown in [5]. In contrast, in residential scenarios, closed subscriber group (CSG) is the typical configuration used. BSR femtocell access is limited to the subscriber identity module (SIM) cards accepted by the node owner under the operator's supervision while any external SIM card is rejected. This aspect introduces a particular interference scenario: a user can be located nearer to a neighbor's node than to his own node, but there is no option to connect to the neighbor's node. In such cases, the neighbor's interference can prevent the user from connecting to his own node in certain areas of his house. This scenario is inconceivable in a macrocell environment since users are always able to connect to the node with the best power reception.

CSG introduces a significant amount of uncertainty in such deployments because it is UMTS Node B for residential use that allows a limited number of simultaneous communications. It connects with the mobile core network through the DSL at the user's home. An overview of this architecture can be found in [5]. The CSG presents another limitation, which is macrocell coverage at the user's home for SIMs not belonging to the CSG. In this case, the BSR femto node may cause interference with the macrocell signal and create a coverage gap for non-CSG users. All radio parameters must be adapted to ensure a grade of coverage robust enough for the user's entire home. Another critical aspect of BSR femtocell nodes is that they are deployed by users—no configuration is performed by the operator, and the whole installation process is automatic. The network should also be adapted to support potentially frequent and random BSR femtocell powering on/off.

The chapter is structured as follows: we begin with an introduction of the different planning and dimensioning procedures for femtocell deployments with Wideband Code Division Multiple Access (WCDMA) R99 and HSPA users. Then, basic radio considerations are explained, and a dimensioning model based on the GoS is introduced. Finally, basic simulator flow is explained, and the results obtained for the femtocell scenario simulations are presented.

4.2 3G Femtocell Planning and Dimensioning Procedures

4.2.1 R99 Planning and Dimensioning General Concepts

In the planning process of CDMA-based system, is capital to take into account that these mobile systems are interference-limited, not coverage-limited, as other mobile systems are. This implies that the maximum number of simultaneous connections to access points (APs) is limited by the maximum allowed interference level. Each connection employs a single channel that is ideally orthogonal to the other connection channels, but, however, there are orthogonally losses that are the reason of the interference-based degradation in these systems.

The interference limitation affects both the uplink and downlink in different ways. In the downlink, where all the base stations' transmitted channels are synchronous, it is possible to make these channels orthogonal while the pseudorandom character of different base stations transmissions is kept. For this purpose, two layers or sequence level codes are employed. In the first layer, named channelization, the base station spreads the signal from each transmission channel using a family of orthogonal sequences. These signals are added after the application of a gain factor, different for each channel spread signal. The final compound signal is randomized by the multiplication by a unique pseudorandom noise (PN) sequence which is typical of each base station. This second stage of randomization assures that the transmissions of different base stations are entailed as pseudorandom signals. This type of system is limited by the interference of adjacent cells while it is orthogonal in its own cell, that is, any channel suffers from interference provoked by the other channels that are transmitted by the same base station. However, in the uplink the improvement obtained from the orthogonalization process is significantly lower due to the fact that this orthogonalization is only applicable to the different channels transmitted by the same mobile terminal.

Because of the interference limitation of this type of systems, power control processes have a strong influence on the network capacity. Therefore, to assess the performance of these kind of networks, it will be crucial to model properly the user association to each base station and the allocation of transmission powers in uplink and downlink cases for mobile terminals and base stations, respectively. These two actions performed by the system network have the objective of fulfilling the required users' QoS levels. If these quality requirements cannot be achieved for a determined network's configuration, the system is considered in degradation situation.

The main constraint for fulfilling the users' QoS requirements in a specific service type is the user capacity to reach the targeted signal to interference ratio (SIR), signal to self interference ratio (SSIR), E_b/N_0, or E_c/N_0 levels. These metrics can be related to each other through the following expression:

$$\text{SIR} = \frac{E_b}{N_0} \times \frac{R_k}{W} = \frac{E_c}{N_0} \times \frac{W}{W} = \frac{\text{SSIR}}{1 - \text{SSIR}} \tag{4.1}$$

For a specific user with a specific service, its SIR value would be able to be calculated as (4.2)

$$\text{SIR} = \frac{P_{\text{RX}}}{I_{\text{int}} + I_{\text{ext}} + N} \tag{4.2}$$

Using this expression, the power that must be transmitted by the femtocell to fulfill SIR requirements in the downlink and by each terminal in the uplink can be assessed. In this chapter, a simplified analysis of the system is presented in order to reduce the necessary computation time for power allocation estimations that will be used in optimization processes. Thus, this simplification implies that in order to calculate transmitted powers and interferences, some factors which could be taken into account in a more exhaustive study will not be evaluated. Among them, effects associated to the power control in closed loop, as the increase of attenuation produced by the correlation among the control in closed loop and the fast variations of the propagation attenuation, produced by the multipath or the power margin, or the effective reduction of the transmitted power limit derived from the use of the closed loop. In the same way, it will be meant that the mobiles are not in transfer situation, and therefore, they are associated to a single base station. So, continuity gain due to the reduction of the objective SIR values with regard to the values for isolated base stations will not be considered.

With these simplifications, the SSIR associated to a user and a service which is received in a base station will be for uplink (4.3)

$$\text{SSIR}(\phi(k),k) = \frac{\alpha(\phi(k),k)P(k)}{\sum_{j=1}^{k} v(j) \times \alpha(\phi(k),j)P(j) + N(k)} \tag{4.3}$$

where

$P(j)$ is the power transmitted by user j
$\phi(k)$ refers to the femtocell that serves user k
$v(j)$ is the activity factor for the service of user j
$\alpha(\phi(k),j)$ is the propagation gain between user j and femtocell $\phi(k)$
$N(k)$ is the noise power
$P(k)$ is the power transmitted by user k
$\alpha(\phi(k),k)$ is the attenuation between user k and the base station $\phi(k)$ the user is associated to

For downlink evaluation, the SSIR will be calculated in a specific mobile terminal as (4.4)

$$\text{SSIR}(m,k) = \frac{\alpha(m,k)P(m,k)}{\sum_{n=1}^{M} \alpha(n,k)\theta(n,k)T(n) + N(k)} \tag{4.4}$$

where

$\alpha(m,k)$ is the attenuation between base station m and user k
$P(m,k)$ is the power transmitted from femtocell m to user k
$\alpha(n,k)$ is the propagation gain between femtocell n and user k
$\theta(n,k)$ is the orthogonality factor between femtocell n and user k, calculated in (4.5)
$T(n)$ is the total power transmitted by femtocell n, and $N(k)$ the noise power

$$\theta(n,k) = \begin{cases} 1 & m = n \\ \rho & m \neq n \end{cases} \tag{4.5}$$

4.2.2 R99 Power Allocation Assessment: Simulation Procedures

The principles which constraint the capacity in CDMA systems are associated to the power limitations, both at the individual mobile stations in the uplink and at the base station in the downlink. As it has been indicated previously, in the uplink a specific value of signal to interference relationship must be maintained at the base station. Thus, if there is a certain number of mobiles placed under the coverage of a cell that provokes a specific level of interference, and the mobiles' location is modified or mobiles go into or out of the coverage area, the interference level can increase or decrease, so the power control mechanism will order the mobiles to change its power level to maintain the SIR value. This procedure will try to ensure that the transmitted power in every link is minimum, since the transmitted power by a mobile terminal influences the global interference and, therefore, the power that the rest of mobiles will have to transmit. If the maximum power that a mobile terminal can transmit should be exceeded to fulfill its quality requirements, the network performance will be degraded because of the impossibility of reaching its objective SIR value.

In the downlink, the power transmitted by the base station is distributed among both the traffic channels associated to the mobiles and the signaling and control channels. Thus, the power control mechanisms will also watch over that the power assigned to the traffic channel of a specific mobile is the strictly necessary one. Besides that, the interference level a mobile suffers in the downlink comes from the own base station as well as from the adjacent ones, so a deficient power control increases the global interference and degrades the overall network operation.

Strict power control in the uplink and downlink as well as the power limitations associated to the maximum transmission power of the mobiles and to the limited and shared power of the base station will determine the simulation procedures of the CDMA network's operation. Therefore, in static simulations of the system a great number of these snapshots will be generated, in a random and independent way, specifying in each one the users' situation (activity of each one in that moment, fadings, etc.). Each one will be analyzed separately obtaining statistics related to the network operation. For that, in each of the time instants simulated, and for a system configuration and femtocell location, a set of transmission powers for the mobiles (uplink) or bases (downlink) will be determined, trying to fulfill the quality objectives for all the users.

4.2.2.1 R99 Uplink Power Allocation Procedure

In the uplink, the transmitted power by a mobile influences the global interference and, therefore, the power levels required for the rest of the mobile terminals to reach a determined E_b/N_0 level. The estimation of the power levels transmitted by every mobile in a time instant will have to be calculated through an iterative process. The iterative algorithm is based on the following idea: Every mobile calculates, using the interference level obtained in the previous iteration, the transmitted power that is required to fulfill E_b/N_0 requirement for the provided service. Once the powers have been calculated, it is checked if these powers are within the mobile power limits, and then the algorithm goes on the following iteration. The algorithm finishes when the power transmitted by the different terminals does not differ on more than a selected threshold. Procedure used for power allocation to each R99 user in the uplink is presented in Figure 4.1. First of all, and taking into account the locations selected of the mobile terminal, each of them will be associated to a base station, following the minimum attenuation criterion. After this, transmitted power values will be initialized in a vector P^0, which will contain random values within the mobile terminal power limits. After this, for every k user, the necessary power to fulfill the quality requirements associated to the service $P^n(k)$ will be calculated (without taking into account power limitations), supposing that the power transmitted by the rest of the mobile terminals are those obtained in the previous iteration. This $P^n(k)$ can be obtained as (4.6)

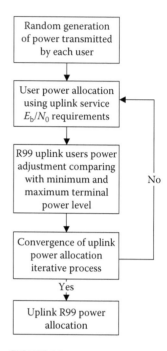

FIGURE 4.1
Procedure used for power allocation to each R99 user in the uplink.

$$P^n(k) = \frac{\mathrm{SSIR}(k)}{\alpha(m,k)} \times \sum_{j=1}^{K} \alpha(m,j) \times P^{n-1}(j) + n(m) \qquad (4.6)$$

Finally, it will be checked that the estimated power for the K users is among the minimum and maximum power values the terminal can transmit, adjusting otherwise (4.7). After that, the algorithm will continue, carrying out the following iteration:

$$P^n(k) = \max\left\{\min\left(P^n(k), P_{\max}(k)\right), P_{\min}(k)\right\} \qquad (4.7)$$

4.2.2.2 R99 Downlink

As in the uplink, in the downlink the powers associated to the traffic channels for every mobile terminal will have to be iteratively calculated, since the power transmitted to a user will influence both the global interference, and, therefore, the power that the different base stations will have to assign to the different mobile terminals to fulfill the quality requirements. Procedure

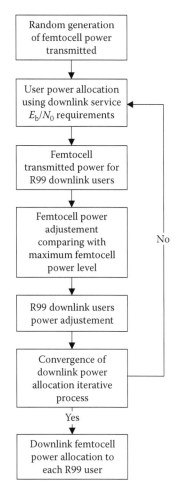

Random generation of femtocell power transmitted

↓

User power allocation using downlink service E_b/N_0 requirements

↓

Femtocell transmitted power for R99 downlink users

↓

Femtocell power adjustement comparing with maximum femtocell power level

↓

R99 downlink users power adjustement

↓

Convergence of downlink power allocation iterative process

No

Yes

↓

Downlink femtocell power allocation to each R99 user

FIGURE 4.2
Procedure used for power allocation to each R99 user in the downlink.

used for power allocation to each R99 user in the downlink is presented in Figure 4.2. The algorithm calculates the power that will be necessary to transmit to a mobile, supposing that the power transmitted to each one of the rest of the terminals is the one obtained in previous iteration. Once the power associated to every terminal has been calculated, the overall power transmitted from every base will be obtained. This power will be compared with the maximum power associated to traffic channels that could transmit every base, that is, the maximum power that can be allocated in each base station taking into consideration the power associated to control and signaling channels. If after the algorithm convergence this limit is exceeded, the associated powers to every terminal would be reduced proportionally to the power that would be necessary to obtain its quality requirement, until the power associated to the traffic channels in each base station will be below the maximum power limit.

First of all, taking into account mobile terminals' locations, power allocated to traffic channels in each base station will be randomly selected and included into a τ^0 vector. After this, for every k user, the necessary power will be calculated to fulfill the quality requirements associated to the service $P^n(k)$ (without taking into account limitations), supposing that the powers of the remaining users are the powers calculated in the previous iteration. This $P^n(k)$ will be able to be obtained as (4.8)

$$P^n(k) = \frac{SSIR(k)}{\alpha(\phi(k),k)} \times \sum_{m=1}^{M} \alpha(m,k)\theta(m,k)\tau^{n-1}(m) + n(k) \tag{4.8}$$

In the same way, the necessary traffic total power will be calculated in every base as (4.9)

$$\tau^n(m) = \sum_{k \in \phi^{-1}(m)} P^n(k) + P_c(m) \tag{4.9}$$

Finally, it will be checked that these necessary powers in the M base stations are found in the maximal value of its transmission powers, adjusting otherwise (4.10):

$$\tau''(k) = \left\{ \min\left(\tau''(m), \tau_{\lim}(m) \right) \right\} \qquad (4.10)$$

WCDMA femtocell deployments present some differences, since they are usually in the coverage area of a macrocell. If there are available frequencies, the ideal situation would be to use different frequencies for macrocells and femtocells placed in the same areas. Usually, since the spectrum is a scarce resource, many deployments of femtocells must share frequencies with macrocells. Because of that, it is important to take into account, when evaluating femtocell deployments, the interference produced by macrocells in the deployment area, because of the considerable difference between power transmitted by macrocells and femtocells. As a consequence, in the procedures presented, a new variable will be included, P_{macro}, which will include the interfering power due to the macrocell in the deployment area. This interference level can be modeled using different values, depending on the estimated distance between the macrocell and the deployment area.

4.2.3 HSPA Planning and Dimensioning: General Concepts

4.2.3.1 HSDPA

High Speed Downlink Packet Access (HSDPA) is deployed with the purpose of increasing downlink packet data throughput as well as reducing round-trip times and latency times. The standard provides new physical channels for data transmission and signaling and includes dynamic adaptive modulation and coding on a frame-by-frame basis, allowing a better usage of the available radio resources. Four new physical channels are introduced in HSDPA. The High Speed Shared Control Channel (HS-SCCH) supports three basic principles: fast link adaptation, fast hybrid automatic repeat request (HARQ), and fast scheduling as result of placing this functionality in the Node B instead of the Radio Network Controller (RNC). Each user to which data can be transmitted on the High Speed Dowlink Shared Channel (HS-DSCH) has an associated Dedicated Channel (DCH) that is used to carry power control commands and the control information necessary to realize the UL-like ARQ acknowledgment and Channel Quality Indicator (CQI). Compared with DCH, the most important difference in mobility is the absence of soft handover for HS-DSCH.

For user data transmission, HSDPA uses a fixed spreading factor of 16, which means that user data can be transmitted using up to 15 orthogonal codes. In order to modulate the carrier, Release 99 uses Quaternary Phase Shift Keying (QPSK). On the other hand, HSDPA can also use 16 Quadrature Amplitude Modulation (16QAM), which in theory doubles the data rate. In HSDPA, the possibility to support the features is optional from the point of view of the mobile terminals. When supporting HSDPA operation, the MT indicates which of the 12 different categories are specified. The achievable maximum data rate varies between 0.9 and 14.4 Mbps in agreement with the category of the MT. The new link adaptation functionality has new metrics to evaluate the performance of HSDPA. Release 99 uses E_b/N_0, but this metric is not appropriate for HSDPA since the bit rate on HS-DSCH is varied every transmission time interval (TTI) using different modulation schemes, effective code rates, and a number of High Speed Physical Downlink Shared Channel (HS-PDSCH) codes. Therefore, the metric used for HSDPA is the average HS-DSCH Signal-to-Interference-plus-Noise-Ratio (SINR) that represents the narrowband SINR ratio after the process of de-spreading of HS-PDSCH. Link adaptation selects the modulation and coding schemes with the purpose of optimizing throughput and delay for the instantaneous SINR.

4.2.3.2 HSUPA

High Speed Uplink Packet Access (HSUPA) uses most of the basic features of WCDMA Release 99. The main changes take place in the way of delivering user data from the user equipment to the Node B. It is based on a dedicated user data channel rather than a shared channel. HSUPA also operates in soft handover because all the Node Bs in the active set are involved. In the uplink, the critical issue is the power control of scheduling. Uplink capacity is limited by the level of interference in the system, which is proportional to each mobile terminal transmission power. The base station can specify the power level used by the MT to transmit HSUPA messages, relative to the power level of the normal data channel for Release 99.

HSUPA introduces five new physical channels. A new uplink transport channel, Enhanced UL Dedicated Channel (E-DCH), supports new features such as fast base station-based scheduling, fast physical layer HARQ with incremental redundancy and, optionally, a shorter 2 ms TTI. Each MT has its own dedicated E-DCH data path to the base station. A comparison between the DCH in Release 99, the HS-DSCH in HSDPA, and the E-DCH in HSUPA is done in Table 4.1.

TABLE 4.1

Comparison between DCH in Release 99, HS-DSCH in HSDPA, and E-DCH in HSUPA

Feature	R99 DCH	HSDPA (HS-DSCH)	HSUPA (E-DCH)
Spreading factor	Variable	16	Variable
Fast power control	Yes	No	No
Adaptive modulation	Yes	No	No
HARQ	No	Yes	Yes
Soft handover	Yes	No	No
TTI length (ms)	10, 20, 40, 80	2	2, 10

Base station-based scheduling has a control signaling that operates faster than RNC-based scheduling with layer 3 control signaling. The performance of the system is improved by the faster adaptation to interference variation and faster reallocation of radio resources among users. HSUPA, by using HARQ and soft combination of HARQ retransmissions, allows a decrease of the necessary E_c/N_0 at the base station comparing with Release 99 for a certain data rate. Therefore, UL spectral efficiency also increases. There are two available TTIs: the 2 ms is appointed to high data rates with good radio channel conditions, and the 10 ms is the default value for cell edge coverage suffering from a high number of retransmissions due to the increase of associated path loss.

4.2.3.3 HSDPA+/HSUPA+

HSPA+ consists of introducing MIMO and high-order modulation, protocols optimization, and optimizations for voiceover Internet protocol (VoIP). The deployment of existing HSPA is easily updated from the point of view of operators, and uses MIMO in order to transmit separately different encoded streams to a mobile terminal, increasing throughput. The link adaptation has two types of components: a spatial one and a temporal one. Release 6 HSPA systems support the use of QPSK and 16QAM in DL and the BPSK and QPSK modulation schemes in the UL. 16QAM and QPSK provide high enough data rates for macrocell environments.

In indoor or small-cell deployments, higher signal-to-noise ratio (SNR) and higher-order modulation can be supported. The best combination of modulation and coding rate for a given SNR is determined by Modulation and Coding Schemes (MCS) tables. In this way, peak rate is limited by the output of the MCS table, in other words, a higher-order modulation with the least amount of coding. Release 7 introduces 64QAM in DL, increasing the peak data rate from about 14 up to 21.6 Mbps. Note that the enhancements inherent to HSPA+ are reflected in the 16QAM modulation for DL, with the need for a smaller SNR value, to achieve the peak data rate, compared to HSDPA Release 5. In UL, the introduction of 16QAM allows for the peak data rate to reach about 11.5 Mbps (per 5 MHz carrier), featuring an increase of 100% compared to the 5.74 Mbps of the enhanced UL in Release 6, with QPSK. In Release 7, MIMO is defined for transmitting up to two streams (2 × 2 MIMO scheme), which for DL, using 16QAM for each stream, leads to peak data rates of approximately 28 Mbps. The combination of MIMO and 64QAM, being considered for Release 8, extends the peak data rate to 43.2 Mbps (per 5 MHz carrier).

4.2.4 HSPA Power Allocation Assessment: Simulation Procedures

4.2.4.1 HSDPA/HSDPA+

In the case of HSDPA, two scenarios must be distinguished. In a dedicated HSDPA carrier, all the power will be allocated for HSDPA users. However, if

R99 and HSDPA users share the same bandwidth simultaneously, the power allocated for HSDPA will be the remaining power not used by R99 users.

In any case, for HSDPA power allocation assessment it is considered that all the power available will be associated to only one HSDPA user in a determined TTI. Procedure used for time allocation to each HSDPA/HSDPA+ user is presented in Figure 4.3.

Therefore, the SIR during the time allocated to the terminal can be obtained as (4.11)

$$SIR(m,k) = 16 \frac{P_{\text{HSDPA}}(m)}{P_{\text{total}}(m)\left(1 - \theta(m,k) + \frac{\sum_{\substack{n=1 \\ n \neq m}}^{M} \alpha(n,k)P_{\text{total}}(m) + N(k)}{\alpha(m,k)P_{\text{total}}(m)}\right)} \tag{4.11}$$

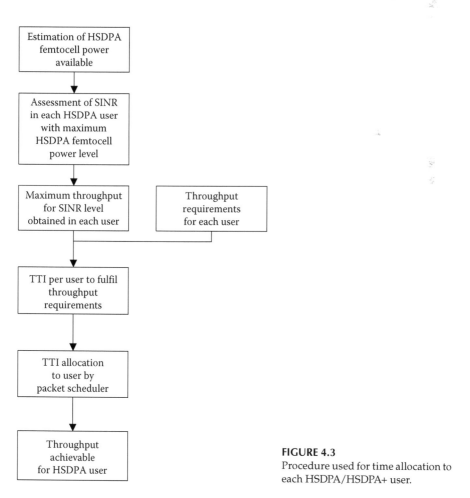

FIGURE 4.3
Procedure used for time allocation to each HSDPA/HSDPA+ user.

Given the mobility model and the characteristics of the terminal (HSDPA or HSDPA+, number of channelization codes, etc.), the maximum throughput for that user can be modeled as a function of SIR. In Figure 4.5, maximum raw throughput values for different SIR values are presented, using as a reference the MCS tables provided by 3GPP in [24] for a pedestrian user channel at 3 km/h. This maximum throughput value achievable by the user must then be compared to the throughput requirements for the service provided to the user, in order to model the time which must be allocated to the user in order to fulfill its throughput requirements.

A scheduler in the femtocell controls how simultaneously active HSDPA users share the shared channel HS-DSCH in time domain. Several scheduling approaches can be adopted, deciding the allocated time only taking into account the number of HSDPA users (Round Robin) or applying scheduling policies trying to benefit HSDPA with best conditions (Proportional Fair, Maximum C/I, Minimum bit rate scheduling, etc.). The policy implemented in this case will be the Round Robin policy, limiting the time allocated to a user to that time that fulfills its throughput requirements. Other users will be able to use the spare time, if they need it to fulfill their requirements.

Therefore, the time allocated to a determined HSDPA user will be

$$\tau_{\text{allocated}}(k) = \left\{ \min\left(\tau_{\text{needed}}(k), \tau_{\text{scheduling}}(k) \right) \right\} \qquad (4.12)$$

where
  $\tau_{\text{needed}}(k)$ is the ratio among the throughput required and the maximum throughput achievable for the SIR value obtained in (4.11)
  $\tau_{\text{scheduling}}(k)$ is the percentage of the total time allocated by scheduler to that user

Finally, using the time allocated and the HSDPA maximum throughput obtained for SIR value calculated, the achievable throughput for each HSDPA user will be assessed.

As for R99 UL resource allocation procedure, and due to the macrocell effect, a variable P_{macro} can be included to take into consideration the interfering power due to the macrocell in the deployment area.

4.2.4.2 HSUPA/HSUPA+

Unlike HSDPA, HSUPA does not use a shared channel for delivering the data calls, following a power allocation policy similar to R99. Therefore, the study of HSUPA performance must, consequently, be based on the study for UMTS R99 uplink model. In this way, the power allocated to a determined HSUPA user can be obtained introducing these users in the iterative process of power transmitted assessment presented for R99 users. Procedure used for power allocation to each HSUPA/HSUPA+ user is presented in Figure 4.4.

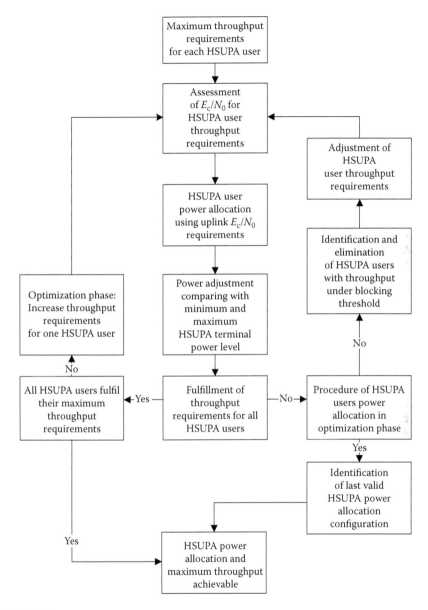

FIGURE 4.4
Procedure used for power allocation to each HSUPA/HSUPA+ user.

For R99 uplink users, if its throughput requirements or E_b/N_0 cannot be fulfilled, the user will be considered in outage. However, the data rate offered to one HSUPA user depends on its E_c/N_0, and, due to that, it is not constant over the time, and the packet scheduler will try to take advantage of this requirement reconfigurability. In this way, the algorithm will try to provide all users

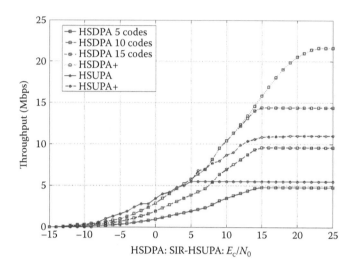

FIGURE 4.5
Maximum raw throughput values for different SIR or E_c/N_0 values and HSPA/HSPA+ configurations.

with the highest data rate possible, equal to their throughput requirements, and, if this is not possible, it will seek for the maximal throughput level all mobile terminals can adopt, taking the load condition into consideration.

For that reason the power allocation assessment must be modified. Let us consider a scenario in which all the uplink users are HSUPA users. First of all, the maximum throughput is considered for all the HSUPA users, and the E_c/N_0 or SIR needed to obtain that throughput is calculated, using for that purpose the MCS tables provided by 3GPP in [25] for a pedestrian user channel at 3 km/h, presented in Figure 4.5.

With those SIR values a process similar to the one in the R99 UL case is carried out, obtaining the transmitted power required for each HSUPA user. If any of them needs to transmit a power higher than its maximum power to fulfill its throughput requirements, the throughput requirements for all the HSUPA users is decreased to a determined level, and the procedure is initiated again.

A HSUPA user is considered in an outage condition when the throughput requirement must be established below a determined threshold defined as minimum acceptable throughput If throughout this process any of the HSUPA users is in an outage condition, this user will be considered blocked, and it will not be taken into account in the assessment procedure, maintaining the throughput level of the remaining users instead of decreasing it.

Once a data rate level is found for all the HSUPA users, the scheduler tries to privilege the largest possible number of the mobile terminals by giving them the throughput level immediately higher, initiating an optimization phase. In this optimization, only one HSUPA user increases its throughput each iteration until any of them needs to transmit a power higher than its maximum power to fulfill its throughput requirements. In that moment, the process is finished, and last valid power allocation and throughput configuration is taken as solution.

4.3 Grade of Service Assessment: Indoor Radio Propagation and Signal Strength Considerations

As it was introduced in [23], to analyze femtocell outage conditions, the behavior of the received signal strength must be understood. In complex propagation scenarios such as indoor areas, small spatial separation changes between femtocells and observation points have great impact, causing dramatic signal amplitude and phase changes. In typical cellular communication systems, the signal strength analysis is based on long-distance outdoor or combined scenarios that experience Rayleigh fading. Several handover studies assume that fading can be averaged to make up a random variable following a lognormal distribution, as described in [8–9,21–22] in the form (4.13)

$$f_i(\hat{s}) = \frac{1}{\hat{s}\sigma_i\sqrt{2\pi}} e^{\frac{-(\hat{s}-\mu_i')^2}{2\sigma_i^2}} \tag{4.13}$$

where
$\hat{s}$ is the received signal amplitude of the envelope
σ_i represents shadowing, which can be reasonably averaged to express slow power variations
μ_i' is the average signal loss received at the mobile node from the wireless access point i, and can be expressed as (4.14)

$$\mu_i' = k_1 + k_2 \log(d_i) \tag{4.14}$$

where
d_i represents the distance from the observation point to the wireless Access Point i, or AP_i
constants k_1 and k_2, respectively, represent the frequency-dependent and fixed attenuation factors, and the propagation constant

Although Equation 4.13 represents the fading probability distribution function for the path losses, as described in Equation 4.14, in complex scenarios, such as indoors, in which many obstacles contribute to the propagation losses, signal strength can be expressed as (4.15)

$$\mu_i = P_{tx} - \left(k_1 + \sum_k \lambda_k + k_2 \log(d_i) \right) \tag{4.15}$$

where
  λ_k is the attenuation of the k passing through walls in the path from the observation point to the wireless access point
  P_{tx} is the transmitted power

Moreover, to take the attenuation into account due to different floors in indoor propagation, one additional term could be added to Equation 4.15, as stated in [10,19].

4.3.1 Extreme Value Signal Distribution

Other possible choices of statistical distributions for envelope modeling have already been described, and detailed studies based on exhaustive measurements have been carried out to explain indoor propagation. The Weibull distribution seems to be one of the statistical models that best describes the fading amplitude and fading power indoor scenarios [1,10,12,19], improving the lognormal distribution in many cases. The Weibull distribution can be expressed as (4.16)

$$f_i(\hat{s}) = ba^{-b}\hat{s}^{b-1}e^{-\frac{\hat{s}^b}{a}} I_{(0,\infty)} \tag{4.16}$$

where
  $\hat{s}$ represents the fading amplitude envelope or the fading power
  a and b, respectively, are the position and shape values of the distribution
  $I_{(0,\infty)}$ indicates that $f_i(\hat{s})$ is defined from 0 to infinite

When the power distribution is represented in dBm, the extreme value distribution function should be used. In fact, if $\hat{s}$ has a Weibull distribution with parameters a and b, $\log(\hat{s})$ has an extreme value distribution with parameters $\mu = \log(a)$ and $\sigma = 1/b$, as shown in [17,18].

The extreme value function for the power probability distribution function (pdf) has been considered a good approximation. An example of fitting is shown in Figure 4.6. As can be seen, the power histogram of an indoor trajectory, modeled by the lognormal pdf function, is sufficiently represented by the extreme value function.

This perfomance has already been observed in scenarios with complex propagation conditions such as vegetation obstacles [15].

Since most of the analysis will be probabilistic, it is interesting to see how the histogram in Figure 4.6 is approximated in terms of cumulative distribution function (cdf). The comparison results are shown in Figure 4.7. After

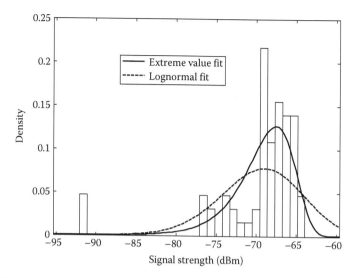

FIGURE 4.6
Comparison of lognormal and extreme value pdf fit for an indoor trajectory power log.

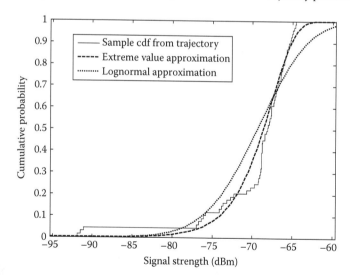

FIGURE 4.7
Comparison of lognormal and extreme value cumulative distribution approximation for an indoor trajectory power log.

integrating the differences between the sample data and the cdf approximations, an overall 2% error occurs for the lognormal function, and a 5.5% error for the extreme value function. Although lognormal is overall better in this scenario, a local analysis shows that the maximum difference between sample data and lognormal cdf is 0.24, while the maximum difference for the extreme value is 0.16. This leads us to consider the extreme value as a

TABLE 4.2

Equivalences between Distributions

	Lognormal	Weibull	Extreme Value
Mean	μ	α	$\mu = \log(a)$
Shape	σ	b	$\sigma = 1/b$

reasonable approximation. The use of this approximation will allow us to derive analytical expressions for the outage probabilities involved. Although all of the mentioned distribution functions can be used, the extreme value distribution presents some advantages; for instance, it has a closed form expression for the cumulative distribution function, which is not possible in lognormal distributions. This in turn makes it possible to present results in more manageable, closed analytical expressions.

Table 4.2 shows the relationship or the equivalences between the different distribution functions.

The extreme value probability distribution function is commonly used in the modeling and analysis of phenomena with low occurrence probabilities, such as in risk analysis or meteorology studies.

The pdf of the extreme value distribution can be expressed as (4.17)

$$f_i(\hat{s}) = \frac{1}{\sigma_i} e^{\frac{\hat{s}-\mu_i}{\sigma_i}} e^{-e^{\frac{\hat{s}-\mu_i}{\sigma_i}}} \qquad (4.17)$$

where
     σ_i and μ_i are as defined above
     the subscript "i" represents the realization for a given femtocell

To analyze the outage probabilities taking place in massive femtocell deployments, it is necessary to assess the probability of exceeding the maximum power limits with the help of signal fading distribution. Assuming that power control is taking place in the femtocell domain, some variations in the propagation path may occur, forcing the transmission power required for the transmitter output to exceed the maximum power, causing an outage.

4.3.2 Grade of Service Assessment

Assuming that femtocell deployments are taking place in a frequency channel different from the one used in the macro network, the dimensioning of massive femtocell deployments in residential environments can be performed based on an estimation of the radio resources outage probability for a given service. The outage probability is in turn giving an estimate of the GoS.

CDMA systems like the universal terrestrial radio access (UTRA) frequency division duplexing (FDD) used in femtocells keep the uplink SINR received at the BSR at a suitable value expressed as (4.18)

$$\text{SINR} = \frac{K_c S_0}{(n-1)S_0 + N_T + S_X} \tag{4.18}$$

where
  K_c is the gain process
  S_0 is the received signal strength
  N_T is the thermal noise
  n is the number of users in the cell
  S_X is the aggregated interference contribution from external femtocells
   located in the same frequency channel in the vicinity

Therefore, a possible approach to estimate the aggregated contribution from other cells can be to take similar power values arriving from other cells and introduce a factor η which depends on the extra propagation losses and the number of neighboring femtocells (4.19).

$$S_X = \eta n S_0 \tag{4.19}$$

S_X depends on the number of users per femtocell (n) and on factor η which relates to the scenario geometry and number of surrounding femtocells. The determination of η can be performed through measurements or scenario simulations.

The signal level required to maintain the system SINR is obtained as follows (4.20):

$$S_0 = \frac{\text{SINR} \cdot N_T}{K_c - \text{SINR}\left[n(1-\eta)-1\right]} \tag{4.20}$$

Taking S_{max} as the maximum power that is received at the femtocell, corresponding to the maximum transmitted power at the terminal end, an outage occurs when $S_0 > S_{max}$.

The probability of experiencing an outage is (4.21)

$$Prob(S_0 > S_{max}) \tag{4.21}$$

Since the cumulative distribution function has a closed form (4.22)

$$F(\hat{s}) = 1 - e^{-e^{\frac{\hat{s}-\mu}{\sigma}}} \tag{4.22}$$

using the complementary probabilities (4.23)

$$Prob(\hat{s} > s_0) = 1 - \left(1 - e^{-e^{\frac{s_0-\mu}{\sigma}}}\right) = e^{-e^{\frac{s_0-\mu}{\sigma}}} \tag{4.23}$$

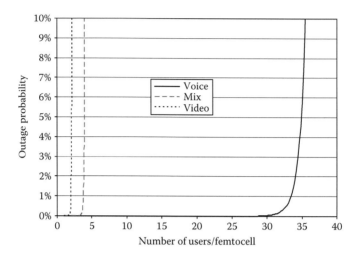

FIGURE 4.8
Grade of service estimation.

and taking the average signal as S_0, the probability of exceeding S_{max} is (4.24)

$$P_O = e^{-e^{\frac{S_{max}-S_0}{\sigma_i}}}. \tag{4.24}$$

Combining above equations, the outage probability is (4.25)

$$P_O = e^{-e^{\frac{S_{max}-\frac{SINR \cdot N_T}{K_C - SINR\left[n(1-\eta)-1\right]}}{\sigma_i}}} \tag{4.25}$$

Taking typical values for SINR and N_T, fixing S_{max} according to signal losses in the coverage area, and $\eta = 0.012$ that corresponds to a geometry in $70\,m^2$ homes with a total of 12 neighboring interferers, the results for voice service, 144 kbps video service and a service mix of 70% voice, 20% video, and 10% 384 kbps data are shown in Figure 4.8. Dimensioning enables us to estimate them as the number of simultaneous users for a given outage probability, e.g., 1%.

Although the analysis is focused on different frequencies in a femtocell and a macrocell, the effect of the macrocell can be introduced as an additional factor in (4.19), providing different GoS results.

4.4 3G Femtocells Dimensioning Study Based on Simulation: Framework Description

As it was commented previously in this chapter, an extensive simulation activity that covered both circuit-type and packet-type services has been carried out in several surveys, evaluating the performance of massive femtocell deployments in several reference scenarios, in order to be able to extract dimensioning rules to be used in the massive deployment of this type of nodes. In this section, a description of the simulation framework used in those studies is presented, describing scenarios selected, the simulation process carried out, and several simulation parameters.

4.4.1 Scenario Description

Different scenarios were designed to cover a wide range of deployment situations. All the scenarios use a four-floor building. Including more floors would increase the processing time while causing difficulties with the global visualization of the scenario. It has been established that due to the high attenuation of cells/floors and the low power used by these types of devices, the influence of BSR femtocells located at a distance of more than two floors is almost negligible. Figure 4.9 provides an example of the full building plot and a second floor layout of a 70 m² apartment.

The apartment distribution is the same for every floor. This assumption means that when a floor distribution is designed, all other floors are obtained by replicating the previous one. In other words, each floor has the same number of apartments with all the walls in the same position as the rest of the floors. Different distribution patterns do not affect the simulation complexity, although this hypothesis is used to facilitate the identification of the factors causing a certain effect.

The results in the study were obtained from the second floor. The first and last floor may present border effects because they have no interferers in any of the existing directions, so the second and third floors are more suitable to study the expected behavior of a BSR femtocell in a real scenario. The second floor was randomly selected, although very similar results were found for the third floor. There is a common zone not belonging to any of the

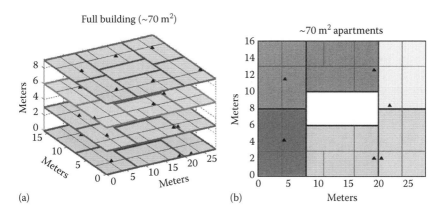

(a)   (b)

FIGURE 4.9
Simulation scenario layout. (a) Full building plot. (b) Second floor layout of 70 m² apartments.

apartments. This zone could include the stairs, a corridor, or an elevator, as in an actual building.

The location of the BSR femtocells was selected in order to simulate a real scenario as closely as possible. All of the BSR femtocells were placed near a wall. The BSR femtocells on different floors are not positioned in the same location as on the selected floor but in a similar one, so as to compare different scenarios. Best- and worst-case situations were avoided. Nevertheless, a specific part of this study was devoted to an analysis of the impact of the BSR femtocell location. A particular scenario was created for this purpose and will be explained in the corresponding section.

The study included scenarios with 70, 120, 150, and 300 m² apartments. For comparison and simplicity reasons, all the internal walls in an apartment are considered to add an attenuation of 2.5 dB while all the external walls which separate two apartments contribute 4 dB to the attenuation. In the case of 2 GHz simulations, these values are updated to 3.5 and 7 dB, respectively. The corresponding floor attenuation data are 10 dB (850 MHz) and 18 dB (2 GHz). All these values are based on [7]. Talking specifically about simulation in Section 4.4.2 (HSPA simulations), the scenario will include six 120 m² apartments, with two different femtocell configurations (in close and open subscriber group (OSG)) and with two different locations of femtocells in each apartment, working in 2 GHz band. All the users will support 384 kbps data services in order to make possible a direct comparison among R99 and HSPA users.

For simulation purposes, it was assumed that the femtocell antennas had a radiation pattern that was similar in all directions. Although actual antennas can have some form of horizontal and vertical directivity, this assumption allows the results to be independent from particular antenna arrangements.

4.4.2 Simulation Process

The simulation process was primarily based on a static simulator that provides results upon request. Each of the Monte Carlo iterations in the simulator performs a number of calculation blocks. As shown in Figure 4.10, the number of user equipment (UE) is set, and then UE positions in the scenario are randomly generated. Each UE is assigned to a corresponding BSR femtocell, depending on its position. Afterward, the path loss value is calculated among each UE and all the BSR femtocells in the scenario. To make the resource allocation, procedures in Figures 4.1 through 4.4 presented in Section 4.2 are used. A number of iterations may be needed to converge. It should be noted that iterations in this step are power control iterations. The number of iterations to converge will depend on the number of users per femtocell and they are included inside each Monte Carlo iteration. Finally, all the other expected data is calculated, e.g., femtocell CPICH E_c/N_0 and macro coverage.

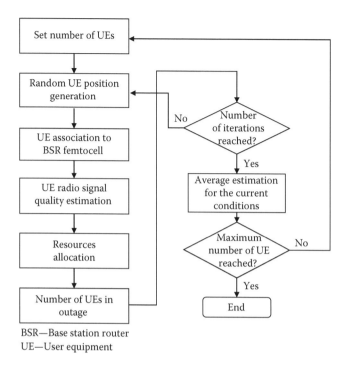

BSR—Base station router
UE—User equipment

FIGURE 4.10
Simulations flow diagram.

Simulations were performed in two channel conditions, i.e., co-channel and adjacent channel. In both cases, the influence of the macrocell is taken into account. In the second case (at a different frequency), the adjacent channel was used as worst case, taking into account the adjacent channel rejection of the receiver. In both cases, the effect of the macrocell was assessed as global interference of a fixed power value, instead of considering the effect of a physical macrocell placed at a certain position. The macrocell power level received at the indoor location can be viewed as the distance to the indoor scenarios.

4.4.3 Simulation Parameters

The simulation process requires the selection of suitable values for a number of parameters used in the calculations. In this section, these values are justified. It should be noted that only the main fixed parameters are included here, as parameters that change in each scenario are clarified in the corresponding sections. Table 4.3 shows a summary of the main simulation parameters. The adjacent channel interference ratio (ACIR) can be obtained as:

$$ACIR = \frac{1}{(1/ACLR) + (1/ACS)} \tag{4.26}$$

TABLE 4.3

Summary of the Main Simulation Parameters

Parameter	Value	Observations
Floor height	3 m	This value is chosen for the sake of simplicity, although it is still realistic. Small variations of around 3 m in the parameter were shown not to lead to important changes in the results.
Number of Monte Carlo iterations	500	
Carrier frequency	850 and 2000 MHz	HSPA simulations are carried out in 2000 MHz only.
Chip rate	3.84 Mcps	This is a standard value in UMTS.
Uplink maximum transmit power	21 dBm	This is the standardized value for UEs with Power Class 4 (3GPP TS 25.101).
UE receiver sensitivity	−118 dBm	The standardized minimum requirement value for the UE receiver sensitivity operating in band V is −115 dBm (3GPP TS 25.101). The selected value includes an improvement margin of 3 dB.
Downlink maximum transmit power	13/20 dBm	HSPA simulations are carried out for 13 dBm configuration.
BSR femto receiver sensitivity	−110 dBm	The standardized minimum requirement value for the BS receiver sensitivity of a Local Area Class BS is −107 dBm (3GPP TS 25.104). The selected value includes an improvement margin of 3 dB.
Control channels power	6 dBm	This value was selected based on the usual configuration of a UMTS macrocell, where the assigned power for control channels (CPICH, SCH, CCPCH, AICH) entails 20% of the maximum power. It should be noted that this value includes a 10% of the maximum power for the primary CPICH power, that is, 3 dBm. For simulations with a maximum transmit power of 20 dBm, the same ratio was used. Therefore, the control channel power is 13 dBm and the primary CPICH power is 10 dBm.
ACIR	30 dB	This value includes the adjacent channel leakage ratio (ACLR = 33 dB) of the transmitter and the adjacent channel selectivity (ACS = 33 dB) of the receiver.
E_b/N_0 Requirements (UL/DL)	12.2 kbps: (5.8/7.8) 64 kbps: (3.1/2.8) 144 kbps: (2.3/1.5) 384 kbps: (2.2/2.2)	
HSDPA terminal codes	15 codes	

Typical values of adjacent channel leakage ratio (ACLR), and adjacent channel selectivity (ACS) are also included in Table 4.3.

4.5 3G Femtocells Dimensioning Study Based on Simulation: Results

The main objective of the simulation process was to obtain a series of general conclusions in order to validate the previously described GoS dimensioning model, which we used in our generic deployment criteria. In this section, for R99 circuit-type services, most relevant results obtained in [23] are presented and discussed. These results have been complemented with posterior studies, which have included the dimensioning of HSPA and HSPA+ in the model.

The analysis of these results will pursue to answer some question. Questions around coverage include: Does the coverage offered by the BSR femtocell comprise user's whole apartment? Are there problems if all neighbors have a BSR femtocell? Is there any relevant interference between them? Regarding GoS, we consider: How many users can the BSR femtocell admit for different services? Are there radio limitations? And regarding the impact of BSR positioning: Is BSR femtocell position important from the radio perspective? Which are the probabilities of a good position? What will be the gain of using HSPA instead of R99 data services? And HSPA+ instead of HSPA? What will be the effect of using a CSG configuration?

4.5.1 R99 Simulation Output

4.5.1.1 BSR Femtocell Coverage

In order to calculate the coverage offered by BSR femtocells, sample scenarios were simulated. Coverage will be measured by the CPICH E_c/N_0 level received by the user. A CPICH E_c/N_0 of −15 dB can be considered as a location with very poor coverage, while a level lower than −20 dB prevents the cell from being detected. Therefore, in this study, a CPICH E_c/N_0 value lower than −17 dB will be considered a point with no coverage.

The ideal situation is a BSR femtocell covering the entire apartment without interfering with service in neighboring apartments. The main coverage problem with BSR femtocells is their uncoordinated deployment. In other words, it is a network without planning. Unlike a standard cellular network, the locations of the nodes cannot be designed because they are chosen by the user, which means that nodes will interfere with each other. Besides, each user must be connected to his own node even if he receives a better signal from a neighboring node.

TABLE 4.4

Parameters Used in Coverage Simulations

Parameter	Value	Comments
Macrocell RSCP	−75 dBm	Macrocell placed to a middle distance from the building
ACIR	30 dB	Macrocell in an adjacent channel
Services ratio	75% voice, 20% video, and 5% data	
Users per BSR femto	2	
Area without coverage	Area where the received CPICH E_c/N_0 is less than −17 dB	

The parameters assumed in coverage simulations are shown in Table 4.4.

Figure 4.11 shows BSR femtocell coverage in different scenarios. Bold triangles symbolize BSR femtocells of a specific apartment. Transparent triangles are BSR femtocells from other floors. Grey areas represent coverage gaps.

These pictures show that coverage problems appear only in some rooms of the apartment, i.e., when a user is close to a neighbor's node and far

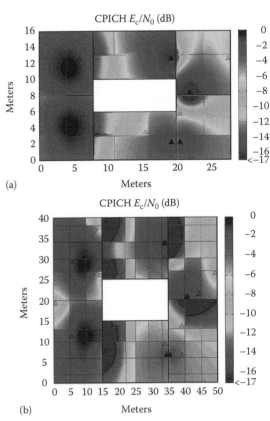

FIGURE 4.11

E_c/N_0 in two types of simulation scenarios. (a) Scenario 2 (70 m^2): 1.38% without coverage. (b) Scenario 4 (300 m^2): 7.85% without coverage.

from his own. This problem is compounded in larger apartments where there are more walls between the user and his node. This situation points out the importance of properly positioning the BSR femtocell within the apartment.

4.5.1.2 Maximum Power Impact

This section analyzes the importance of maximum power in BSR femtocell coverage. To make a comparison, all sample scenarios have been simulated with 13 and 20 dBm of maximum power. Table 4.5 shows the results for an area without coverage.

As shown in the table, the results are practically the same for both powers. This occurs when the power of the BSR femtocell is increased because users not only receive more signal power from their own node but from the other BSR femtocells as well. That is, the interference increases as the signal strength increases, which means the coverage area remains the same.

4.5.1.3 R99 Grade of Service

This section analyzes the GoS offered by a BSR femtocell, i.e., the number of calls that it can admit, and the capacity to offer different services.

In GoS simulations, parameters assumed are the same as those included in Table 4.3. Figure 4.12 shows, for different services, the percentage of users that cannot get service from a BSR femtocell within a 70 m² apartment.

In this scenario, accepting that 5% of users will be without service, up to 40 users are supported for voice service, 5 are supported for video service, between 1 and 2 are supported for 384 kbps data service, and 11 are supported for mixed services. These results are consistent with the theoretical model described and the results shown in Figure 4.8.

Table 4.6 compares simulations and models described in the "Grade of Service" Section 4.3.2.

TABLE 4.5

Areas Without Coverage

		Node Max Power (dBm)	
Area Without Coverage (%)		13	20
Apartment size (m²)	70	0.54	0.54
	120	1.38	1.39
	150	2.27	2.24
	300	7.85	7.82
	Mixed (2 of 300, 2 of 150, and 2 of 120)	5.45	5.47

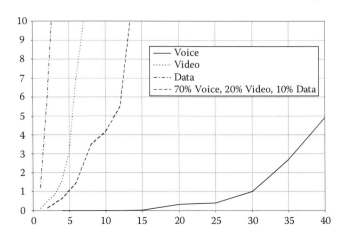

FIGURE 4.12
Grade of service simulation results.

TABLE 4.6

Simulation and Model Results Comparison

Users for 1% Outage	Voice	Video	Mix
Estimated	33	2	4
Simulated	30	3	5

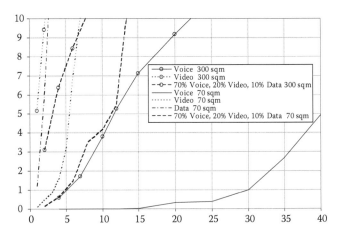

FIGURE 4.13
Comparison of scenario simulation.

The influence of scenario geometry on the results is shown in Figure 4.13, comparing scenario 1 (70 m² apartments) and scenario 4 (300 m² apartments). As with coverage, the GoS is better in smaller apartments.

In scenario 4, again accepting that 5% of the users will be without service, up to 12 users are supported for voice service, 1 is supported for video service, between 0 and 1 are supported for 384 kbps data service, and 3 are supported for mixed services.

4.5.1.4 BSR Femtocell Positioning Impact

In a BSR femto network, the position of the nodes cannot be decided because the user can place the node anywhere in the house. If a node is placed near a neighbor's wall, it will generate strong interference, and if the neighbor's own BSR femtocell is far from that wall, he will probably not be able to connect to his own node from that room.

This effect turns node position into a very important issue.

This section studies the influence of BSR femtocell positioning on coverage and GoS. The idea is to consider all possible positions to determine the probability of obtaining good or bad coverage and GoS.

Due to the complexity of this simulation, a particular scenario was created for this purpose. This configuration reduces processing time, which enables results to be generalized.

This scenario includes four 100 m² apartments with five possible BSR femtocell positions in each apartment. Figure 4.14 shows the floor layout.

With 5 possible positions in each apartment, there are 625 possible combinations of BSR femtocell positions. The objective is to determine how many of these combinations generate good coverage or GoS scenarios and how many of them produce bad results.

The covered area in an apartment will be affected by both the position of the user's own node and the positions of the neighbor's nodes. Table 4.7 shows the percentage of possible positions that generate a coverage area with a specific signal strength.

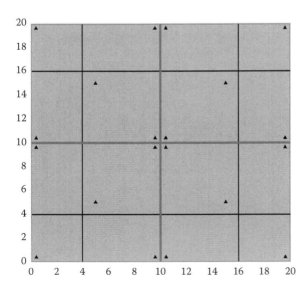

FIGURE 4.14
Femtocell positions in the layout.

TABLE 4.7

Effect of Femtocell Positions

BSR Femtocell Position Combinations (%)		Covered Area (%)			
		100	98	95	90
CPICH E_c/N_0 received (dB)	>−15	2.72	35.17	83.33	99.84
	>−17	4.29	56.91	98.56	100
	>−20	9.01	92.19	100.00	100

Notes: BSR—Base station router.
 CPICH—Common pilot channel.

These results show the mean percentages of coverage obtained with several simulations: four users of voice service, two users of video service, one user of 384 kbps data service, two users of 144 kbps data service, two users of 64 kbps data service, and four users of mixed services (70% voice, 20% video, and 10% 384 kbps data).

It can be observed that, for instance, 98.56% of the combinations produce a 95% coverage area, if the area covered is assumed to be one that can provide reception (e.g., CPICH E_c/N_0 > −17 dB).

It is almost impossible to get 100% coverage due to neighbor interference at particular points of the home, but for most of the combinations, it is feasible to obtain area coverage of more than 95%.

As with coverage, the GoS is also affected by the positioning of the BSR femtocell: if the user is in a low coverage area, more power will be needed from the node for essentially the same level of service.

As an example, a mixed service scenario was considered, with a 70% voice, 20% video, and 10% data 384 kbps service distribution.

As shown in Table 4.8, assuming that 5% of users will be without service, all the possible position combinations (100%) will support four users, and 73.76% of the combinations will support seven users. If the level is reduced to 2% of users without service, 81.76% of the possible combinations will support four users and 45.76% of them support seven users.

TABLE 4.8

Femtocell Position Performance

BSR Femtocell Positions Combinations (%)		Number of Users per BSR Femtocell				
		4	7	10	12	15
Users without service (%)	=0	3.04	1.92	0.48	0	0
	<2	81.76	45.76	25.76	11.2	0
	<5	100	73.76	56.32	28	1.76
	<10	100	93.12	68.8	53.6	9.92

Note: BSR—Base station router

TABLE 4.9

Frequency Band Comparison

		Frequency (MHz)	
Area Without Coverage (%)		**850**	**2000**
Apartment size (m²)	70	0.54	0.28
	120	1.38	0.96

For comparison purposes, some of the coverage and GoS simulations were carried out in 2000 MHz. Scenario 1 (70 m²) and scenario 2 (120 m²) were used in these simulations, as shown in Table 4.9.

Regarding GoS, the results obtained with 2 GHz simulations are better than the results obtained at 850 MHz. This is because in massive deployments, network behavior becomes capacity-limited instead of coverage-limited. Therefore, the better the propagation, the higher the interference, and the lower the capacity for the same GoS. This result is not a contradiction of the fact that from a macrocell point of view, 850 MHz will produce better performance than 2 GHz; it simply solves the issue of coverage limitations. As an example, the number of users supported, assuming 5% of users without service, obtained in scenario 1 (70 m²) is shown in Table 4.10.

To analyze the impact of the macrocell on the adjacent channel, instead of using a fixed position, it has been considered that the macrocell introduces received signal code power (RSCP) interference in the building according to the following criteria: −60 dBm for a macrocell placed near the building, −75 dBm for a macrocell placed at a middle distance from the building, and −90 dBm for a macrocell placed far from the building. The corresponding results are shown in Table 4.11.

TABLE 4.10

Frequency Band Impact on the 70 m² Scenario

		Frequency (MHz)	
Number of Users Supported		**850**	**2000**
Service	Voice	40	65
	Video	5–6	10
	Data	2	4
	Mixed (70% voice, 20% video, 10% 384 kbps)	12	13–14

TABLE 4.11

Macrocell Position Impact on the Performance

		Macrocell Position	
Area Without Coverage (%)		**Near (−60 dBm)**	**Far (−90 dBm)**
Frequency (MHz)	850	0	49.97
	2000	0	10.65

4.5.2 HSPA/HSPA+ Simulation Output

4.5.2.1 HSPA Impact

First of all, when talking about HSPA, it is necessary to compare throughput and GoS offered by a BSR femtocell with regard to WCDMA R99, both in the UL and in the DL. It is important to remark, that, due to the characteristics of HSDPA, HSUPA, HSDPA+, and HSUPA+, the definition of GoS can be slightly varied. In the case of WCDMA R99 users, a user was considered in outage when its service-dependent E_b/N_0 requirement could not be fulfilled. All the users in that situation will be considered as calls blocked, and, therefore, as non-served when evaluating GoS. Although for HSPA users the metric used to determine when a call is blocked can be the same, i.e., when a user cannot be served with the throughput required, this metric can be modified, taking into account that the data rate offered to one HSPA user depends on its E_b/N_0 or E_c/N_0, and, due to that, can be configurable. Therefore, GoS for HSPA users can be defined as a ratio among throughput achieved and throughput required. In this case, a call will be considered as blocked taking into account the maximum reduction of throughput that is allowed for each type of service. Below this throughput threshold, HSPA users will be considered in outage, and therefore power or time will not be allocated to them.

In this section, and in order to make a fair comparison between R99 and HSPA users, the metric to assess GoS will be the same when comparing them, i.e., a percentage of users that cannot fulfills their maximum throughput requirements. Nevertheless, when comparing HSPA and HSPA+ throughput throughout this section, throughput ratio will be the metric selected. For comparison purposes, the scenario used is presented in Figure 4.15, and the location of the femtocells will be correspondent to configuration A.

The results of this comparison are presented in Figures 4.16 and 4.17, and Tables 4.12 and 4.13. for R99, HSPA, and HSPA+ users. As it can be seen, in the DL, there will be a remarkable gain when using HSDPA instead of WCDMA R99. 384 kbps users who can be simultaneously active in the network can be increased from three to four times, depending on the different GoS requirements. HSDPA takes advantage of adaptive modulation and coding to use the higher modulation which will be possible in short periods of time, sharing the channel more efficiently and, therefore, increasing considerably the mean throughput in the network, and, in that way, the maximum number of simultaneous users. As HSDPA+ introduces higher modulation and coding schemes, there will be a gain as well when using HSDPA+ instead of HSDPA, because shorter time allocation will be needed to ensure a determined throughput, and, therefore, more users will be able to be served. However, in the uplink, there will not be a great gain when comparing to WCDMA, because HSUPA terminals, due to power control, will only transmit the exact power needed to achieve the throughput required, and, therefore, they do

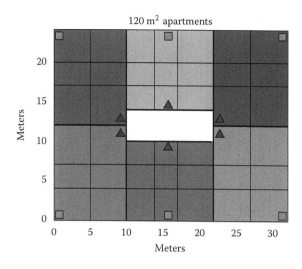

FIGURE 4.15
HSPA simulation scenario: configuration A (triangles) and B (squares).

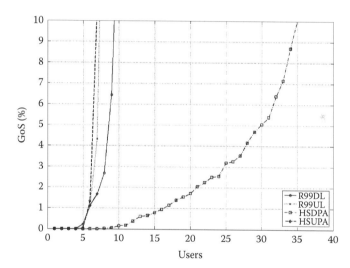

FIGURE 4.16
HSPA impact on the performance.

not take advantage of their capabilities to provide higher throughputs. For the same reason, there will not be any gain using HSUPA+ instead of HSUPA.

Therefore, HSDPA and HSDPA+ will allow to achieve higher throughputs and increase the number of simultaneous users, even for low throughput services, because of the characteristics of the HS-DSCH channel and their time-sharing procedure, whereas HSUPA will only allow to achieve higher throughputs per user, but, if high throughputs are not required, there will not be a remarkable gain with regard to WCDMA R99.

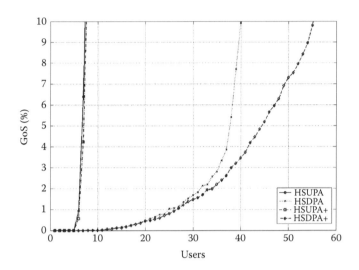

FIGURE 4.17
HSPA+ impact on the performance.

TABLE 4.12

HSPA Impact on the Performance (Normal Interiors)

GoS Users Without Service (%)	Number of Users per BSR Femtocell			
	WCDMA R99 UL	WCDMA R99 DL	HSUPA	HSDPA
<1	5	5	5	16
<2	6	7	6	20
<5	6	8	7	29
<10	6	9	7	34

TABLE 4.13

HSPA+ Impact on the Performance (Cap, Interior y CSG)

GoS Ratio Throughput (%)	Number of Users per BSR Femtocell			
	HSUPA	HSUPA+	HSDPA	HSDPA+
<1	6	6	26	26
<2	6	6	34	34
<5	6	6	44	44
<10	7	7	55	55

4.5.2.2 BSR Femtocell Positioning Impact

As in the R99 case presented in Section 4.5.1, a particular scenario has been created for this purpose, in order to reduce processing time, trying to model the best and worst case which could occur, and, therefore, the maximum and minimum capacity for a determined GoS threshold. This scenario includes in each one of the four floors six $120\,m^2$ apartments with six femtocells, one located in each apartment. Two femtocells configuration will be evaluated, as presented in Figure 4.15. In configuration A, the femtocells will be very close to each other and, therefore, they will generate strong interference, and more power will be needed from the node for essentially the same level of service, affecting overall capacity and GoS. On the other hand, configuration B deals with the situation in which femtocells are farther from each other, while maintaining the coverage in the whole apartment.

The objective will be to determine, for configurations A and B, respectively, which will be the minimum and maximum number of 384 kbps users in UL and DL. Results are presented in Figure 4.18 and Table 4.14.

Results show that when femtocells are located very close to each other, the number of active users for a determined GoS requirement decreases considerably. This effect is caused because of the overall increase of the interference in all the system, which will affect both overall throughput and GoS. As it has been previously commented throughout the text, in these types of deployments, the femtocell location in each apartment will be decided by the

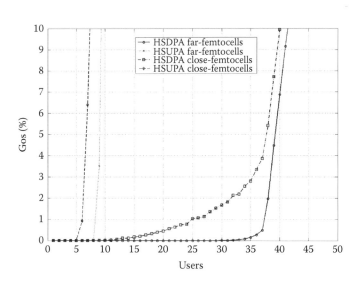

FIGURE 4.18
Femtocell distribution impact on the performance.

TABLE 4.14

Femtocell Distribution Impact on the Performance

GoS Ratio Throughput (%)	Number of Users per BSR Femtocell										
	Close Femtocells				Far Femtocells						
	HSUPA	HSUPA+	HSDPA	HSDPA+	HSUPA	HSUPA+	HSDPA	HSDPA+			
<1	6	6	24	26	8	8	6	46			
<2	6	6	31	34	8	8	6	51			
<5	6	6	37	44	9	9	6	56			
<10	7	7	40	55	9	9	7	59			

user, and, therefore, the coverage and GoS of these types of networks will have to deal with this effect.

4.5.2.3 Close Subscriber Group Impact

When a BSR femtocell is deployed, it can be configured in two modes: open or closed subscriber group. In a residential scenario, a typical configuration is the closed subscriber group (CSG), where the femtocell access is limited to the subscriber identity module (SIM) cards accepted by the node owner under the operator's supervision, rejecting all the other users. This aspect will provoke that even when a user can have better propagation conditions from a neighbor's node than to his own, it will not be able to connect to the neighbor's femtocell, being obliged to connect to its own one. This situation will increase unnecessarily the interference level in the system and, therefore, will affect the capacity and the GoS.

Results associated to femtocells in our scenario in open or CSG configuration are presented in Figure 4.19 and Table 4.15.

Two different situations must be distinguished to comment on the effect of using close or open groups. When femtocells are located, as presented in Figure 4.15, for configuration B, i.e., very far from each other, both close and OSG configuration show similar results. This situation can be explained, because, for all or most part of the users inside an apartment, the femtocell in that apartment will be their best server. Therefore, association between user and femtocell for both configurations will be the same, and, evidently, results with regard to GoS, throughput, and number of active users will be the same as well. On the other hand, when femtocells are located, as

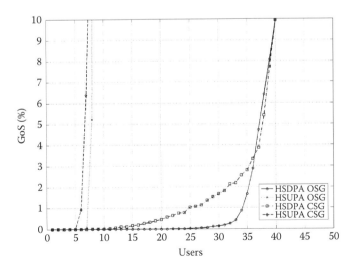

FIGURE 4.19
Close subscriber group configuration impact on the performance.

TABLE 4.15

Close Subscriber Group Configuration Impact on the Performance

GoS Ratio Throughput (%)	Number of Users per BSR Femtocell							
	Close Subscriber Group				Open Subscriber Group			
	HSUPA	HSUPA+	HSDPA	HSDPA+	HSUPA	HSUPA+	HSDPA	HSDPA+
<1	6	6	24	26	7	7	34	43
<2	6	6	31	34	7	7	35	47
<5	6	6	37	44	7	7	37	51
<10	7	7	40	55	8	8	40	55

presented in Figure 4.15, for configuration A, i.e., very close to each other, situation changes. In this case, OSG performs better than CSG. For OSG, in each moment, best server will be selected for each user, and that server can be the femtocell in its same apartment, or a femtocell in any other apartment, depending on building structure and propagation conditions. Therefore, the overall interference in the system will be reduced with respect to the close case, increasing overall throughput, and, as a consequence, maximum number of active users per apartment for a determined GoS required.

In Figure 4.19, for a determined number of users per apartment, a curious effect can be detected. From that moment, CSG configuration outperforms OSG configuration. This effect will be associated to an excess of load in some of the femtocells due to the open association. While in CSG all the femtocells serve the same number of users (supposing an equal number of users in each apartment), for the OSG case, some femtocells, due to building structure of propagation conditions, can serve more users than the others because of its coverage. When the overall throughput to fulfill the requirements of all users associated to a determined femtocell is exceeded, the GoS will be reduced, although some femtocells could have some remaining unused resources. This unbalanced performance of the network due to coverage differences between femtocells will be the cause of the effect commented.

4.6 Conclusions

In this chapter, a dimensioning method for massive 3G femtocell deployments has been shown and validated by means of simulation. The model was used only in the adjacent channel case, considering femtocell and macrocell operation in different frequency channels. Simulations also show that the performance of dense femtocell networks with respect to the BSR femto position has a significant influence on its performance. Specifically, this impact is mainly associated with the relative positioning of a BSR femtocell among all its neighbor nodes. A situation where femtocells are not too close is a favorable scheme from a global point of view, while a region in which a neighbor's BSR femtocell is closer than one's own is likely to involve interference problems. In any case, it should be noted that under certain constraints, in this deployment model, the user ultimately decides the location of the BSR femtocells. Moreover, global optimization in massive deployments based on the users placing femtocells on a trial-and-error basis is not likely to happen; therefore, unfavorable scenarios should be considered as a reference for coverage and GoS evaluation. According to the simulation output, a BSR femtocell is able to provide voice service to a high number of users even in a high-density deployment. For video and 3GPP R99 data services, the number of simultaneous users can be quite limited, so an acceptable value for the maximum percentage of non-admitted calls must be carefully studied. For this reason, HSPA and HSPA+ have been evaluated in order to quantify the increase of this limited number of simultaneous users with relatively high throughput requirements. In the

DL, an increase of the number of simultaneous users will be achieved, even for low throughput requirements, because of the characteristics of the HS-DSCH channel, whereas HSUPA will only allow achieving higher throughputs per user, but if high throughputs are not required, there will not be a remarkable gain with regard to WCDMA R99. Apartment size has a significant impact on the maximum number of simultaneous users, decreasing this value considerably in the case of larger apartments. Additionally, the simulation results suggest that using the 2 GHz frequency band is slightly more beneficial than the 850 MHz band for the BSR femtocell modeling. The main reasons are the higher obstacle attenuation and less favorable propagation conditions in the 2 GHz band, which contribute in adapting coverage to the required area and reducing the level of interference. In the same way, the effect produced by using CSG configuration instead of OSG configuration has been taken into account, because of the important limitations in terms of throughput that introduce for unfavorable cases of femtocell locations.

Acknowledgments

The authors are grateful for the support of the Spanish Ministry of Education and Science within the framework of the TEC2008-02148/TEC project. The authors are also thankful for the support of CELTIC Project EasyWireless II TSI-020400-2008-56 of the Ministry of Industry, Tourism and Commerce.

References

1. F. Babich and G. Lombardi, Statistical analysis and characterization of the indoor propagation channel, *IEEE Trans. Commun.*, 48(3), 2000, 455–464.
2. C. C. Chan and S. V. Hanly, Calculating the outage probability in a CDMA network with spatial poisson traffic, *IEEE Trans. Veh. Technol.*, 50(1), 2001, 183–204.
3. V. Chandrasekhar and J. G. Andrews, Uplink capacity and interference avoidance for two-tier cellular networks, *Proceedings of the IEEE Global Telecommunications Conference (GLOBECOM '07)*, Washington, DC, 2007, pp. 3322–3326.
4. H. Claussen, Performance of macro- and co-channel femtocells in a hierarchical cell structure, *Proceedings of the IEEE 18th International Symposium on Personal, Indoor and Mobile Radio Communications (PIMRC '07)*, Athens, Greece, 2007.
5. H. Claussen, L. T. W. Ho, and L. G. Samuel, An overview of the femtocell concept, *Bell Labs Tech. J.*, 13(1), 2008, 221–245.
6. H. Claussen, L. T. W. Ho, and L. G. Samuel, Self-optimization of coverage for femtocell deployments, *Proceedings of the Wireless Telecommunications Symposium (WTS '08)*, Pomona, CA, 2008, pp. 278–285.

7. E. Damosso and L. M. Correia (eds.), Digital mobile radio towards future generation systems. COST 231 Final Report, European Commission, Brussels, Belgium, 1999.

8. P. Dassanayake, Dynamic adjustment of propagation dependent parameters in handover algorithms, *Proceedings of the IEEE 44th Vehicular Technology Conference (VTC '94)*, vol. 1, Stockholm, Sweden, 1994, pp. 73–76.

9. M. Gudmundson, Analysis of handover algorithms, *Proceedings of the IEEE 41st Vehicular Technology Conference (VTC '91)*, St. Louis, MO, 1991, pp. 537–542.

10. H. Hashemi, M. McGuire, T. Vlasschaert, and D. Tholl, Measurements and modeling of temporal variations of the indoor radio propagation channel, *IEEE Trans. Veh. Technol.*, 43(3) 1994, 733–737.

11. L. T. W. Ho and H. Claussen, Effects of user-deployed, co-channel femtocells on the call drop probability in a residential scenario, *Proceedings of the IEEE 18th International Symposium on Personal, Indoor and Mobile Radio Communications (PIMRC '07)*, Athens, Greece, 2007.

12. M. H. Ismail and M. M. Matalgah, On the use of padé approximation for performance evaluation of maximal ratio combining diversity over Weibull fading channels, *EURASIP J. Wireless Commun. Netw.*, 2006(3), 2006, 62–68, Article ID 58501, doi: 10.1155/WCN/2006/58501.

13. S. B. Kang, Y. M. Seo, Y. K. Lee, M. Z. Chowdhury, W. S. Ko, M. N. Irlam, S. W. Choi, and Y. M. Jang, Soft QoS-based CAC scheme for WCDMA femtocell networks, *Proceedings of the 10th International Conference on Advanced Communication Technology (ICACT '08)*, vol. 1, Phoenix Park, Korea, 2008, pp. 409–412.

14. S. Kishore, L. J. Greenstein, H. V. Poor, and S. C. Schwartz, Uplink user capacity in a CDMA macrocell with a hotspot microcell: Exact and approximate analyses, *IEEE Trans. Wireless Commun.*, 2(2), 2003, 364–374.

15. S. Kourtis and R. Tafazolli, Evaluation of handover related statistics and the applicability of mobility modelling in their prediction, *Proceedings of the IEEE 11th International Symposium on Personal, Indoor and Mobile Radio Communications* (PIMRC '00), vol. 1, London, U.K., 2000, pp. 665–670.

16. X. Lagrange, Multitier cell design, *IEEE Commun. Mag.*, 35(8), 1997, 60–64.

17. A. Murase, I. C. Symington, and E. Green, Handover criterion for macro and microcellular systems, *IEEE 41st Vehicular Technology Conference (VTC '91)*, St. Louis, MO, 1991, pp. 524–530.

18. S. Perras and L. Bouchard, Fading characteristics of RF signals due to foliage in frequency bands from 2 to 60 GHz, *Proceedings of the Fifth International Symposium on Wireless Personal Multimedia Communications (WPMC '02)*, vol. 1, Honolulu, HI, 2002, pp. 267–271.

19. T. K. Sarkar, Z. Ji, K. Kim, A. Medouri, and M. Salazar-Palma, A survey of various propagation models for mobile communication, *IEEE Antennas and Propagation Mag.*, 45(3), 2003, 51–82.

20. J.-S. Wu, J.-K. Chung, and M.-T. Sze, Analysis of uplink and downlink capacities for two-tier cellular system, IEE Proc. Commun., 144(6), 1997, 405–411.

21. N. Zhang and J. M. Holtzman, Analysis of handoff algorithms using both absolute and relative measurements, *Proceedings of the IEEE 44th Vehicular Technology Conference (VTC '94)*, vol. 1, Stockholm, Sweden, 1994, pp. 82–86.

22. M. Zonoozi and P. Dassanayake, Handover delay and hysteresis margin in microcells and macrocells, *Proceedings of the IEEE Eighth International Symposium on Personal, Indoor and Mobile Radio Communications (PIMRC '97)*, vol. 2, Helsinki, Finland, 1997, pp. 396–400.

23. A. Fernandez and G. García, UMTS femtocell performance in massive deployments: Capacity and GoS implications, *Bell Labs Tech. J.*, 14(2), (2009), 185–202.
24. Ericsson, 3GPP TSG-RAN WG1, 64QAM for HSDPA—Performance with Realistic Algorithms. R1-063499, Meeting 47, November 2006.
25. R1-062266 3GPP TSG-RAN WG1, 16QAM for HSUPA—Link-Level Simulation Results. Ericsson, 46.

5

Deployment Modes and Interference Avoidance for Next-Generation Femtocell Networks

İsmail Güvenç, Mustafa E. Şahin, Hisham A. Mahmoud, and Hüseyin Arslan

CONTENTS

5.1 Introduction

Femtocell networks, which are also known as personal-use base stations, have recently received considerable attention from industry and academia due to their tremendous potential for capacity improvements of next-generation wireless systems [1,2]. The data rate requirements for the next-generation wireless networks have been specified under the wireless standard referred to as the IMT-Advanced, where peak rates on the order of 1 Gbps are targeted for low-mobility scenarios. Such high data rate requirements can be achieved through deploying smaller cells for improving the wireless link quality as in the case of femtocell networks.

As discussed in [3], the wireless capacity has grown about one million times over the last 50 years, which can be factorized into three components: better transmitter/receiver efficiency, larger spectrum, and larger number of cells. While the growth due to efficiency and spectrum correspond to a 20 time and a 25 time growth in the capacity, respectively, growth due to the number of cells amounts to a 2000 time increase in the capacity. Forecasts until the year 2015 show that small cell deployments will continue to be the major source of growth in the capacity [3], where femtocells are expected to play an important role.

Even though operators have been trying to benefit from the capacity gains due to small-cell deployments, there are several environmental problems with the previous solutions. For example, picocells are a different type of small cells that may support on the order of 32 users. However, planned deployment by the operator and site availability are critical requirements. Femtocell networks do not face these challenges; they are deployed by the users themselves, and they use the existing broadband connection of the users, introducing very little overhead on the operators. Some other factors explaining why femtocells suddenly became a viable element in the evolution of mobile networks include (1) widespread wired broadband usage (i.e., DSL, cable, and fiber optic broadband), (2) low-cost processors (i.e., FPGAs, DSPs, and ASICs), and (3) flat architectures and the collapsing of network software stacks at femtocells due to powerful processing [4].

An example architecture for a femtocell network is illustrated in Figure 5.1. A femtocell BS (fBS) is connected to the mobile operator's core network through the existing broadband Internet connection of the user. A macrocell-associated mobile station (mMS) does not have to be a dual mode terminal in order to communicate both with an fBS and a macrocell BS (mBS).*

The advantages of femtocell networks for mobile operators and users are numerous. For example, existing handsets may easily be used with femtocell deployments, as opposed to UMA-based solutions that require WiFi capability in the handsets. Some other benefits of femtocells include improvements

---

* Note that the abbreviations H-UE, HNB, M-UE, and NB are used in 3GPP documents for referring to femtocell mobile station (fMS), fBS, mMS, and mBS, respectively.

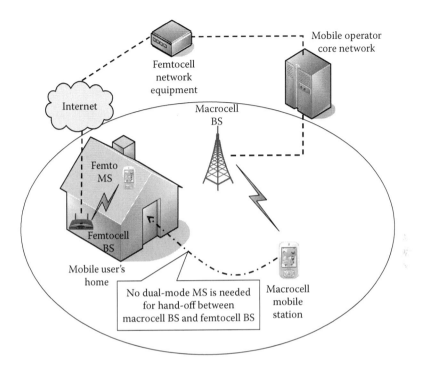

FIGURE 5.1
Femtocell deployment at homes. Connection between the macrocell and the femtocell is established through the Internet.

in cellular network coverage, reduced capital and operational expenditure (CAPEX/OPEX) for the operators, longer battery lives for cell-phones due to shorter communication ranges, increased average revenue per user, deployment in operator-owned spectrum, and enhanced emergency services. The benefits of femtocells from the perspective of the mobile operator and the user are summarized in Figure 5.2.

Due to the infancy of the technology, femtocell networks face several research challenges, such as radio interference mitigation, spectrum assignment, power control, cell selection and hand-off, security, scalability, access control, and activation and management. Among these several challenges, as acknowledged by [5], interference between the femtocell and macrocells, and between neighboring femtocells, remain a key problem that needs to be addressed.

In this chapter, interference problems related to next generation femtocell networks are discussed, and possible solutions for improvements are proposed. The chapter is composed of two main sections. In Section 5.2, different femtocell deployment options are overviewed, and several capacity trade-offs are discussed. Simulation results are provided at the end in order to compare the cumulative distribution functions (CDFs) of the user capacities corresponding to different scenarios. In Section 5.3, an orthogonal frequency division multiple access-(OFDMA) based technique

Benefits for Mobile Operators	Benefits for Users
• Counters the threat from wireless voice over IP (VoIP) service providers • Improves customer loyalty, reducing churn • Increases network coverage and capacity • Works with existing handsets (unlike UMA) • Increases average revenue per unit • Could increase the speed of adoption of new technologies such as 3G, LTE or WiMAX • Lower backhaul costs, reduced CAPEX/OPEX	• Increased indoor coverage and elimination of "dead spots" • High quality, high speed signal in the home, giving excellent quality voice calls and fast downloads • Works with existing handsets (unlike UMA) • Mobile phone battery lasts longer due to shorter distances to BS • Mobile operator may offer incentives to have a femtocell (unlimited minutes etc.) • Enhanced emergency services

FIGURE 5.2
Benefits of femtocells for mobile operators and users.

is considered, and it is proposed to avoid a co-channel interference (CCI) between the femtocell and the macrocell by not using the resource blocks occupied by closely located mMSs. The availability of macrocell frequency scheduling information is assumed, and this information is effectively utilized in conjunction with spectrum sensing. The impact of inter-carrier interference (ICI) from the macrocell UL to the femtocell UL is demonstrated via simulations, and how this affects the decisions about spectrum opportunities at a femtocell is discussed. Section 5.4 provides some concluding remarks on interference problems and related solutions for femtocells.

5.2 Femtocell Deployment Scenarios and Capacity

How interference affects the system performance is directly related to the deployment configurations of femtocells. In [6], three different deployment criteria have been specified for femtocell networks: (1) dedicated channel vs. co-channel, (2) open access vs. closed subscriber group (CSG), and (3) fixed downlink (DL) transmit power vs. adaptive DL transmit power. The prior-art works related to femtocells typically investigate the throughput/capacity performance of femtocells and/or macrocells under one of these three deployment criteria. For example, a comparison of the public (open) access and the private (CSG) access femtocells has been presented in [1,7,8] through computer simulations. Co-channel deployments of femtocells have been preferred compared to dedicated channel deployments in [7–10]. The spectrum allocation for femtocells has been discussed in [11,12], while different power control techniques have been proposed in [8,13,14]. A comprehensive collection of simulation results for several important interference scenarios of interest are documented in [15].

Despite several studies, to our best knowledge, a comparative study that captures several trade-offs among different femtocell deployment criteria is not available in the prior-art. The goal of this section is to provide a comparative study of the three different femtocell deployment modes through the help of channel capacity formulations, and to discuss various related trade-offs. Moreover, a comprehensive set of simulation results in realistic macrocell/femtocell settings and channel models will be presented to discuss the impact of key design parameters.

5.2.1 Capacity of Macrocell Users without any Femtocells

Consider a macrocell network, where there are $\tilde{M}$ macrocell mobile stations (mMSs) communicating with a macrocell base station (mBS). Let B_{mac} denote the bandwidth of the spectrum available for the macrocell network. Also, consider a simple scheduler that assigns an equal bandwidth $B_{m,i} = B_{mac}/\tilde{M}$ to each of the mMSs. Then, in the absence of any femtocells, the capacity of an mMS can be written as

$$C_{m,i}^{(nf)} = \frac{B_{mac}}{\tilde{M}} \log\left(1 + \frac{\tilde{M}P_{m,i}}{B_{mac}N_0}\right), \tag{5.1}$$

where
  (nf) refers to a no-femtocell scenario
  $P_{m,i}$ is the received power for the ith mMS
  N_0 denotes the noise spectral density

As apparent from (5.1), the capacity of mMSs will improve with a smaller $\tilde{M}$, a larger B_{mac}, and a larger $P_{m,i}$. Therefore, the capacity of the indoor users will be relatively lower compared to outdoor users; they suffer from wall penetration loss, which results in lower $P_{m,i}$ values. This problem may be handled through deploying femtocells within indoor locations, which will result in larger received powers (and in typical settings, larger bandwidths) for indoor users, yielding higher channel capacities. In the next three subsections, we will review three different femtocell deployment modes and investigate how they affect the channel capacity for macrocell and femtocell users.

5.2.2 Dedicated Channel vs. Co-Channel

5.2.2.1 Dedicated Channel Deployment

For dedicated channel assignments, femtocells are assigned a separate spectrum (of bandwidth $\tilde{B}_{fem}$) than that of the macrocell, as illustrated in Figure 5.3a. Even though this mostly eliminates a potential interference from the

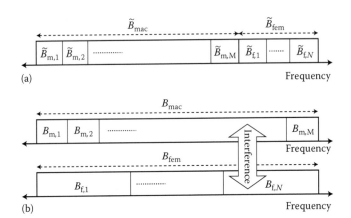

(a)   Frequency

(b)   Frequency

FIGURE 5.3
(a) Dedicated channel operation, vs. (b) co-channel operation of femtocell and macrocell networks. Dedicated channel deployment offers lower bandwidth per femtocell user, while co-channel deployment suffers from interference limitations. (From Mahmoud, H. and Guvenc, I., A comparative study of different deployment modes for femtocell networks, in *Proceedings of the First IEEE International Workshop on Indoor and Outdoor Femto Cells (IOFC),* *in conjunction with IEEE PIMRC 2009,* Tokyo, Japan, Sept. 2009. With permission.)

macrocell,* frequency resources are not efficiently utilized. The capacity of an mMS with a dedicated channel assignment can be written as

$$C_{m,i}^{(dc)} = \frac{B_{mac} - \widetilde{B}_{fem}}{M} \log\left(1 + \frac{MP_{m,i}}{(B_{mac} - \widetilde{B}_{fem})N_0}\right), \qquad (5.2)$$

where
   (dc) refers to a dedicated channel deployment scenario
   i is the index for the mMS
   $\widetilde{B}_{fem}$ denotes the bandwidth dedicated to the femtocell networks
   $M < \widetilde{M}$ is the number of mobile stations (MSs) associated with the
      macrocell[†]

Comparing (5.2) with (5.1), it is observed that with the introduction of femtocells, there is a less available spectrum for the macrocell network. However, also, the $\widetilde{M} - M$ of the users are shifted to the femtocell networks, and they no longer use the macrocell's frequency resources. Hence, in gen-

---

* There may still be co-channel interference between neighboring femtocells, and adjacent channel interference between the macrocell and the femtocells. Even though we neglect both in capacity formulations within this section, the impact of inter-femtocell interference will be taken into account during simulation results.
† After deploying femtocells in the macrocells, some of the MSs that were originally associated with the macrocell are assumed to make hand-off to femtocells due to better signal quality.

eral, the capacity of macrocell users may improve for smaller values of $\widetilde{B}_{\text{fem}}$ and M.

On the other hand, the capacity of a femtocell mobile station (fMS) with a dedicated channel assignment can be written as

$$C_{f,i}^{(\text{dc})} = \frac{\widetilde{B}_{\text{fem}}}{N} \log\left(1 + \frac{NP_{f,i}}{\widetilde{B}_{\text{fem}}N_0}\right),\tag{5.3}$$

where

N is the number of users per femtocell

$P_{f,i}$ is the received signal power from the femtocell base station (fBS)

Comparing (5.3) with (5.1), even though the bandwidth per user ($\widetilde{B}_{f,i} = \widetilde{B}_{\text{fem}}/N$) may be similar to the bandwidth of indoor macrocell users without any femtocell deployment, the received powers $P_{f,i}$ would typically improve significantly with femtocell deployments, hence improving the capacity of indoor users.

5.2.2.2 Co-Channel Deployment

Co-channel deployment of femtocells enables a more efficient utilization of the available spectrum. As illustrated in Figure 5.3b, both the macrocell and the femtocell will have larger bandwidths available per user with co-channel deployments ($B_{m,i} > \widetilde{B}_{m,i}$ and $B_{f,i} > \widetilde{B}_{f,i}$, respectively). Moreover, the cell-search process for an mMS becomes easier since it will not have to search for cells in different frequency bands (e.g., for handover purposes). However, femtocells and the macrocell will observe co-channel interference from each other.

The channel capacity of an mMS with co-channel femtocell deployment can be written as

$$C_{m,i}^{(\text{cc})} = \frac{B_{\text{mac}}}{M} \log\left(1 + \frac{P_{m,i}}{I_{\text{fem}} + B_{\text{mac}}N_0/M}\right),\tag{5.4}$$

where

(cc) refers to a co-channel deployment scenario

I_{fem} is the total interference observed from all the nearby femtocell networks

Comparing (5.4) with (5.2), we observe that the bandwidth available per user improves with co-channel deployment (i.e., $B_{m,i} > \widetilde{B}_{m,i}$). However, the mMSs also observe an interference I_{fem} from nearby femtocell networks, which may degrade the capacity if it is significant. Hence, whether the capacity improves or not with respect to a dedicated channel scenario depends jointly on $\widetilde{B}_{\text{fem}}$ and I_{fem}. Similarly, comparing (5.4) with (5.1), whether the capacity improves or not with respect to a no-femtocell scenario depends jointly on M and I_{fem}.

On the other hand, the capacity of an fMS with a co-channel deployment can be written as

$$C_{f,i}^{(cc)} = \frac{B_{fem}}{N} \log\left(1 + \frac{P_{f,i}}{I_{mac} + B_{fem}N_0/N}\right), \qquad (5.5)$$

where $B_{fem} = B_{mac} \gg \tilde{B}_{fem}$, which implies a significant increase in the available bandwidth per femtocell user. This comes at the expense of the interference I_{mac} observed from macrocell users and the mBS. Since the bandwidth affects the channel capacity linearly, and interference affects the channel capacity through the logarithm function, as also discussed by several other work in the literature [2,7,8,10] co-channel deployments of femtocells typically result in better overall capacities compared to dedicated channel deployments. This will be verified through computer simulations in Section 5.2.5.

5.2.3 Open Access vs. Closed Subscriber Group

For open access femtocell networks, any mMS is allowed to join a particular femtocell network. For CSG type of femtocells, on the other hand, the mMSs that may join a particular femtocell network are restricted to a certain group. Therefore, for the CSG mode, a particular femtocell may receive significant interference from (and cause significant interference too) a close-by co-channel mMS since it will not be granted admission to the femtocell [1,7,8].

How the channel capacity changes for CSG and open access modes can be interpreted using Equations 5.4 and 5.5. For the open access mode, a femtocell will serve a larger number of fMSs since some close-by mMSs will make a hand-off to the femtocell. Therefore, based on (5.5), the bandwidth available per fMS user (B_{fem}/N) will be smaller. However, those mMSs that join the femtocell would typically be the ones that were causing the strongest interference to the femtocell. Hence, interference term I_{mac} will also decrease significantly for the open-access mode and only the far-away mMSs will still be causing interference. This may typically compensate for the bandwidth reduction per fMS and improve the femtocell capacity. From the macrocell network's perspective, based on (5.4), it is easy to see that the open access mode will increase the available bandwidth per macrocell user (B_{mac}/M), improving the capacity of the remaining users associated with the macrocell.

Example simulation results in [1,7,8] show that the open access mode yields a better overall system throughput and coverage, while [1] also shows that CSG results in larger areal capacity gains in general (defined as the ratio of the system capacity with femtocells to the system capacity without any femtocells). Note that compared to the CSG mode, the open access operation may also have several concerns such as privacy issues, extra burden on the backhaul of a femtocell's owner, etc. In Section 5.2.5, we will investigate when CSG may be preferable over the open access mode through a simple example.

5.2.4 Fixed DL Transmit Power vs. Adaptive DL Transmit Power

For fixed transmission power, the fBS fixes its maximum transmission power to a pre-determined value, which is typically set as 13 or 20 dBm. With adaptive transmission power, on the other hand, the fBS may adjust its transmit power based on the interference caused/received to/from the macrocell and other neighboring femtocells. For example, femtocells closer to the mBS may transmit at the maximum power level (due to significant interference from the mBS), while the femtocells closer to the cell edge may decrease their transmit powers since the mBS interference will be weaker. Based on (5.4), the reduction of the fBS transmit power would also reduce the interference I_{fem} to the macrocell network. This results in a larger capacity for macrocell users at the expense of some decrease in the capacity of certain femtocell networks with lower transmit powers. A typical approach for power control is to set the fBS transmit power so that signal to interference ratio (SIR) is equal to 0 dB at the borders of the femtocell [16,18], which is also adopted in this chapter.

5.2.5 Simulation Results

Computer simulations are made to evaluate several trade-offs discussed in Sections 5.2.2 through 5.2.4. Simulation parameters are selected based mostly on [5] and a key set of parameters are summarized in Table 5.1. The downlink of a macrocell/femtocell scenario as in Figure 5.4 is considered with an mBS located at the center of the cell (shown with a triangle). The buildings (squares) and 150 mobile users (circles*) are uniformly scattered over a 600 × 600 grid, within the (hexagonal) borders of the macrocell, where 18% of the area is inside the buildings and 82% of the area is outdoors. Cell selection is based on the SIR metric, where MSs join the femtocell/macrocell that has the best SIR.

In all simulations, idle femtocells with no users are detected and disabled (i.e., their transmit powers are set to zero) to minimize interference. Note that active femtocells are determined in the following manner. All femtocells are initially enabled and the SIR of all femtocells and the macrocell are compared at each user location. Then, users make a hand-off to femtocells with a higher SIR, and femtocells with no users are disabled. Since this process initially enables all femtocells, the interference between femtocells is significant. Therefore, most users prefer not to hand-off to a femtocell, which results in the majority of femtocells being disabled. This is especially true for adjacent channel assignment where interference between femtocells is not observed by macrocell users. A more efficient way to determine idle femtocells (which is not considered in this chapter) would be to turn on femtocells sequentially, so that the interference between femtocells is increased incrementally until a stable point[†] is reached.

---

* Circles with darker color for indoor users and circles with lighter color for outdoor users.
[†] Where, increasing the number of active femtocells would degrade the system performance rather than enhancing it.

TABLE 5.1

Simulation Parameters

Parameter	Value
Central frequency	2.1 GHz
Bandwidth	5 MHz
Coverage (radius) (mBS, fBS)	0.5 km, 8 m
Maximum transmit power (mBS, fBS)	41.75 dBm, 20 dBm
Thermal noise density	−174 dBm/Hz
Wall penetration loss (WL)	10 dB, 20 dB
Antenna gain (mBS, fBS)	17 dBi, 2 dBi
Feeder/cable loss (mBS, fBS)	3 dB, 1 dB
Antenna heights (mBS, fBS, MS)	15 m, 1.5 m, 1.5 m
House size	15 m × 15 m
Street width	10 m
Distance between grid points	1.7 m
Number of users per macrocell	150
Indoor area vs. outdoor area	18% vs. 82%
Scheduling strategy	Equal bandwidth per user
Indoor to indoor path loss model	ITU P.1238
Indoor to outdoor path-loss model	ITU P.1411 + Wall Loss
Outdoor to outdoor path-loss model	ITU P.1411
Outdoor to indoor path-loss model	ITU P.1411 + Wall Loss

Source: Mahmoud, H. and Guvenc, I., A comparative study of different deployment modes for femtocell networks, in *Proceedings of the First IEEE International Workshop on Indoor and Outdoor Femto Cells (IOFC), in conjunction with IEEE PIMRC 2009*, Tokyo, Japan, Sept. 2009. With permission.

Based on the fixed building/macrocell layout in Figure 5.4 and the simulation parameters in Table 5.2, the cumulative distribution functions (CDFs) of the capacities of indoor users, outdoor users, and all users are plotted for different scenarios. First, Figure 5.5 includes the capacity plots in the absence of any femtocells and for two different wall penetration losses 10 and 20 dB, where the results are averaged over 50 different realizations of user locations. Results show that compared to outdoor users, the capacity of indoor users suffers significantly from the wall penetration loss, which can be addressed through deploying femtocells. In Sections 5.2.5.1 through 5.2.5.3, capacity improvements due to introducing femtocells with different deployment modes will be demonstrated through simulations.

5.2.5.1 Dedicated Channel vs. Co-Channel

Figures 5.6 and 5.7 compare the capacity CDFs of the outdoor and the indoor users for dedicated channel and co-channel deployments, respectively, using an open access mode of operation and without any power control at femtocells. In Figure 5.6, a dedicated channel femtocell deployment

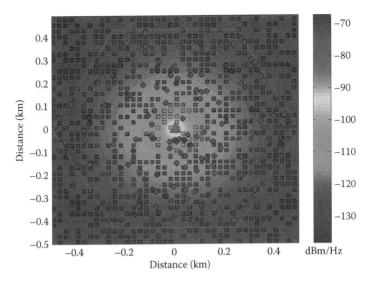

FIGURE 5.4
Macrocell/femtocell simulation scenario in consideration. Received power from the mBS (in dBm/Hz) is also illustrated. (From Mahmoud, H. and Guvenc, I., A comparative study of different deployment modes for femtocell networks, in *Proceedings of the First IEEE International Workshop on Indoor and Outdoor Femto Cells (IOFC), in conjunction with IEEE PIMRC 2009*, Tokyo, Japan, Sept. 2009. With permission.)

TABLE 5.2

Comparison of Median Capacities with and without Femtocells

	No Femtocells (Mbps)	With Femto (dc) (Mbps)	With Femto (cc, no pc) (Mbps)	With Femto (cc, with pc) (Mbps)
All users (WL = 20 dB)	1.15	4.13	14.2	5.11
Indoor users (WL = 20 dB)	0.80	16.2	72.2	21.7
Outdoor users (WL = 20 dB)	1.23	1.35	1.23	1.4
All users (WL = 10 dB)	1.19	1.23	8.60	4.74
Indoor users (WL = 10 dB)	1.03	1.70	42.9	20.9
Outdoor users (WL = 10 dB)	1.23	1.12	0.98	1.15

Source: Mahmoud, H. and Guvenc, I., A comparative study of different deployment modes for femtocell networks, in *Proceedings of the First IEEE International Workshop on Indoor and Outdoor Femto Cells (IOFC), in conjunction with IEEE PIMRC 2009*, Tokyo, Japan, Sept. 2009. With permission.

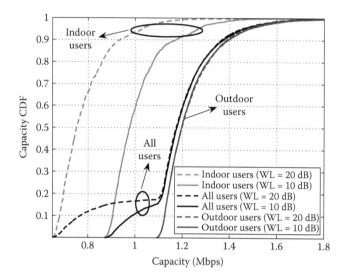

FIGURE 5.5

Comparison of indoor and outdoor capacity CDFs with no femtocells. The capacity of indoor users suffer from wall penetration loss, which may be improved through deploying femtocells. (From Mahmoud, H. and Guvenc, I., A comparative study of different deployment modes for femtocell networks, in *Proceedings of the First IEEE International Workshop on Indoor and Outdoor Femto Cells (IOFC), in conjunction with IEEE PIMRC 2009*, Tokyo, Japan, Sept. 2009. With permission.)

scenario is considered, where it is assumed that all the buildings have femtocells (100% femtocell penetration), there is no interference between femtocells and the macrocell, while there occurs interference among neighboring femtocells. Macrocell users use 90% of the available bandwidth B_{mac}, while 10% of the bandwidth is assigned to femtocell users.* The bandwidth is equally distributed among the users of the femtocells and the macrocell based on the number of users associated with each cell. An open access deployment is considered, where, each MS chooses the cell with the best signal-to-interference-ratio (SIR). Results in Figure 5.6 show a significant capacity improvement of indoor users compared to Figure 5.5 (no femtocells), especially when the wall penetration loss is large. This is because for a larger WL, interference from neighboring femtocells is lower, yielding larger SIRs at femtocells. Note that for WL = 10 dB, most of the indoor MSs still observe better SIRs when they are connected to the mBS. However, since the bandwidth available per user is smaller at the mBS, the capacity of most of the indoor users is significantly lower for WL = 10 dB when compared with the indoor user capacity of WL = 20 dB.

In Figure 5.7, on the other hand, a co-channel deployment scenario is considered for femtocells. In addition to inter-femtocell interference, we

---

* Note that optimization of the dedicated bandwidth size of femtocells may further improve the capacity performance of users, which is not specifically investigated in this chapter.

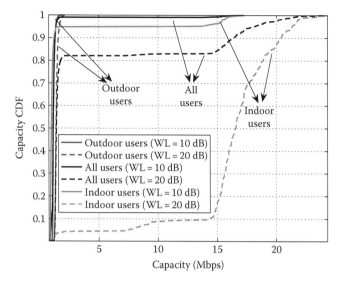

FIGURE 5.6
Comparison of indoor and outdoor capacity CDFs for dedicated channel femtocell deployments. (From Mahmoud, H. and Guvenc, I., A comparative study of different deployment modes for femtocell networks, in *Proceedings of the First IEEE International Workshop on Indoor and Outdoor Femto Cells (IOFC), in conjunction with IEEE PIMRC 2009*, Tokyo, Japan, Sept. 2009. With permission.)

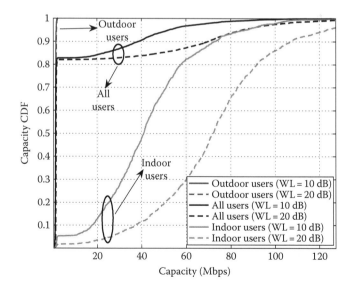

FIGURE 5.7
Comparison of indoor and outdoor capacity CDFs for co-channel femtocell deployments. (From Mahmoud, H. and Guvenc, I., A comparative study of different deployment modes for femtocell networks, in *Proceedings of the First IEEE International Workshop on Indoor and Outdoor Femto Cells (IOFC), in conjunction with IEEE PIMRC 2009*, Tokyo, Japan, Sept. 2009. With permission.)

also consider interference between the femtocells and the macrocell for co-channel deployment. However, the available bandwidth per femtocell user is significantly improved, since femtocells have access to the whole spectrum that is used by the macrocell (see, e.g., (5.5)). Results in Figure 5.7 show significant gains compared to dedicated channel deployment results in Figure 5.6. This verifies our discussion in Section 5.2.2 that in general, the large bandwidth gain compensates for the increase in interference, making co-channel deployments preferable over dedicated channel deployments.

5.2.5.2 CSG vs. Open Access

A simple single-femtocell simulation scenario as in Figure 5.8a is considered for the comparison of channel capacities with open access and CSG modes of co-channel femtocells. The uplink (UL) transmission of an fMS at a distance of 7 m to the fBS is considered, with fBS being located at the center of a 15 × 15 m apartment. The UL signal received at the fBS is interfered by N_m mMS signals, located at a distance d_2 to the fBS. If the femtocell operates in the open access mode, these mMSs are accepted and served by the femtocell, sharing its bandwidth. Otherwise, with the CSG mode of operation, they are served by the macrocell, causing interference to the femtocell users.

Mean capacities for the users associated with the femtocell are plotted in Figure 5.8b for open access and CSG modes, for $N_m = 1, 2, 3, 4$ and $d_2 \in [3, 6, ..., 39]$ meters. When the mMS(s) are located within the borders of the apartment, there is significant UL interference in the CSG mode, which results in poor channel capacities for the femtocell users. On the other hand, in the open access mode, the channel capacity with indoor mMSs is acceptable even with large numbers of mMSs. For the mMSs located outdoors, the CSG mode of operation becomes preferable for longer distances between the fBS and the mMSs, and for a larger number of interfering mMSs.

5.2.5.3 Impact of Power Control

To test the performance of femtocells under adaptive power control, the following setup was employed. The femtocell transmit power is adjusted to maintain an SIR level of 0 dB (with respect to the macrocell signal) at the edge of the coverage area. The minimum femtocell range is set to 8 m with the assumption that the coverage area is indoors. Thus, the SIR just outside the building is −20 or −40 dB for a WL of 10 or 20 dB, respectively.* This setup

---

* Note that even though wall loss in a realistic scenario is not homogenous due to windows, doors, etc., simplified homogenous model that we consider in this chapter still provides useful insights about associated trade-offs for different wall loss values.

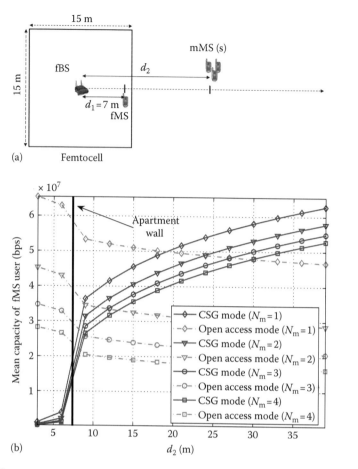

FIGURE 5.8
Simulation results comparing the mean femtocell UL capacities for CSG and open access femtocell deployments. Different numbers of mMSs located at a distance d2 to the fBS are considered. (a) Simulation scenario in consideration; (b) mean capacity of an fMS user, as a function of the distance of an mMs user to a femtocell. (From Mahmoud, H. and Guvenc, I., A comparative study of different deployment modes for femtocell networks, in *Proceedings of the First IEEE International Workshop on Indoor and Outdoor Femto Cells (IOFC), in conjunction with IEEE PIMRC 2009*, Tokyo, Japan, Sept. 2009. With permission.)

reduces interference to macrocell users, especially users in the proximity of femtocells located near the edge of the macrocell area.

The impact of the power control on the capacity CDFs of macrocell and femtocell users is illustrated in Figure 5.9. When compared with the results in Figure 5.7, we note that the adaptive power control can, as expected, improve the macrocell/outdoor user capacity due to reduction in the interference levels from femtocells. However, the degradation to femtocell users, represented by indoor users in this case, is clearly more significant. This is explained by the fact that macrocell users only use a small fraction of the

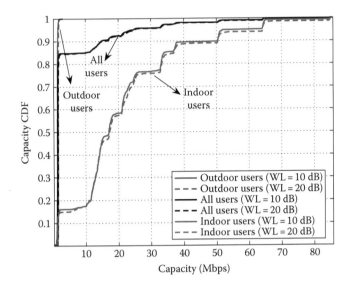

FIGURE 5.9
Comparison of indoor and outdoor capacity CDFs for co-channel femtocell deployments with power control (pc). (From Mahmoud, H. and Guvenc, I., A comparative study of different deployment modes for femtocell networks, in *Proceedings of the First IEEE International Workshop on Indoor and Outdoor Femto Cells (IOFC), in conjunction with IEEE PIMRC 2009*, Tokyo, Japan, Sept. 2009. With permission.)

total bandwidth when compared to femtocell users who typically utilize 50%–100% of total bandwidth (1–2 users per femtocell). As such, the reduction in capacity in femtocells is more sensitive to the SIR level when compared to macrocell users.

The median capacities for the three simulation scenarios in Figures 5.5 through 5.7 and in Figure 5.9 are summarized in Table 5.2 for indoor users, outdoor users, and all users. Results show that co-channel femtocell deployments result in tremendous capacity gains for indoor users, which comes at the expense of possible capacity degradation for outdoor users due to interference from femtocells. It is observed that wall loss becomes a critical factor that determines the capacity gains with femtocell deployments. In order for the indoor users to have larger capacities, larger wall loss is desirable due to better isolation between the femtocells and the macrocell. For larger wall loss, it is also observed that deploying femtocells in essence may improve the median capacity of macrocell users; this is due to the off-loading of macrocell users to femtocells, which results in larger bandwidths for the remaining macrocell users. When the wall loss is lower, on the other hand, the bandwidth gain due to off-loading of users to femtocells does not compensate for the interference observed from the femtocells, and the median capacity of outdoor users degrades due to interference from femtocells.

5.3 Interference Scenarios and Avoidance Techniques in Femtocells

Femtocell networks have been studied extensively under the 3GPP standard (see e.g., [5,6], and the references therein), where wideband code division multiple access (W-CDMA) is used as the physical layer technology. Even though the system capacity and the performance have been analyzed for macrocells and femtocells in the presence of CCI in [2,10,17], it can be said that these studies are more specific to CDMA-based systems. There are only limited studies on OFDMA-based femtocell networks (e.g., based on the LTE or WiMAX standards*), which offer greater flexibility in terms of allocation of frequency resources. An efficient utilization of spectrum resources that causes minimum interference to other networks is a critical requirement for the successful deployment of next-generation femtocell networks, which requires intelligent scheduling techniques. Some operators summarized their views regarding LTE femtocells in the 3GPP standard as "femtocell can actively attempt to configure its cell resources to minimize impacts to the operator-deployed cells, and to avoid interference interactions with other home cells in its vicinity" [18], and "in order to maximize system capacity and throughputs in the uplink, the fBS could decide to schedule its users to transmit on those resource blocks that experience the lowest other cell/channel interference" [19].

Some related work on avoiding interference in OFDMA networks through spectrum sensing [20] and through intelligent radio resource allocation [21,22] is available in the literature in the context of cognitive radio systems; however, these works do not consider system-specific issues related to femtocells. Significant improvements in throughput per unit area have been demonstrated in [1] when OFDMA-based WiMAX femtocells are used (on the order of 15 times throughput improvement for dense-deployments in large cells); however, co-channel interference was stated as an important factor that limits the overall network performance. In [7], trade-offs between public access and private access were compared for WiMAX femtocells through realistic system-level simulations, and public access was shown to yield considerably larger throughput due to reduced interference. It was discussed in [11] that the throughput of a naïve co-channel WiMAX femtocell deployment may suffer a lot from the interference; hence, an interference avoidance technique that uses a dynamic frequency-planning technique was introduced, which considerably improves the throughput performance.

In the following section, first, CCI and ICI issues in co-channel OFDMA femtocells will be discussed. Then, a frequency scheduling method for

---

* WiMAX uses OFDMA both in downlink and uplink, while LTE employs OFDMA in downlink and single-carrier frequency division multiple access (SC-FDMA) in uplink.

avoiding CCI will be described and the impact of ICI on co-channel femto-cells will be analyzed.

5.3.1 CCI and ICI Issues in Femtocell Deployments

As discussed in Section 5.2, co-channel deployments have the potential of providing significantly better throughputs compared to dedicated chan-nel deployments. Hence, we consider co-channel deployments of femtocell networks in this section, primarily due to significant bandwidth benefits for the femtocells. However, co-channel interference still remains a critical problem, especially for scenarios where there are mMSs in the vicinity of a femtocell. In the access mode referred to as the CSG mode [6] discussed in the Section 5.2, close-by mMSs are not allowed to make hand-offs to the femtocell network. Especially for femtocells on the edge of the macrocell, this implies significant interference concerns between the mMSs and the femtocell: during a macrocell downlink, the mMSs will be interfered with significantly by the femtocell, while during a macrocell uplink, the mMSs will cause significant interference to the femtocell (see e.g., the discussion on *dead zones* in [2]). In [7,17], it has been shown that due to such interfer-ence problems, open access femtocells (which allow the hand-off of close-by mMSs to the femtocells) will provide better throughput compared to the CSG type of femtocells. However, an open-access approach may also have some concerns such as privacy issues, reduction of available bandwidth per femtocell user, extra burden on the femtocell owner's backhaul connection etc. One of the contributions of the present work is an alternative method for handling the interference from close-by mMSs, by avoiding use of their frequency resources at the femtocell network, which will be discussed in more detail in Section 5.3.2.

A different type of interference that may be observed in next generation femtocell deployments is the inter-carrier-interference (ICI) [2,23]. It occurs during the uplink transmission of mMSs and it is caused by the timing mis-alignment of mMS signals at the femtocell. Typically, the mMSs get synchro-nized to the mBS in the uplink through a ranging process, where the mBS determines and announces the instant to start transmission for each mMS. Since such a synchronization mechanism is not applicable for the fBS, the mMS signals arrive at the fBS with different delays due to their different dis-tances to the fBS, as illustrated in Figure 5.10b. Assuming that the femtocell is synchronized to the first arriving mMS signal in the uplink, if the delays of some of the other mMSs exceed the cyclic prefix (CP) of the femtocell symbol as in Figure 5.10c, orthogonality between the subcarriers is lost. This causes inter-carrier interference in the resource blocks that can be reused by the fBS or fMS, as illustrated in Figure 5.10d. Since the spread of the mMS arrival times will be larger, this is especially a critical problem for femtocells that are located closer to the macrocell edge [23]. An analysis of ICI in femtocell networks will be discussed in more detail in Section 5.3.3.

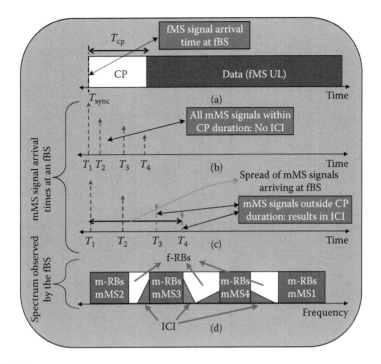

FIGURE 5.10
The relation between the arrival times of mMS signal delays, CP-size, and ICI: (a) structure of a femtocell symbol, (b) signal arrival times from four different mMSs at an fBS that is within the CP, (c) mMS signal arrival times that exceed the CP, and (d) illustration of ICI due to delays larger than the femtocell CP size. (m-RB: macrocell resource block, f-RB: femtocell resource block.) (From Sahin, M.E. et al., *IEEE Trans. Consumer Electron.*, 55(4), 1936, 2009. With permission. ©2009 IEEE.)

5.3.2 Frequency Scheduling for Avoiding the CCI

In CSG femtocell deployment, the potential co-channel interference problem between the femtocell and the mMSs can be prevented by ensuring that the femtocell avoids using the macrocell resource blocks that belong to nearby mMSs. Two exemplary methods that can be used in CSG femtocell deployment are outlined in the framework in Figure 5.11a. One of these methods is based on macrocell-centered scheduling, while the other is based on autonomous scheduling at a femtocell network.

The steps illustrated in Figure 5.11a for the two different scheduling methods for avoiding CCI can be summarized as follows. First, the fBS performs spectrum sensing for finding the occupied parts of the spectrum, which are supposedly the resource blocks of nearby mMSs. If macrocell-centered scheduling is employed, the spectrum sensing results and fBS scheduling decisions are communicated with the mBS. After collecting these results from all fBSs, the mBS schedules or re-schedules its own users in an effort to minimize the potential CCI to the femtocells. The availability of fBS and mMS geo-location information can considerably increase the efficiency of scheduling decisions made by the mBS. If autonomous scheduling is employed, the fBS receives the mMS scheduling information from the mBS and determines the spectrum opportunities by comparing the sensing results with this scheduling information. In both methods, the fBS schedules its fMSs over the spectrum opportunities determined as illustrated in Figure 5.11b for the macrocell UL and in Figure 5.11c for the macrocell DL.

In the remainder of this section, certain critical steps of the proposed framework will be discussed in detail.

5.3.2.1 Obtaining Scheduling Information from the mBS

We consider two possible methods for the fBSs to obtain the scheduling information of mMSs from the mBS. These two methods are to receive the scheduling information through the backhaul, or to capture the scheduling information over the air. In the first method, upon initialization, fBS establishes a secure and a stable backhaul connection to the operator network

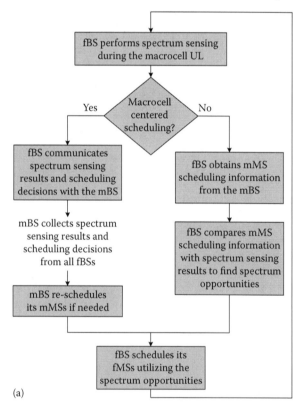

(a)

FIGURE 5.11
Framework for handling co-channel interference between the femtocell and the macrocell. (a) Flowchart for the proposed scheduling method; (b) macrocell uplink; (c) macrocell downlink. (From Sahin, M.E. et al., *IEEE Trans. Consumer Electron.*, 55(4), 1936, 2009. With permission. ©2009 IEEE.)

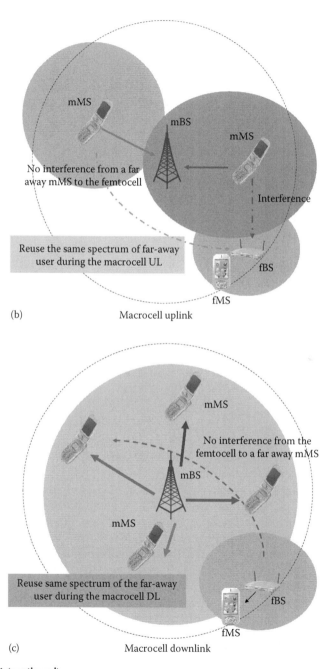

(b)   Macrocell uplink

(c)   Macrocell downlink

FIGURE 5.11 (continued)

and obtains information about the mMS resource blocks. Depending on the availability of geo-location information, the mMS scheduling information provided to an fBS can be limited to the mMSs that are at a certain neighborhood of that fBS. The mBS needs to make and then deliver the scheduling decision well ahead of the scheduled transmission, the *ahead time* being greater than the latency of the backhaul. Because the large ahead time makes it difficult for a channel dependent scheduler to capture instantaneous channel variation, the efficiency of such a scheduler is likely to be degraded. Also, the delivery of the scheduling information consumes precious bandwidth of the backhaul, which otherwise could be used to deliver the actual data. One possible improvement to address this efficiency problem is to make an ahead scheduling decision (and accompanied sensing and comparison) only when it is necessary. For example, it can be done only in the following cases: initialization of fBSs, handoff of mMSs from/to neighboring mBSs, significant change of channel status, interference level, or resource requests, etc. Most of the other times, the channel-dependent scheduler could make scheduling decisions considering the instantaneous channel status. Resource sharing between the mBS and the fBS can be made semi-static based on the most recent scheduling that was accompanied by sensing and comparison. In the second method, the mMS scheduling information is received from the mBS over the air. For this purpose, the fBS may connect to the mBS as an mMS, and use this connection in order to obtain the scheduling information.

5.3.2.2 *Jointly Utilizing Scheduling Information and Spectrum Sensing Results*

In a well designed OFDMA system, it is expected that almost all resource blocks are allocated to users. Therefore, scheduling information that the fBS obtains from the mBS would normally indicate that the spectrum is mostly occupied. However, since many of the mMSs are far away from the fBS, their resource blocks can still be utilized at the femtocell. Through spectrum sensing, an fBS can detect resource blocks that are either not used by the macrocell network, or belong to far away users. Two methods that can be employed for spectrum sensing by the fBS are energy detection and ESPRIT algorithms (a detailed analysis of UL spectrum sensing by femtocells can be found in [24]).

Once the spectrum sensing results are available, one of the two methods outlined in Figure 5.11a can be used. The fBS may send the spectrum sensing results to the mBS for macrocell-centric scheduling, or, the fBS may compare the spectrum sensing results with the scheduling information to decide about spectrum opportunities. Spectrum sensing results are impaired with missed detections (MD) and false alarms (FA) due to additive noise. In the macrocell uplink, another reason for these impairments is the inter-carrier interference that is caused by timing misalignment. In the presence of ICI, the resource blocks that are adjacent to the nearby mMSs' resource blocks may also observe strong interference.

Given the existence of spectrum sensing impairments, it is not reasonable to expect all subcarriers in a supposedly unoccupied block to be detected as empty. Empirically, it is found that a resource block should be considered unoccupied as long as the number of subcarriers detected as non-empty does not exceed 25% of the total number of subcarriers in a resource block.

As a simple example, consider the macrocell UL scenario illustrated in Figure 5.12, where a resource block size of 12 subcarriers as in LTE is considered. The fBS compares the spectrum-sensing results (shown in Figure 5.12c) with the scheduling information (shown in Figure 5.12b) to determine the resource blocks that constitute spectrum opportunities. Energy detection applied to the received signal (Figure 5.12a) yields some FAs due to the noise and due to ICI as shown in Figure 5.12c. These false alarms may or may not cause a resource block to be considered as occupied based on their number within each resource block. The spectrum opportunities found for the illustrated scenario are shown in Figure 5.12d, where scheduling information is plotted again for visual comparison. Note that even though the scheduling information indicates otherwise, the resource block associated with the mMS2 may be utilized by the fBS in this scenario. This is because the mMS2 is a far-away mMS to the fBS, and the spectrum sensing results indicate that the received signal power is relatively weak.

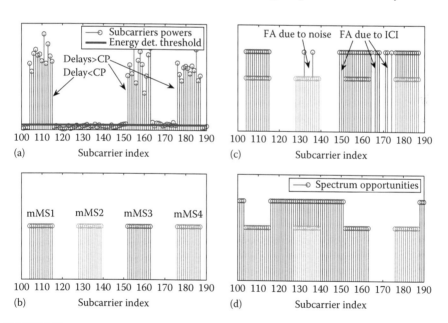

FIGURE 5.12
Combining scheduling information with spectrum sensing results: (a) received subcarrier powers and the energy detection threshold; (b) scheduling information for multiple mMSs, where an LTE resource block size of 12 subcarriers is considered; (c) energy detection results matched with scheduling information, false alarms occur due to noise and due to ICI; (d) spectrum opportunities detected. (From Sahin, M.E. et al., *IEEE Trans. Consumer Electron.*, 55(4), 1936, 2009. With permission. ©2009 IEEE.)

5.3.3 Analysis of the ICI

The distance between the mMS and the fBS has two opposing effects on the inter-carrier interference. While a longer distance leads to a larger delay and hence a stronger ICI, since path loss increases exponentially with increasing distance, a weaker ICI should be observed at large distances. A simulation is performed in order to quantify the variation of the ICI with an increasing mMS to fBS distance. In the simulation, two mMSs (mMS1 and mMS2) are considered, and it is assumed that they are both synchronized to the mBS. A worst case scenario is considered where round-trip-delays (RTDs) need to be taken into account. Consider that the mMSs, the mBS, and the fBS are on the same line, where mMSs are in between the mBS and the fBS. mMS2 has a constant location that is very close to the mBS, while the mMS1 is very close to the femtocell. The distance d between the two mMSs is increased by moving the femtocell (and mMS1) away from the mBS. In order to synchronize to the mBS, the mMS1 needs to start its transmission d/c seconds earlier than the mMS2, where c is the speed of light (this arrangement is done via initial ranging). Therefore, the difference between the arrival times of the mMS2 and the mMS1 signals at the fBS is $2d/c$ seconds, which corresponds to the round-trip-delay (see, e.g., Figure 5.10).

In the simulation results shown in Figures 5.13 and 5.14, the resource blocks allocated to the mMS2 are randomly spread around the given spectrum, and the ICI is measured by determining the total energy in the unused resource

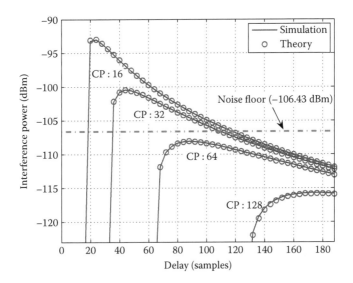

FIGURE 5.13
Variation of the ICI power depending on round-trip-delay values for various cyclic prefix sizes (in samples), where the FFT size is 512 samples. (From Sahin, M.E. et al., *IEEE Trans. Consumer Electron.*, 55(4), 1936, 2009. With permission. ©2009 IEEE.)

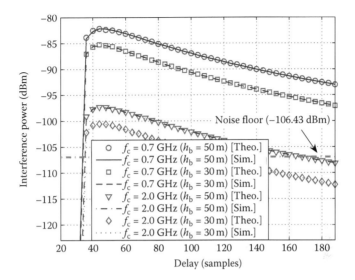

FIGURE 5.14
Variation of the ICI power with respect to RTD for different center frequencies and femtocell
BS heights.

blocks. The theoretical curves are plotted using the expressions for the ICI
given in [24] to verify the simulation results.

The path-loss model used in the simulations, which is derived from the
Okumura–Hata model, is obtained from [5]. The model, which is applicable
in small to medium cities, yields the path loss (in dB) as follows:

$$L = 46.3 + 33.9\log(f_c) - 13.82\log(h_b) + (44.9 - 6.55\log(h_b))\log(d) - F(h_M),$$
(5.6)

where
  f_c is the center frequency
  h_b is the base station height above ground
  h_M is the mobile station height above ground, and $F(h_M)$ is given by

$$F(h_M) = (1.1\log(f_c) - 0.7)h_M - (1.56\log(f_c) - 0.8).$$
(5.7)

Concerning the scenario at hand, h_b should be considered as the height
of a femtocell BS, which is found in a high-rise building. The parameters
related to wave propagation and path loss used in the simulations are mainly
selected according to the values given in [5,17]. All simulation parameters are
listed in Table 5.3.

TABLE 5.3

Simulation Parameters

Parameter	Value
Center frequency (f_c)	700 MHz, 2 GHz
Bandwidth (B)	5.714 MHz
FFT size	512
Symbol duration	89.9 μs
CP sizes	1/32, 1/16, 1/8, 1/4
BS height (h_b)	30 m, 50 m
MS height (h_M)	2 m
MS transmit power	27 dBm
Antenna gain	16 dB
Wall penetration loss	15 dB
Number of Walls	1 (external)
Noise floor of fBS	-174 dBm/Hz $+ 10 \log_{10}(B) = -106.43$ dBm

Source: Sahin, M.E. et al., *IEEE Trans. Consumer Electron.*, 55(4), 1936, 2009. With permission. ©2009 IEEE.

In Figure 5.13, the variation of the ICI power depending on the RTD is plotted (for $f_c = 2$ GHz, $h_b = 30$ m), where the largest delay corresponds to a distance of 5 km. From the curves plotted for various CP sizes, it can be concluded that

- ICI is typically close to zero for delays smaller than the CP size. Once the CP is exceeded, there is a sharp rise in the ICI, but at larger distances the ICI decreases due to increased path loss.
- For larger CP sizes, the ICI is always lower compared to smaller CP sizes, regardless of the delay.
- Under certain conditions, the ICI power might be lower than the noise floor, i.e., the effect of the ICI can become negligible.

As apparent from Figure 5.13, for some mMSs whose signals arrive at the femtocell after the CP, the interference level may be higher than the noise level as much as 13 dB. Hence, certain resource blocks, which would have normally been used by the femtocell, might not be utilized due to their high interference level.

To investigate how the ICI power is affected by the changes in certain important system parameters, the ICI versus delay analysis is performed for two different center frequencies and two BS heights (employing a CP size of 32 samples). The results demonstrated in Figure 5.14 show that employing a lower center frequency or having the BS at a higher location might considerably increase the received interference power.

Another analysis is performed on the error probability in detecting the occupied and unoccupied subcarriers in the received UL signal via energy detection. The error probability is computed as the sum of the probability of missed detection (PMD) and the probability of false alarm (PFA). The PMD is defined as the ratio of the number of subcarriers detected as unused although they are used to the total number of subcarriers N. PFA, on the other hand, is the ratio of the number of subcarriers detected as used although they are unused to N.

Two different subcarrier assignment schemes are considered. The first one is a blockwise assignment, where the two schemes used in the simulations are an LTE resource block with 12 subcarriers and 7 symbols, and a WiMAX UL PUSC tile with 4 subcarriers and 3 symbols. These schemes will be shortly denoted as 12×7 and 4×3, respectively. The other assignment scheme considered is a randomized assignment, where each individual subcarrier may be assigned to a different user. The two randomized assignment schemes employed in the simulations are 1×7 and 1×3. Although not used in any standard, this scheme is included in our simulations to investigate the effect of using a small number of subcarriers as an assignment unit. The maximum RTD that the latest-arriving user signal can have (τ_{max}) is considered to be between 0 and 60 µs, where all other user RTDs are between 0 µs and τ_{max}. Note that τ_{max} values greater than 11.2 µs exceed the CP duration.

Figure 5.15 demonstrates the related simulation results, where the signals of all 12 users are considered to have the same power. An optimum energy detection threshold is used in all cases (see [24] for a detailed analysis of obtaining the optimum threshold in different scenarios). It is observed that in a randomized assignment, the ICI has a more destructive effect on the detection performance. This is because in a randomized assignment, each occupied subcarrier affects its adjacent subcarriers, some of which may be

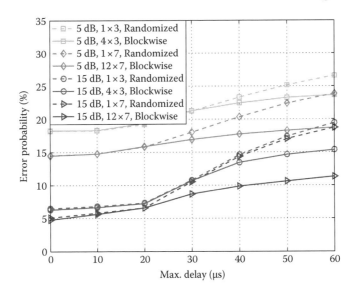

FIGURE 5.15
Error probability versus τmax for energy detection with blockwise and randomized assignments. (From Sahin, M.E. et al., *IEEE Trans. Consumer Electron.*, 55(4), 1936, 2009. With permission. ©2009 IEEE.)

unoccupied. In the blockwise assignment, however, the subcarriers that are strongly affected are limited to the ones that are adjacent to each block. The two reasons for the error rates being higher for block size 4×3 compared to block size 12×7 are that the number of subcarriers is smaller leading to a higher number of affected empty subcarriers, and the number of symbols is smaller resulting in a worse noise averaging.

5.4 Conclusion

In this chapter, different deployment modes for femtocell networks are compared through the help of capacity formulations and computer simulations. Bandwidth and interference are shown to be two critical parameters inversely affecting the capacity. However, due to the low transmission powers of femtocells, it is typically desirable to have co-channel operation that favors bandwidth gains over interference-free communications. In cases where interference is intolerable (e.g., for mMSs inside a femtocell), the open access mode is preferable over the CSG type of operation. Femtocell power control can be utilized for limiting the amount of interference caused to the macrocell users, which comes at the expense of some degradation in the femtocell capacity.

In the second part of the chapter, the CCI and the ICI problems and possible interference mitigation approaches for OFDMA-based co-channel femtocells are considered. The CCI is avoided by determining the spectra of closely located users through spectrum sensing and by not using their resource blocks. Moreover, the impact of ICI is taken into account, which occurs due to the asynchronous arrival of mMS signals to an fBS. In order to improve the spectrum sensing results, the scheduling information obtained from the mBS is utilized. The proposed framework may help in solving the interference problems that may be observed when using the CSG mode of OFDMA-based femtocell networks.

Acknowledgment

The authors would like to thank Moo-Ryong Jeong and Fujio Watanabe from DOCOMO USA Labs for fruitful discussions.

References

1. S. P. Yeh, S. Talwar, S. C. Lee, and H. Kim, WiMAX femtocells: A perspective on network architecture, capacity, and coverage, *IEEE Commun. Mag.*, 46(10), 58–65, Oct. 2008.
2. V. Chandrasekhar, J. G. Andrews, and A. Gatherer, Femtocell networks: A survey, *IEEE Commun. Mag.*, 46(9), 59–67, Sept. 2008.

3. M. Rumney, IMT-advanced: 4G wireless takes shape in an olympic year, White Paper, Sept. 2008 [Online]. Available: http://cp.literature.agilent.com/litweb/pdf/5989–9793EN.pdf

4. Airvana, Femtocells: Personal base stations, 2007, white Paper [Online]. Available: http://www.airvana.com/files/FemtoOverviewWhitepaperFINAL12-July-07.pdf

5. FemtoForum, Interference management in UMTS femtocells, White Paper, Dec. 2008 [Online]. Available: http://www.femtoforum.org/femto/Files/File/Interference Management in UMTS Femtocells.pdf

6. 3rd Generation Partnership Project, Technical specification group radio access networks, 3G Home NodeB Study Item Technical Report (Release 8), 3GPP, 3GPP TR 25.820, Valbonne, France, Mar. 2008.

7. D. L. Perez, A. Valcarce, G. D. L. Roche, E. Liu, and J. Zhang, Access methods to WiMAX femtocells: A downlink system-level case study, in *Proceedings of the IEEE International Conference on Communications System (ICCS)*, Guangzhou, China, Nov. 2008, pp. 1657–1662.

8. H. Claussen, Performance of macro- and co-channel femtocells in a hierarchical cell structure, in *Proceedings of the IEEE International Symposium on Personal, Indoor and Mobile Radio Communications (PIMRC)*, Athens, Greece, Sept. 2007, pp. 1–5.

9. L. T. W. Ho and H. Claussen, Effects of user-deployed, co-channel femtocells on the call drop probability in a residential scenario, in *Proceedings of the IEEE International Symposium on Personal, Indoor and Mobile Radio Communications (PIMRC)*, Athens, Greece, Sept. 2007, pp. 1–5.

10. V. Chandrasekhar and J. G. Andrews, Uplink capacity and interference avoidance for two-tier cellular networks, in *Proceedings of the IEEE Global Telecommunications Conference (GLOBECOM)*, Washington, DC, Nov. 2007, pp. 3322–3326.

11. D. L. Perez, G. D. L. Roche, A. Valcarce, A. Juttner, and J. Zhang, Interference avoidance and dynamic frequency planning for WiMAX femtocells networks, in *Proceedings of the IEEE International Conference on Communications Systems (ICCS)*, Guangzhou, China, Nov. 2008, pp. 1579–1584.

12. V. Chandrasekhar and J. Andrews, Spectrum allocation in tiered cellular networks, *IEEE Trans. Commun.*, 57(10), 3059–3068, Oct. 2009.

13. Nokia Siemens Networks, LTE Home Node B downlink simulation results with flexible Home Node B power, in *3GPP TSG-RAN Working Group 4 (Radio) Meeting*, Shangai, China, Oct. 2007 [Online]. Available: ftp://ftp.3gpp.org/tsgran/WG4Radio/TSGR444bis/Docs/R4–071540.zip

14. V. Chandrasekhar, J. G. Andrews, T Muharemovic, Z. Shen, and A. Gatherer, Power control in two-tier femtocell networks, *IEEE Trans. Wireless Commun.*, 8(8), 4316–4328, Aug. 2009.

15. H. Mahmoud and I. Guvenc, A comparative study of different deployment modes for femtocell networks, in *Proceedings of the First IEEE International Workshop on Indoor and Outdoor Femto Cells (IOFC), in conjunction with IEEE PIMRC 2009*, Tokyo, Japan, Sept. 2009.

16. I. Guvenc, M. R. Jeong, F. Watanabe, and H. Inamura, A hybrid frequency assignment for femtocells and coverage area analysis for co-channel operation, *IEEE Commun. Lett.*, 12(12), 880–882, Dec. 2008.

17. H. Claussen, L. Ho, and L. Samuel, An overview of the femtocell concept, *Bell Labs Tech. J.*, 13(1), 221–245, 2008.

18. Orange, Telecom-Italia, T-Mobile, and Vodafone, Requirements for LTE Home eNodeBs, 3GPP Document R4–070209, Lemesos, Cyprus, Mar. 2007.
19. Vodafone-Group, Home eNodeB considerations for LTE, 3GPP Document R4–070456, Valbonne, France, Apr. 2007.
20. S.-Y. Tu, K.-C. Chen, and R. Prasad, Spectrum sensing of OFDMA systems for cognitive radios, in *Proceedings of the IEEE International Symposium on Personal, Indoor, and Mobile Radio Communications (PIMRC)*, Athens, Greece, Sept. 2007, pp. 1–5.
21. F. S. Chu and K. C. Chen, Radio resource allocation in OFDMA cognitive radio systems, in *Proceedings of the IEEE International Symposium on Personal, Indoor, and Mobile Radio Communications (PIMRC)*, Athens, Greece, Sept. 2007, pp. 1–5.
22. T. H. Kim and T. J. Lee, Spectrum allocation algorithms for uplink sub-carriers in OFDMA-based cognitive radio networks, in *Proceedings of the IEEE International Conference on Innovations in Information Technology*, Nashville, TN, Nov. 2007, pp. 51–54.
23. I. Guvenc, Statistics of macrocell-synchronous femtocell-asynchronous users' delays for improved femtocell uplink receiver design, *IEEE Commun. Lett.*, 13 (4), 239–241, Apr. 2009.
24. M. E. Sahin, I. Guvenc, M. R. Jeong, and H. Arslan, Opportunity detection for OFDMA systems with timing misalignment, in *Proceedings of the IEEE Global Telecommunications Conference (GLOBECOM)*, New Orleans, LA, Nov. 2008, pp. 1–6.
25. M. E. Sahin, I. Guvenc, M. R. Jeong, and H. Arslan, Handling CCI and ICI in OFDMA femtocell networks through frequency scheduling, *IEEE Trans. Consumer Electron.*, 55(4), 1936–1944, 2009.

6

Conversational Quality and Wireless Network Planning in Fixed Mobile Convergence

Mariano Molina-García, Alfonso Fernández-Durán, and José I. Alonso

CONTENTS

6.1 Wireless Local Area Networks in Fixed Mobile Convergence Scenarios

In recent times, wireless local area networks (WLAN) technology based on the 802.11 family has become a resource in solving communication needs in multiple applications. Some of the applications that have gained momentum are those related to the communications that not only transport LAN data or web type Internet access, but also mainly IP-based voice and video. This is possible because of the availability of proprietary and standardized (IEEE802.1le) techniques for traffic prioritization. This circumstance has enabled the possibility to have not only WLAN-based handhelds, but also dual-mode terminals including cellular and WLAN.

There are many scenarios in which wireless convergent voice and video over IP are being considered for both private and public applications. In private communications, companies already using WLAN for LAN data-only are considering the replacement of their obsolete TDM PBXs with new communications methods capable of handling VoIP, and therefore merging the data and voice networks. In this scenario, the availability of dual-mode cellular and WLAN voice handsets enables the possibility to migrate the cordless communications to the WLAN domain, to take advantage of the lower rates in the fixed networks with respect to the cellular.

There are different approaches to define reference architectures for a communication convergence solution based on dual-mode handsets. The objective of these kinds of solutions is to provide a seamless mobility user experience, whether the user is under cellular (GSM or CDMA) or WLAN coverage, ensuring service continuity for both, voice and data, when roving between GSM and WiFi areas. These solutions offer high value to the end-user in terms of convenient improvements by providing unique number and unique handset.

Although still in an early phase for adoption and waiting for a wide availability of dual-mode handsets, there are already several initiatives being carried out by main service providers around the world, some of them with commercial deployments. Three different architecture alternatives have been conceived to support cellular/wireless convergence.

6.1.1 Enterprise-Based Architecture

Some PBX manufacturers are implementing "wireless extensions" over WiFi. These are proprietary solutions based on dual-mode handsets, running a specific voice client in charge of interwork with the communication server at the PBX structure. As a matter of fact, this is the same principle that is applied earlier to pure WLAN-based handsets. From a user experience standpoint, the handset not only allows mobility around the campus, but enables access to the PBX supplementary services as well.

6.1.2 Unlicensed Mobile Access

This is a mobile oriented solution promoting basically an alternative access to GSM/GPRS network. From a commercial point of view, it can be seen as a complement to low-quality coverage areas, a lower cost alternative access network, or simply as a way of accelerating fixed mobile substitution. Dual-mode handsets with specific Unlicensed Mobile Access (UMA) clients are linked to WiFi access points which, in turn, are connected throughout a broadband link (ADSL or cable-modem) to a specific UNC (UMA Network Controller), working very much like a BSC (Base Station Controller) in a cellular network. In fact, it is connected to the core network using the same interfaces, A for GSM circuit-switching network, and Gb for GPRS data network.

The main advantages of this architecture are total service continuity (voice, SMS, and data sessions) and handover between cellular and wireless areas. However, it lacks flexibility for local switching and horizontal handover in campus deployments. UMA is being specified by 3GPP body on GAN (Generic Access Network) [1] and supported by the FMCA (Fixed Mobile Convergence Alliance) [2].

6.1.3 SIP/WiFi Architecture

The third alternative for fixed mobile convergence is a network-based architecture, aligned with IMS (IP Multimedia Subsystem) standards and specified by 3GPP in the VCC (Voice Call Continuity) Technical Report [3] and supported by the FMCA [4].

It is a pure SIP-based solution, with a multimedia SIP client installed on the dual-mode handset. Presence, numbering, and routing decisions are taken at a network application Server, connected to a VoIP network and, optionally, to a cellular network via a CAMEL signaling interface. At the access layer, a variable number of WiFi access points are deployed at customer premises (or at public hotspot) depending on the area to be covered, the number of users, and voice traffic. These access points are controlled by a WiFi Network Controller, implementing Fast Roaming capabilities for horizontal handover. This handover is achieved by means of break-before-make and acceleration of authentication using an authentication proxy. Moreover, the WiFi access network support mechanisms to separate Virtual WLANs and differentiate voice and data packets. Vertical handover, from WiFi to GSM, and the other way around, uses a "make-before-break" procedure, which is controlled and instructed by the Application Server, and synchronized with the SIP client running on the dual-mode handsets. This handover is based on signal quality measurements and uses hysteresis between two thresholds.

6.2 Conversational Services and Wireless Network Planning

The planning and dimensioning of IEEE802.11 [5] wireless local area networks to be deployed in large buildings or areas that require multiple access points is very complex. In these kinds of networks, frequently oriented to provide best effort data services, the main tasks to fulfill in the design process will first be the estimation of the number of access points needed to handle the capacity required by the users, secondly the selection of the location of each access point in the zone in which services are going to be provided, and thirdly the assignment of one of the nonoverlapping different frequency channels to each access point.

There are many determining factors behind these design decisions. The number of access points in a WLAN is always a trade between the desired coverage in which the service is going to be provided, the capacity the network wants to be able to handle, and the interference between the different access points. On one hand, to guarantee a required coverage and a determined capacity per user demands to deploy enough access points to cover the zone. On the other hand, concentrating on the access points in a zone can provoke high levels of interference in the network that can lead to the degradation of the performance of the network. The location of the access points plays a very relevant role as well. An adequate distribution of the access points can ensure the fulfillment of coverage and interference conditions, as well as maximize the overall capacity in the network, whereas a bad selection of the location of the access points can cause a global degradation of the performance of the network. Finally, the assignment of radio frequencies to each access point has to be made, ensuring that as adjacent access points function by using frequencies as separated as possible, in order to minimize co-channel interferences.

Therefore, the calculation of these parameters would not be trivial at all, and, to increase the complexity of the planning process, many different criteria can be established to determine their values. Requirements such as maximizing the percentage of locations in which a signal strength above a determined threshold that guarantees coverage is obtained, and maximizing capacity that the deployed network is able to handle or balance the statistical distribution of the load of each access point that forms part of the wireless network are some of the most used strategies used by WLAN designers.

The most commonly used approach for WLAN planning and optimization has been the manual and intuitive location of access points by reasonably spacing them in the deployment area, followed by a trial and error procedure. However, due to the increasing complexity of the deployments in terms of number of access points, new approaches based on automatic optimization models and algorithms to calculate the optimum placement of access point in WLANs have been proposed.

Several studies have analyzed processes of automatic planning and dimensioning WLANs, using different criteria as targets, and different algorithms to obtain the optimum results. According to Wertz et al. [6], a method to optimize WLANs taking into account required coverage, capacity, and interference situation is presented. According to Bosio et al. [7], mathematical models to tackle the WLAN planning with the purpose of maximizing network efficiency evaluation inter-AP domain interference and the access mechanism are proposed. According to Jaffres et al. [8], a tabu multiobjective algorithm is implemented to evaluate coverage, interference, and average throughput per user. According to Lee et al. [9] and Gondran et al. [10], a new approach where location selection and frequency assignment are tackled together is introduced. According to Liang et al. [11], an automatic base station planning approach trying to assess coverage and mean available bandwidth per user using Markovian chains is proposed. According to Vanhatupa et al. [12], a novel genetic algorithm to explore the design space and an IEEE 802.11 rate adaptation aware QoS estimation functionality are presented. According to Prommak et al. [13], a WLAN design approach concentrating on maximizing the network capacity, and covering user population density, traffic demand, and required coverage area is introduced. According to Amaldi et al. [14], a hyperbolic and quadratic formulation of the coverage planning problem aiming to maximize the capacity of the network is provided. According to Kamenetsky and Unbehaun [15], a combination of two approaches has been developed, using pruning to achieve an initial position for the access points and refining these using simulated annealing and neighborhood search, using maximizing coverage area and signal quality as optimization targets. According to Kouhbor et al. [16], a discrete gradient algorithm has been used to solve the problem of deciding number of access points and their locations.

Nevertheless, most optimization procedures for IEEE802.11 WLANs have been developed under the perspective of networks that was originally intended to transport best-effort data traffic, and that were not able to handle different types of traffic in different ways. The incorporation of new standards like IEEE802.11e [17] has brought about the opportunity of deploying delay and bandwidth sensitive services. Voice and conversational video over IP are two applications that have experienced a greater growth over the last few years. These facts, together with the massive deployment of WLAN IEEE802.11g networks and the appearance of some new fixed mobile convergent networks based on IMS and UMA have caused voice and video over WLAN to emerge as key applications in WLAN.

Although these new services have many advantages, they raise some challenges as well, since the characteristics of physical and MAC layer are not adapted to provide real time services. Due to this, if it is desired to provide real time services over WLAN with some special quality of service requirements, it will be necessary to carry out different tasks in relation to the network architecture, the network planning and dimensioning, control admission, or QoS provision. Many efforts have been carried out in this way. According to Cai et al. [18], a survey of recent advances in MAC layer quality of service enhancement mechanisms is provided. According to Saliba et al. [19], the concept of user-perceived QoS is introduced and linked to specific wireless data networks parameters. According to Skyrianoglou et al. [20], they address the potential issues that emerge when using WLANs as access systems to an IP core and focuses on defining a consistent mechanism that supports IP QoS and resource reservation over an IEEE 802.11e WLAN. According to Diwakar and Iyer [21], a new scheme is proposed for better QoS guarantees to the end-user in a WLAN supporting real time applications.

The quality of voice and conversational video over IP are mainly determined by delay, jitter, and packet losses rate. In WLAN VoIP and conversational video over IP applications, delay must include delay associated with codec, packetizing process, and networking, including network delay and the delay on account of being transmitted through the backbone and through the WLAN. Jitter depends on dynamic fluctuations of the performance of the different network an IP packet is transmitted through, and, because WLAN can be considered the bottleneck in terms of performance, jitter due to the WLAN IEEE802.11 network can be considered as the dominant part. Assuming a traffic model with a constant bit rate, jitter will be caused mainly by the time that the network interface takes to transmit successfully a packet, and that time depends on MAC protocol and transmission data rate. Jitter can be compensated using a buffer in the receiver with an efficient algorithm to deliver packets, discarding packet when a limit time is exceeded. Buffer size must be carefully selected in order to avoid including a delay that can produce degradation in the overall quality. Finally, the packet loss rate will represent the greater source of quality degradation for voice and conversational video over IP services. The packet loss rate is made up by packet losses because of congestion, link failures, routing problems, and transmission errors, as well as packets discarded because of excessive delays.

So, because of the peak of voice and conversational video over IP services and, with the aim of adapting WLAN deployments oriented to provide real time services to specific requirements in terms of delay, jitter, and packet losses rate will be necessary, not only to distinguish between services with different requirements and, therefore, priorities, but to develop new planning and dimensioning concepts that take into account special requirements of these multimedia real time services.

Taking into account IEEE802.11 WLAN planning procedures, and special requirements of WLANs if real time services are considered, the following sections introduce the framework for a simple method to estimate voice and conversational video over IP quality, as well as concepts associated to traditional WLAN planning methods to maximize coverage and capacity, evaluating the effect of using these methods on the quality of voice and conversational video services. After that, a procedure that adapts planning strategies to those WLAN IEEE802.11 network oriented to provide real time services, and, therefore, to deal with voice and conversational video quality concepts is introduced.

6.3 Subjective Quality Assessment for Conversational Services

The concept of user-perceived QoS, and its estimation for real time services, especially voice and conversational video services has been a key issue in recent years. To estimate voice quality in a network, several objective methods have been developed through the years. Objective measurement of voice quality in modern communication networks can be intrusive or nonintrusive. Intrusive, objective speech quality measurement systems are referred to as intrusive due to the injection of test signals and the need to utilize the network. They normally use two input signals, namely a reference signal and the degraded signal measured at the output of the network or system under test. Within these systems the perceptual domain measures, based on models of human auditory perception, have been shown to be the most successful objective speech quality measures. Typical perceptual measure methods are Perceptual Speech Quality Measure (PSQM), adopted as ITU-T Recommendation P.861 [22], Perceptual Assessment of Speech Quality (PAMS) [23], Measuring Normalizing Blocks (MNB) [24], Enhanced Modified Bark Spectral Distortion (EMBSD) [25], and Perceptual Evaluation of Speech Quality (PESQ) [26], which is the latest ITU standard for assessing speech quality for communication systems and networks. However, these intrusive methods are accurate to measure end-to-end perceived speech quality, but their application is not useful for designing, planning, and dimensioning WLAN IEEE802.11, because they do not provide mathematical expression to relation parameters derived from planning and dimensioning WLANs and voice quality associated with them. For this purpose, parameter-based nonintrusive speech quality prediction methods, which predict speech quality directly from network and/or non-network parameters, can be used.

Many efforts have been taken up in this way. According to Markopolou et al. [27], an assessment of subjective voice quality based on delay and loss measurements is carried out. According to Clark [28], a nonintrusive monitoring technique for voice over IP networks using an extended version of ITU G.107 is proposed. According to Choudhury et al. [29], an evaluation of voice quality in frequency selective multipath fading for different combination of codecs, payload size, and data rates is presented. According to Narbutt and Davis [30], the performance of different audio codecs under varying load conditions on a VoWLAN is studied. The model specifically used in this survey to model the quality of voice for different network parameters will be a modification of the well known E-Model, extracted from Recommendation ITU T-G107 [31], and widely accepted to model the performance of quality of voice in telephony networks.

Talking about the estimation of conversational video quality, this quality will be obviously dependent on video quality and audio quality of voice and video integrated in the videocall. There are many widely accepted techniques to measure video quality, divided into subjective methods [32,33] that require human viewers to rate the video quality either looking at a single clip or both the original and the distorted video, and the objective methods [34,35] that estimate video quality is based on the error signal, defining as an error the absolute difference between the original and transmitted signal evaluated, using metrics as Mean Square Error (MSE) or Peak Signal-to-Noise Ratio (PSNR).

Many surveys have tried to relate quality of video and codec and network parameters to characterize multimedia performance. According to Pace and Viterbo [36], a technique for accurate estimation of the perceived streaming quality delivered within a wireless network is proposed. According to Yamagishi and Hayashi [37], a computational opinion model for estimating video quality of videophone services is presented. According to Winkler and Faller [38], the quality of multimedia content at very low bit rates is assessed, and the interaction between audio and video in terms of perceived quality is explored. According to Ries et al. [39], quality assessment for H.264/AVC codec low rate and low resolution video sequences is evaluated based on content class, frame rate, and bit rate. In this survey, the model used to estimate video and multimedia quality will be an adaptation of the one presented in recommendation ITU T-G.1070 [40], which defines quality of video in relation to network parameters for MPEG4 codec and the relation between multimedia, video, and audio quality, taking as a reference the speech quality estimation model presented in recommendation ITU-T G.107.

In Sections 6.3.1 and 6.3.2, speech and video quality estimation models presented in [41] will be used. After that, a multimedia quality estimation, which uses parameters provided by speech estimation and video estimation will be proposed.

6.3.1 Voice Over IP Quality Estimation Model

VoIP quality depends deeply on delay, jitter, and packet loss rate. To characterize the quality of voice perceived by the user using these parameters, the procedure proposed in ITU-T G.107 has been used calculating a rating factor R that is a additive combination of five factors as (6.1):

$$R = R_0 - I_s - I_d - I_e + A \tag{6.1}$$

where
    R_0 is the basic signal to noise ratio
    I_s is the simultaneous impairment factor function of the SNR impairments associated with the switched circuit network paths
    I_d is the delay impairment factor which includes all delay and echo effects
    I_e is the equipment impairment factor which models impairments caused by low-bit-rate codecs and the expectation factor
    A is a parameter of the model known as the advantage factor

For the purpose of the survey, the most important contributions are those which are associated to delay and packet loss. Due to this fact, parameters I_d and I_e will be taken into account, and the rest of the parameters will be considered as constants, taking their default values from ITU-T G.114 [42]. These values are presented in Appendix 6.A. Parameter I_d can be approached by

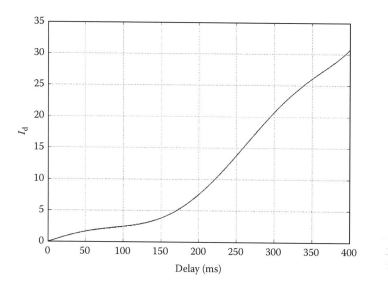

FIGURE 6.1
Relation between delay and I_d.

means of a sixth order polynomial (6.2), taking the tabular values from ITU-T G.107. The relation between I_d and delay is presented in Figure 6.1.

$$I_d = 1.62 * 10^{-13} d^6 - 1.76 * 10^{-10} d^5 + 6.45 * 10^{-8} d^4$$

$$- 8.22 * 10^{-6} d^3 + 23.15 * 10^{-5} d^2 + 3.52 * 10^{-2} d - 0.024 \qquad (6.2)$$

ITU-T G.114 assures that, if delays can be kept below 150 ms, most applications, both speech and non-speech, will experience essentially transparent interactivity. If this range of values is assumed for one-way transmission delay, transmission rating factor will be determined by (6.3)

$$R = 93.2 - I_e \qquad (6.3)$$

And, therefore, perceived voice can be calculated from the specific voice codec together with the packet loss rate associated. The main difficulty is to determine a relationship between the packet loss P_{pl} and the equipment impairment factor I_e. ITU-T G.113 [43] can be used to determine a relationship between the packet loss P_{pl} and the equipment impairment factor I_e, providing values for different codecs and for several values of packet losses. The codecs characterized are the most commonly used in VoIP, i.e., G729, G723.1, and G711, the latter is characterized according to the presence or absence of PLC (Packet Loss Concealment), which introduces an enhancement in performance at the expense of extra processing. Taking the tabular values from ITU-T G.113, it is possible to see that all of them can be approached by means

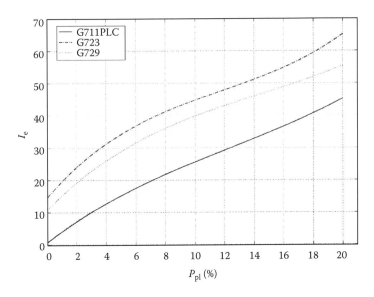

FIGURE 6.2
Relation between P_{pl} and I_e for difference VoIP codecs.

of third order polynomials (6.4) through (6.6) in their respective ranges of use, with a correlation factor better than 0.99. Relation between packet loss and I_e for different VoIP codecs is presented in Figure 6.2.

$$I_e^{G729} = 0.0051P_{pl}^3 - 0.2197P_{pl}^2 + 4.5869P_{pl} + 10.788 \quad P_{pl} < 20\%, \quad (6.4)$$

$$I_e^{G723.1} = 0.0087P_{pl}^3 - 0.3094P_{pl}^2 + 5.2348P_{pl} + 14.598 \quad P_{pl} < 20\%, \quad (6.5)$$

$$I_e^{G711-PLC} = 0.0037P_{pl}^3 - 0.1376P_{pl}^2 + 3.4978P_{pl} + 0.6663 \quad P_{pl} < 20\%, \quad (6.6)$$

From the above expressions, it is possible to obtain the transmission rating factor R as a function of the packet loss for each of the voice codecs. Although the R factor represents the quality of the transmission, the common way to represent the user perceived quality is the MOS (Mean Opinion Score), which comes from statistical surveys of quality graded from 1 to 5 or bad quality to excellent quality. ITU G.107 provides an expression (6.7) to relate R with MOS, where MOS_{CQE} is referenced as CQE (Conversational Quality Estimated), to distinguish it from the measured one. The values obtained could be used to characterize the performance of the system in different conditions and take decisions on design and deployment, considering MOS = 4 as excellent quality, MOS = 3.6 as good quality, and MOS = 3 as acceptable quality.

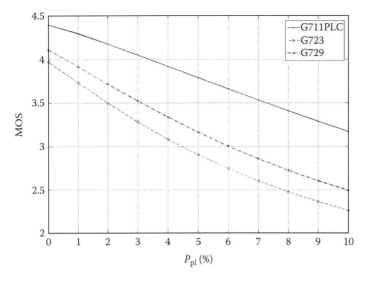

FIGURE 6.3
Audio MOS for different P_{pl} (%) and VoIP codecs.

$$\text{MOS} = \begin{cases} 0 & R < 0 \\ 1 + 0.035R + R(R-60)(100-R)*7*10^{-6} & 0 \leq R < 100 \\ 4.5 & R > 100 \end{cases} \quad (6.7)$$

In Figure 6.3, the relation between packet loss rate and perceived quality of voice for different VoIP codecs is presented.

6.3.2 Video Over IP Quality Estimation Model

Talking about video quality, there are many widely accepted techniques to measure video quality, divided into subjective methods, that require human viewers to rate the video quality either looking at single clip or both the original and the distorted video, and objective methods that estimate video quality based on the error signal, defining an error as the absolute difference between the original and transmitted signal evaluated using metrics as Mean Square Error (MSE) or Peak Signal-to-Noise Ratio (PSNR). However, these methods do not allow quality of experience planners to help to ensure that users will be satisfied with end-to-end service quality, due to the fact that it would not be possible to define quality of video in relation to packet loss and delay experienced by a user for different video codecs and configurations. In order to predict video and multimedia quality from IEEE802.11 network parameters, a model is proposed by ITU in its recommendation ITU T-G.1070. Combining this video quality assessment with the speech quality estimation model presented in recommendation ITU-T G.107, conversational video over IP performance can be predicted, which can be used in planning and dimensioning processes. The model assumes some specific evaluation conditions for terminals, environments, and evaluation contexts, leading to a video over IP quality of experience assessment mainly dependent on the video codec, the bit rate and the frame rate transmitted, the packet losses, delays, and video display size.

In order to model the quality perceived by a user who is visualizing a videocall in a handheld, the procedure proposed in Recommendation ITU-T G.1070 express the subjective quality of video over IP perceived as a function of Icoding, which represents the basic video quality affected by the coding distortion and the packet loss robustness factor Dpplv, which in turn expresses the degree of video quality robustness due to packet loss. The values of these parameters will be dependent on the codec type, the video format, the key frame interval, and the video display size, and this dependence will be introduced in the model by means of coefficients v. The default values for these coefficients is presented in Appendix 6.A. Apart from those considerations, the basic video quality affected by coding distortion Icoding will depend on IOfr which is the maximum video quality at each video bit rate (Brv), Ofr which is an optimal frame rate that maximizes the video quality at each video bit rate (Brv), and DFrv which represents the degree of video quality robustness due to frame rate (Frv). Dpplv represents the degree of video quality robustness against packet loss and will depend on the video bit rate (Brv) and frame rate (Frv).

Talking about the estimation of conversational video quality, this quality will be obviously dependent on video and audio quality of voice and video integrated in the videocall. The multimedia quality MMq can be calculated taking into account the audiovisual quality MM_{SV}, the audiovisual impairment factor MM_T and coefficients m dependent on display size and conversational task. The default values for these coefficients is presented in Appendix 6.A. Audiovisual quality will be calculated from speech quality Sq and video quality Vq values, whereas audiovisual impairment factor will be obtained from delays associated to speech TS, and video TV, which will affect the absolute audiovisual delay AD and the audiovisual media synchronization MS. Both audiovisual quality and audiovisual impairment will depend on video display size and conversational task.

The procedure that will be used to estimate voice perceived quality (Sq), video perceived quality (Vq), and multimedia perceived quality from packet loss rates and delays is presented in Figure 6.4. In Figures 6.5 and 6.6, the relations between the packet loss rate and the perceived quality of video for MPEG4 codec and different bit and frame rate configurations are presented, and in Figures 6.7 and 6.8, multimedia quality values for different MPEG4 frame and bits rates using G711PLC for speech coding are presented, assuming that voice and video, which are part of the videocall, are perfectly synchronized. Finally, in Table 6.1, quality of experience values for different voice codecs and videoconference configurations obtained using this

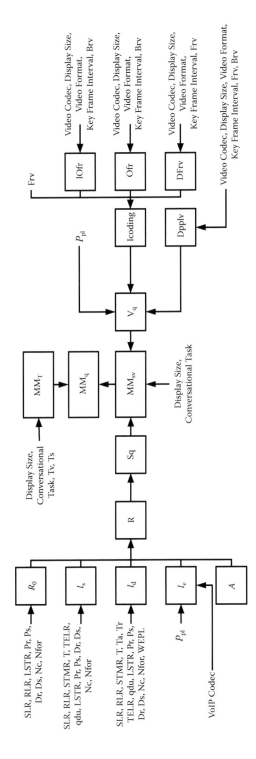

FIGURE 6.4
Procedure followed to estimate multimedia perceived quality.

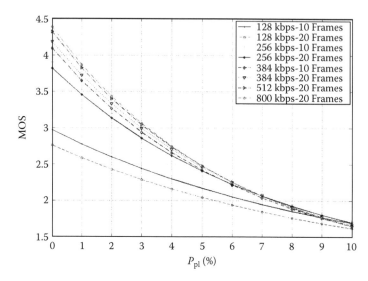

FIGURE 6.5
Video MOS for different P_{pl} (%) and configurations for MPEG4 QVGA, 2.1 in. screen.

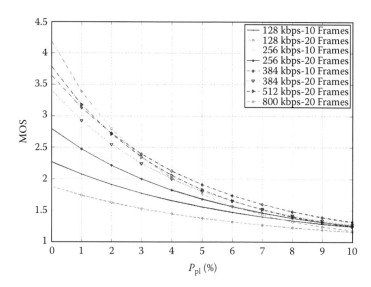

FIGURE 6.6
Video MOS for different P_{pl} (%) and configurations for MPEG4 QQVGA, 4.2 in. screen.

procedure are presented, assuming 150 ms end-to-end delay. Several conclusions can be extracted from the results obtained. At the outset, and talking about VoIP services, G711PLC will have the best performance, supplying an acceptable quality for packet loss up to 10%. For the rest of the codecs evaluated, packet losses around 5% will ensure an acceptable quality perceived

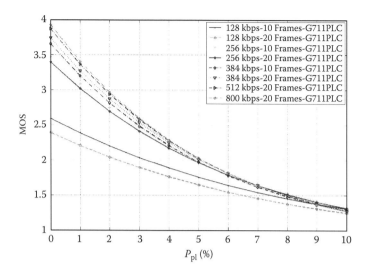

FIGURE 6.7
Multimedia MOS for different P_{pl} (%) and configurations for MPEG4 QQVGA, 2.1 in. screen and G711PLC.

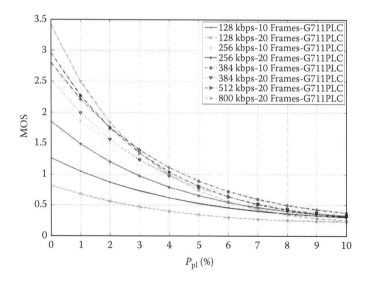

FIGURE 6.8
Multimedia MOS for different P_{pl} (%) and configurations for MPEG4 QQVGA, 4.2 in. screen and G711PLC.

by the user. In videoconference services, these packet loss thresholds will be stricter, due to the combined effect of audio and video degradation as a consequence of the process in Figure 6.4. Using G711PLC as audio codec, for mobile terminals screen size, a minimum bit-rate of 256 kbps combined with a frame rate of 20 frames per second will be necessary in order to fulfill

TABLE 6.1

Maximum Packet Loss Rates to Obtain Different Audio and Multimedia Quality of Experience Values

Type of Service	Configuration	MOS Requirements		
		Excellent (%)	Good (%)	Acceptable (%)
VoIP	G711PLC	3.5	6	>10
	G723	0	1.5	4.5
	G729	0.5	2.5	6
Videoconference over IP	MPEG4, 256 kbps-20 Fr/s, 2.1 in., G711PLC	NA	NA	1
	MPEG4, 384 kbps-20 Fr/s, 2.1 in., G711PLC	NA	0.25	1.75
	MPEG4, 512 kbps-20 Fr/s, 2.1 in., G711PLC	NA	0.6	1.9
	MPEG4, 800 kbps-20 Fr/s, 2.1 in., G711PLC	NA	0.75	2
	MPEG4, 800 kbps-20 Fr/s, 4.2 in., G711PLC	NA	NA	0.5

minimum quality requirements. Strict requirements of packets loss percentages, from 1% to 2%, depending on video configurations, must be ensured. For PDA terminals screen size, these requirements are even stricter. In this case, a minimum bit-rate of 800 kbps combined with a frame rate of 20 frames per second will be necessary, combined with a 0.5% packet loss threshold.

6.4 WLAN Planning Principles Oriented to Data Services

To ensure the satisfactory performance of an IEEE802.11 wireless network, meticulous planning and dimensioning procedures previous to the deployment must be carried out. As it was pointed out before, due to the increasing complexity of the deployments in terms of number of access points, new approaches, based on automatic optimization models and iterative algorithms to calculate the optimum number of access points and their placement in WLANs have been proposed. In each iteration, the structure of the WLAN and different parameters of the access points will be defined. In this phase, the number of access points and their location, radiation patters of the antennas associated to each access point, orientation of the antennas in elevation and azimuth, frequency, and isotropic power radiated are defined. With this configuration, the performance of the network is checked using schemes based on static simulations, which consider independent snapshots of the system that represent the situation of the network in a determined moment. Static simulations will need to define a group of active users for that determined moment, each one with different characteristics, in terms of service type, throughput required, or WLAN standard supported by the terminal. The generation of users will be carried out enough times to be statistically relevant. For each group of users generated, the signal level received in each user from each access point will be calculated from propagation losses. With that information, each user will be associated to the access point with a higher signal level. Finally, this signal strength received in each user, due to the adaptive modulation schemes used in IEEE802.11 WLANs, will be mapped to a determined combination of modulation and coding schemes. After the required number of iterations, the performance of the network configuration can be evaluated, assessing its capacity of handling the overall quality requirements established by the network designer in a statistical way.

The raw throughput that can be handled by a handheld is determined by the signal strength that is received from the base station it is associated to and the modulation mode used. Current stations are able to commute between different modulation modes using a mechanism known as ARS (Automatic Rate Selection), following different strategies not defined in the standards. Therefore, it is possible to decide changing the modulation mode taking into account the throughput associated to the modulation mode used, the packet loss rate, the signal noise rate, the number of packets retransmitted, or any other policy decided by the network manager. Several studies have evaluated the importance of ARS in the performance of WLAN IEEE 802.11 networks. According to Haratcherev et al. [44], an advanced hybrid control algorithm that uses SNR information to achieve fast responses has been implemented. According to Pavon and Choi [45], a novel link adaptation algorithm, which aims to improve the system throughput by adapting the transmission rate to the current link condition, is presented. According to Qiao et al. [46], a novel MPDU (MAC Protocol Data-Unit) based link adaptation scheme for the 802.11a systems is proposed.

For simulation purposes, all these different strategies must be finally translated into a SNR-configuration lookup table that defines low and high SNR thresholds of use for different modulation and coding schemes. These algorithms are typically used in static simulation processes to characterize the performance of wireless networks with adaptive modulations, in order to select the configuration of a determined user depending on the signal level received. Therefore, it will be crucial to characterize accurately the performance of the automatic rate selection algorithm implemented in the terminal, in order to assess correctly the network performance to be used in the IEEE802.11 WLAN design process, planning, and dimensioning procedures.

For data users, a typical approach of automatic rate selection algorithm to be used is the throughput-based rate control. In this approach, a constant small fraction of the data is sent at the two adjacent rates to the current one (an adjacent rate is the next higher or lower one available). Then, at the end of a specified decision window, the performance of all three rates is determined by dividing the number of bytes transmitted at each rate by their cumulative transmission times. Finally, a switch is made to the rate that provided the highest throughput during the decision window.

To simulate the performance of this type of algorithms, and, therefore, to evaluate the optimum threshold to commute between different modulation schemes, it is possible to use some values provided in IEEE 802.11 standard that relate throughput and the power value in the receiver. In this way, the signal strength can be evaluated at each point to determine if the signal strength exceeds a certain value to offer a throughput for a defined modulation mode. However, this procedure is very imprecise, because of the fact that the throughput supported is affected by the probability of error, that varies for different values of signal noise rate implying that throughput is continuous, and can take any value lower than the maximum for that modulation scheme, and, therefore, it is possible to obtain a higher throughput value admitting some packet losses rather than switching to another modulation mode.

Due to this, it will be necessary to analyze each IEEE 802.11 standard, and obtain, for each modulation supported by the standard, the expression that define the error rate as a function of the signal noise rate in the receiver. Using that expression, the throughput could be estimated from the raw throughput and the packet error rate, as expressed in the following:

$$C = C_0(1 - \text{PER}) \tag{6.8}$$

This is because IEEE802.11 standard does not permit any scheme to correct errors, but the retransmission of the damaged packets takes place. Beside that, as soon as a bit is received in the damaged form, the whole packet is discarded, and the increase of retransmissions causes a proportional loss of throughput. In this way, given a signal noise rate, if a terminal supports ARS, and the strategy defined is maximizing the throughput, the terminal should select the modulation scheme to achieve the best throughput.

Nowadays, most part of the access points and terminals are based on IEEE802.11g standard, which essentially consists of running 802.11a in the ISM frequency band (2.4 GHz). It will support data rates from 6 to 54 Mbps, using OFDM and different modulation schemes and coding rates. The mandatory modes in 802.11g are slight modifications of existing physical layers, with a few minor alterations for backward compatibility. This will imply that access points and terminals based on IEEE 802.11g are compatible with IEEE 802.11b equipments, and therefore, are able to support differential modulations (802.11) at 1 and 2 Mbps and complementary code keying modulation schemes (802.11b) at 5.5 and 11 Mbps as well as OFDM with binary and quadrature modulations. Thus, throughput for each signal noise rate can be expressed as (6.9)

$$C_{802.11g} = \max\left[C_i(1 - \text{PER}_i)\right] \tag{6.9}$$

where
   C_i is the maximum rates supported by the different modulation schemes
   (6.10)
   PER_i are the packet error rate probabilities for a modulation mode and the
   signal noise rate received at that moment

$$C_i = \begin{cases} 1, 2\,\text{Mbps} & \text{DBPSK, DQPSK} \\ 5.5, 11\,\text{Mbps} & \text{CCK} \\ 6, 9, 12, 18, 24, 36, 48, 54\,\text{Mbps} & \text{OFDM} \end{cases} \qquad (6.10)$$

Therefore, to calculate the throughput associated to a modulation mode for each possible value of signal noise rate, the packet loss rate must be estimated. PER values will be worked out, for equipments working in 802.11 and 802.11b modes, or, in other words, using differential modulations (DBPSK, DQPSK) and Complementary Code Keying modulations as (6.11)

$$PER \leq 1 - \left[(1-BER)^{N\text{bits/data packet}} (1-BER)^{N\text{bits/ACK packet}} \right] \qquad (6.11)$$

where
   $N_{\text{bits/data packet}}$ are the number of bits which form each data packet
   $N_{\text{bits/ACK packet}}$ are the number of bits which form each acknowledgement
   packet

It is considered erroneous for packet that has an error both in its own transmission and in its ACK transmission.

In the same way, to obtain PER for 802.11g equipments using OFDM modulations, it is necessary to calculate the probability of having an error in each subcarrier SER subcarrier, which will depend on the modulation used. Then, PER is the complementary probability of transmitting correctly all the subcarriers that form an OFDM symbol.

$$PER \leq 1 - \left[(1-SER_{\text{subport}})^{48 \circ N\text{symb.OFDM.data}} (1-SER_{\text{subport}})^{48 \circ N\text{symb.OFDM.ACK}} \right] \qquad (6.12)$$

As can be seen from the above expression, it will be required to estimate the OFDM symbols transmitted in each packet. According to IEEE802.11, symbols for packet are (6.13):

$$N_{\text{symb}} = \left\lceil \frac{16 + 8 \circ \text{data} + 6}{N_{\text{DBPS}}} \right\rceil \qquad (6.13)$$

NDBPS values for each modulation are included in Table 6.B.1 in Appendix 6.B. The total number of subcarriers can be obtained by multiplying $N_{\text{symb,OFDM}}$

by 48, which is the number of subcarriers in each OFDM symbol. Eventually, bit error rate (BER) for differential and Complementary Code Keying modulations and SER for OFDM modulations for each value of signal noise rate will be estimated using expressions included in Table 6.B.2 in Appendix 6.B. Expressions to relate SNR and BER or PER can be extracted from [47]. $Q(x)$ can be expressed from the complementary error function as (6.14):

$$Q(x) = \frac{1}{2} \cdot \text{erfc}\left(\frac{x}{\sqrt{2}}\right) \tag{6.14}$$

Throughput in a WLAN depends on the packet loss rate, and this rate can be estimated from bit error rate or subcarrier error rate for a determined size of data and acknowledgment packets. The probability of a packet being corrupted depends on the length of the transmission, and, consequently, long packets should be transmitted at a lower rate than short packets. Due to this, the typical distribution of packet size corresponding to the main applications used in an IEEE802.11 network must be taken into account. Several studies have been carried out for this purpose [48]. The typical distribution of packet size used in this survey has been taken from, and is shown in Figure 6.9.

Distribution supplies the probability of having a determined packet size in a generic WLAN that supports different applications. Different non–real time kind of applications will have different packet size distributions, each one with certain frequent packet sizes. In Table 6.2, typical most frequent packet sizes for different applications are presented.

Using the expressions previously presented to estimate throughput for each modulation, the trade between throughput and signal-noise rate for each

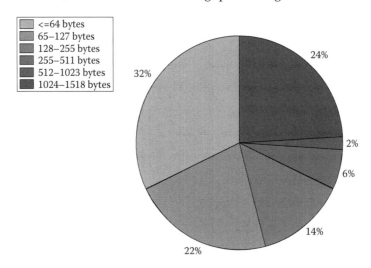

FIGURE 6.9
Typical packet distribution in WLAN IEEE802.11 deployments.

TABLE 6.2

Frequent Packet Size for Different Non Real Time Applications

Application Name	Class	Frequent Packet Size (bytes)
BitTorrent	P2P	377
eMule	P2P	1180
Apache	HTTP	1216
zFTP server	FTP	688

modulation scheme used in IEEE802.11, IEEE802.11b, and IEEE802.11g can be modeled as it is shown in Figures 6.10 through 6.12. In these figures, a packet size corresponding to the typical maximum MTU size of 1500 bytes has been selected. In Table 6.3, optimum SNR transition values between modulation schemes to maximize capacity working under different standards are presented for different frequent packet sizes. These will be the values used to determine the modulation used by a user located in a determined location of an scenario.

To justify the need for modifying design criteria when planning a WLAN in which real time voice and video services are supported, the effect over VoIP, video over IP, and conversational video of using throughput-based rate selection schemes are provided. From the analysis shown previously in this section, the relation modulation scheme-signal to noise rate that ensure the maximization of the capacity in the network can be derived. Using the values presented in Table 6.3 for the typical maximum MTU size of 1500 bytes,

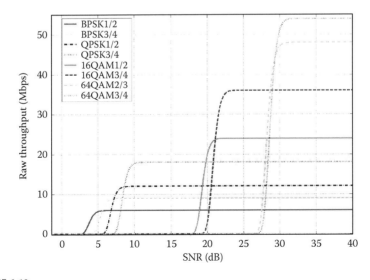

FIGURE 6.10

Throughput associated to different modulation schemes and SNR for equipment working under IEEE802.11b standard.

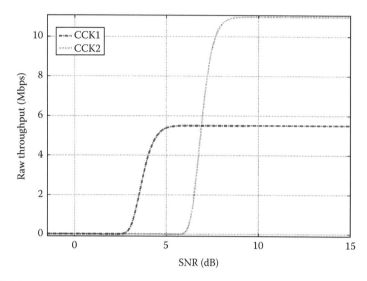

FIGURE 6.11

Throughput associated to different modulation schemes and SNR for equipment working under IEEE802.11b standard.

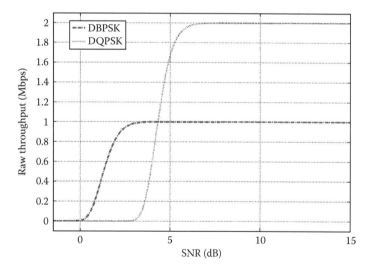

FIGURE 6.12

Throughput associated to different modulation schemes and SNR for equipment working under IEEE802.11 standard.

the curve that relates modulation scheme and SNR to maximize capacity for terminals working under IEEE802.11g standard is presented in Figure 6.13.

In spite of the fact that the design of the WLAN IEEE802.11 network using these thresholds to commute between modulation modes will maximize overall capacity, the selection of the modulation used by a determined user

TABLE 6.3

Optimum SNR Transition Values between Modulation Schemes to Maximize
Capacity Working under Different Standards

Standard	Modulation	Optimum SNR Transition (dB) for Different Packet Sizes (bytes)			
		1500 bytes	1216 bytes	688 bytes	377 bytes
IEEE802.11g	64QAM3/4 to 64QAM2/3	29.4	29.3	28.9	28.6
IEEE802.11g	64QAM2/3 to 16QAM3/4	28.7	28.6	28.28	27.9
IEEE802.11g	16QAM3/4 to 16QAM1/2	21.1	21.0	20.6	20.3
IEEE802.11g	16QAM1/2 to QPSK3/4	19.8	19.7	19.4	19.0
IEEE802.11g	QPSK3/4 to QPSK1/2	8.6	8.5	8.2	7.9
IEEE802.11g	QPSK1/2 to BPSK3/4	7.2	7.1	6.8	6.5
IEEE802.11g	BPSK3/4 to BPSK1/2	5.6	5.5	5.2	4.9
IEEE802.11g	No coverage	2.3	2.1	1.7	1.2
IEEE802.11b	CCK2 to CCK1	6.9	6.8	6.5	6.3
IEEE802.11b	No coverage	2.4	2.3	1.9	1.5
IEEE802.11	DQPSK to DBPSK	4.3	4.2	3.9	3.5
IEEE802.11	No coverage	−0.2	−0.3	−0.8	−1.3

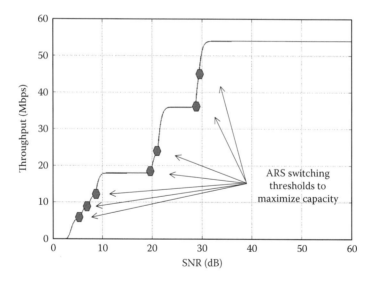

FIGURE 6.13
Relation between throughput and SNR using ARS switching values to maximize capacity.

obtained as a result of this procedure will provoke that real time services
in the WLAN deployed could have a poor performance. This performance
could be explained by analyzing the relation between packet loss rate and
signal-to-noise rate obtained as a consequence of using the strategy of maximizing capacity when selecting the policy of the ARS mechanism. This

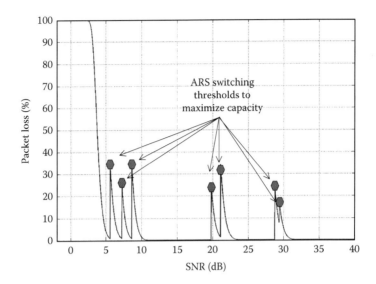

FIGURE 6.14
Relation between packet losses rate and SNR using switching values to maximize capacity.

relation between packet loss rate and signal-to-noise rate is presented in Figure 6.14. The figure shows that, for many SNR values, the value of the packet loss rate will be considerably higher, exceeding the maximum rate allowed to ensure a user experience of a minimum quality. Using the models presented in the previous section to estimate the perceived quality of VoIP, video over IP, and conversational video, it is possible to calculate values of audio MOS, video MOS, and multimedia MOS with regard to signal-to-noise rate. In Figures 6.15 through 6.17, correspondence between SNR values and quality for voice, video, and multimedia services assuming ARS switching thresholds for best-effort data services are presented.

Because of the problems with VoIP and teleconference quality presented previously when using a throughput-based rate control algorithm, other approaches are used for real time services rate selection, based on packet error rate. This rate can easily be determined since under 802.11, all success-fully received data frames are explicitly acknowledged by sending an ACK frame to the sender, and a missing ACK is a strong indication of a lost data frame. By counting the number of received ACK frames and the number of transmitted data frames during a rather short time window, the packet error rate can be computed. If the packet error rate exceeds some threshold and the current rate is not the minimal rate, then the control algorithm switches to the next lower rate. If the packet error rate is below a determined threshold, a few frames are transmitted at the adjacent higher rate frames and, if all of them get acknowledged, the control algorithm switches to that rate. To pre-vent the control algorithm from oscillating between two adjacent rates, the upscale action may be prohibited for some time after a downscale decision.

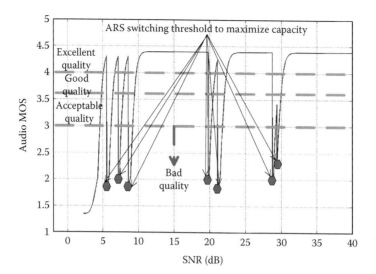

FIGURE 6.15

Relation between audio MOS values and SNR using ARS switching values to maximize capacity.

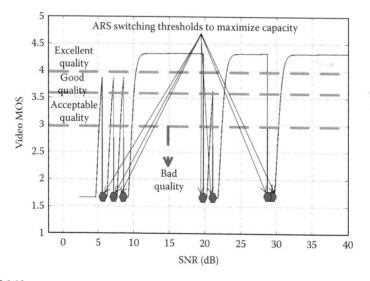

FIGURE 6.16

Relation between video MOS values and SNR using ARS switching values to maximize capacity.

Using these PER based control algorithms, real time services quality, very dependent on packet loss rates and delay, can be optimize. These algorithms must be modeled, in order to be used when simulating and evaluating the performance of a WLAN IEEE802.11e network supplying real time services. In Section 6.5, the performance of PER-based control algorithms is modeled,

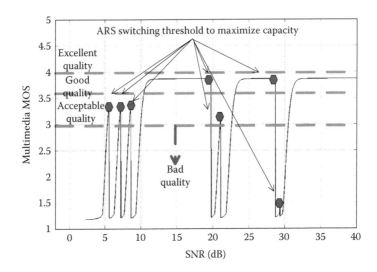

FIGURE 6.17
Relation between multimedia MOS values and SNR using SNR transition values to maximize capacity.

using as a reference for that purpose subjective quality assessment procedures proposed in Section 6.3.

6.5 WLAN Planning Based on Conversational Services Quality of Service

As mentioned earlier, some concepts concerning WLAN IEEE802.11 planning oriented to maximize capacity shared in a coverage area have been revised. However, with the appearance of the concepts of quality of service in IEEE802.11e, these networks have stopped being used only to support data services, in which the priority is the network capacity, to deal with other types of services, as voice over IP, video over IP, or conversational video services, not only affected by network capacity, but by delay, jitter, and packet loss rates, making it compulsory to ensure strict quality requirements. Due to these new services and requirements, techniques to be applied in software tools oriented to automatically planning and dimensioning WLAN IEEE802.11 networks should be adapted to encounter these new challenges.

In a WLAN planning oriented to maximize coverage and capacity, designers try to optimize the deployment area in which signal strength above a fixed threshold can be received and carry out an analysis of the global capacity of the network for different locations for the access points. For that purpose, signal-to-noise rate values to commute between different modulation schemes are obtained and used to define network areas covered for each type of modulation, and, therefore, to calculate overall capacity in the network if determined locations of access points are assumed.

On the other hand, if a WLAN is conceived to support real time voice, video and conversational video services, the analysis must be completely different, since as was shown modeling the subjective quality of VoIP, video over IP, and conversational video, the perceived quality is highly dependent on the delay and the packet loss rate. Because of this fact, software tools will have to vary the strategies when trying to assess the performance of WLAN terminals using these kinds of services, modifying the design criteria to ensure that the quality perceived by the user is above a quality threshold, indicating not exceeding a rate of packet losses and a determined delay.

6.5.1 WLAN Quality of Experience Design Process: Contention Mechanism and Delay

As it has been pointed out earlier, end-to-end delay is one of the main factors that have influence on the quality perceived by real time services users. Due to that fact, it will be necessary to assess the delay undergone by each service class in order to carry out an accurate performance estimation of a WLAN oriented to provide real time services. In an IEEE802.11 WLAN, delay can be evaluated modeling the performance of its medium access control mechanism. This mechanism is based on the Carrier Sense Multiple Access with Collision Avoidance protocol (CSMA/CA). When a station with a packet to transmit senses the channel and it is busy, the station waits until the channel becomes idle for a DIFS time. Later, it starts a backoff process, generating a random value chosen from a uniform distribution between 0 and a parameter known as Contention Window (CW), which depends on the number of transmissions failed for the packet. At first, CW is set equal to a value CW_{min}, and it is doubled after each unsuccessful transmission, up to a maximum value of CW_{max}. The backoff time counter is decremented once every time interval Te for which the channel is detected empty, frozen when a transmission is detected on the channel, and reactivated when the channel is sensed empty again for a DIFS time or an EIFS time, for a successful or unsuccessful transmission, respectively. The station transmits when the backoff time counter reaches zero. If the packet is correctly received, the receiving station sends an ACK frame after a SIFS time. If the ACK frame is not received within an ACK timeout time, a collision is assumed to have occurred and the packet transmission is rescheduled or discarded depending on a predefined Retry Limit. Upon completing the backoff process (either with a success or with a discard), the transmitting station resets the Contention Window to its initial value.

As it has been mentioned earlier, 802.11g operates in the same frequency band as 802.11 and 802.11b, and is required to remain backward-compatible. When only 802.11g terminals are present in a WLAN, by using the CSMA/CA mechanism, all the terminals will be able to sense when one of them is transmitting, and, therefore, collisions will be avoided as required. However, the situation will change when 802.11 or 802.11b terminals are presents in the network. In that case, modulations used by 802.11g will not be recognized by 802.11 or 802.11b stations, so protection mechanisms must be defined to limit the cross-talk in mixed b/g environments. Essentially, the protection mechanisms require that 802.11g stations operating at high rates reserve the radio medium by using slower 802.11b-compatible reservation mechanisms. The minimal protection contemplated by the standard is that 802.11g stations will protect the fast 802.11g frame exchange with a slow Clear To Send (CTS) frame that prevent other stations from access to the medium. In the CTS-to-self protection, when a station has a frame for transmission, it will transmit a CTS frame with a receiver address of its own MAC address, which means that the destination of the CTS frame will be the station itself. In the CTS it will update the network allocation vector (NAV) to tell other stations using the physical medium that it will be using the radio link for the time necessary to transmit the CTS, the data frame and the corresponding ACK. By means of this CTS frame, all the stations in the network will be aware of the status of the medium and will update the NAV. Thus, the CTS frame must be transmitted at the highest rate understood by all stations attached to the access point, which will be at the most 11 Mbps, when only 802.11b stations are present or 2 Mbps, if 802.11 stations are present. Using only a CTS frame to reserve the medium is the minimum requirement, but it may fail in some cases where there are so-called "hidden nodes" that do not see the CTS. To fully reserve the medium, a protection mechanism known as RTS/CTS is used. This mechanisms consists of a two-frame exchange that would fully announce the impending transmission composed of a Request To Send (RTS) frame followed by the CTS frame. Although the standard requires only a CTS-to-self, using the full RTS/CTS will better protect the inner exchange from interference. As in the case of CTS-to-self mechanism, RTS and CTS frames will be transmitted at the highest rate understood by all stations.

To deal with different service classes with different quality requirements, in standard 802.11e standard, the medium access control mechanism was modified by means of the Enhanced Distributed Channel Access (EDCA). In EDCA was provided a probabilistic quality of service support classifying the traffic into four access categories (AC): voice, video, best effort, and background, from highest to lowest priority. Each access category has its own transmit queue and its own set of medium access parameters, stimulating that high priority traffic has a higher chance of being sent than low priority traffic, and, therefore, giving priority to real time traffic with severe delay requirements. Each access category will have its own DIFS time, known as AIFS, and its own values of minimum and maximum contention window. Lower values of AIFS and CW defined for real time services will reduce the average waiting time for these services against non–real time services. In addition to that, each priority level is assigned a transmit opportunity (TXOP). A transmit opportunity is a bounded time interval during which a station can send frames, as long as the duration of the transmissions does not extend beyond the maximum duration of this time. The use of TXOPs reduces the problem of low rate stations taking up the medium too much time, increasing delays remarkably.

In the IEEE802.11 WLAN design process, planning and dimensioning procedures, EDCA procedures must be taken into account to assess the performance of real time services users in terms of quality of experience perceived by the users. For a determined snapshot of the static simulation under a WLAN configuration, a group of active users will be positioned in the scenario, each one with a different service type, throughput, and quality required. Using different criteria, each user will be associated to an access point, an users under the control of the same access point will have to compete to use the medium. The definition of active user entails that all the stations in the WLAN have always packet to transmit. Therefore, to model the average delay for each user, a backoff delay analysis in saturation conditions must be made. Therefore, a procedure as the one presented in [49] must be implemented, defining different EDCA parameters for each service class to assess delays for different users. The delay parameter used in the quality of experience models for real time services presented in Section 6.2 is the end-to-end latency, which includes WLAN and backbone delays. The goal commonly used in designing networks to support VoIP is the target specified by ITU standard G.114, which states that 150 ms one-way, end-to-end delay ensures user satisfaction in voice applications, so this value will be defined as an upper threshold to comply with. Talking about videoconference services, these services will have similar one-way latency requirements as voice services, because it is a voice component. So, taking that value as a reference, in the design process of an IEEE802.11 network some delay threshold for each network must be established. A typical approach consists of assuming that delay in each network will be similar, and, in that conditions, for a VoIP call initiated and terminated in a WLAN, WLANs delay for real time services should not exceed 50 ms. This will be the reference value to compare with the value obtained from the backoff delay procedure, in order to estimate if the IEEE802.11 network configuration is able to ensure an acceptable quality of experience for real time services. In Table 6.4, delay assessment for an IEEE 802.11e network using EDCA is presented, with the following Access

TABLE 6.4

Saturation Backoff Delay for Different Access Category Users

Users of Each Class	Voice Users Delay (ms)	Videoconference Users Delay (ms)	Best Effort Users Delay (ms)	Background Users Delay (ms)
$N = 2$	5	7	12	38
$N = 4$	7	14	23	110
$N = 6$	9	21	43	250
$N = 8$	10	38	68	470
$N = 10$	12	50	95	700
$N = 12$	14	62	125	1000
$N = 14$	16	82	170	1450
$N = 16$	18	104	230	1950

Class Parameters: CW_{min} = {32, 32, 64, 128}, CW_{max} = CW_{min} * {32, 32, 16, 8}, AIFS = DIFS + TimeSlot * {0, 2, 2, 4} and Retry limit = 6. It has been assumed that all the stations present in the network are working under 802.11g standard, and, due to this fact, CTS-to-self and RTS-CTS protection mechanism have not been taken into account. According to the values in Table 6.4, 10 users of each type of service will determine the limit of users associated to an access point, for the 50 ms delay requirement established.

6.5.2 WLAN Quality of Experience Design Process: Automatic Rate Selection and Packet Losses

Packet loss will be the other parameter that will have a notable influence on the quality perceived by real time services users. The performance of a user in terms of packet loss will be determined by the modulation scheme used and the delivery probability at each bit rate, which will have a close correlation with the signal-to-noise rate. The Standard IEEE 802.11g, which made possible the convergence of IEEE802.11a and IEEE802.11b, uses DSSS, OFDM, or both, through the provision of different physical layers, defined as extended rate physical or ERP. With this extended physical layers, IEEE802.11g introduced the possibility of supporting both data rates of IEEE802.11b (1, 2, 5.5, and 11 Mbps) and IEEE802.11a (6, 9, 12, 18, 24, 36, 48, and 54 Mbps) in 2.4 GHz band. Different rates in ERP-OFDM are supported through the use of an adaptive modulation and coding scheme (BPSK, QPSK, 16 QAM, and 64 QAM), being able to adapt parameters of a determined user depending on propagation conditions. This multi-rate characteristic of IEEE 802.11g makes the choice of frame transmission rate a key issue with an enormous impact on the performance of the networks, and, for that purpose, a control algorithm needs information about the current link conditions, selecting this rate in an adaptive manner, since the wireless channel varies its conditions over time due to factors as station mobility, time varying interference, and location dependent errors. As it was commented previously, SNR-based link adaptation algorithms the selection of the switching thresholds used to define the SNR lookup-table will have a great influence on the accuracy of the assessment of the performance of WLAN IEEE802.11g networks, and, therefore, it will be crucial in the planning and dimensioning optimization processes of this kind of networks to define them properly. For users with best-effort data services, without packet loss or delay constraints, these rate selection procedures will be completely focused on maximizing throughput supplied, as it was exposed before. Throughput can be modeled as continuous, and it will depend on the maximum data rate for the modulation scheme evaluated and the packet loss rate, which varies for different values of signal noise rate. For users with voice and video services, rate selection algorithms must be modeled in a different way, since, as it was shown modeling the quality of experience of VoIP and conversational video, the perceived quality is highly dependent on the packet loss rate. Because of that, strategies when trying to estimate the performance of rate selection algorithms must be varied, modifying the design criteria to ensure that the quality perceived by the user is above a quality threshold. The main objective of the procedure proposed is fulfilling the subjective perceived quality requirements, trying to keep the overall capacity of the network as high as possible. As it was previously commented, for a determined snapshot of the static simulation under a WLAN configuration, a group of active users will be positioned in the scenario and associated to an access point, modeling the expected delay with EDCA backoff delay procedure presented. For a determined signal-to-noise rate and a mean delay value assessed, packet error rate is calculated for each modulation supported by the standard used to establish the communication between terminal and access point. For terminals working under IEEE802.11b mode, packet error values will be worked out, for differential modulations (DBPSK, DQPSK) and Complementary Code Keying, from bit error rate, while for terminals working under IEEE802.11g mode, which means using OFDM modes, it is necessary to calculate the probability of having an error in each subcarrier assessing packet error rate as the complementary probability of transmitting correctly all the subcarriers that form an OFDM symbol. Expressions to relate SNR and BER or PER can be extracted from [8]. Using packet loss value, and end-to-end delay obtained from the EDCA backoff delay procedure, both throughput and quality of experience can be estimated, using the models presented in Section 6.3. Then, a pair throughput-Mean Opinion Score for each modulation is obtained. Finally, and depending on the quality threshold selected by the designer, the decision module will select an optimum modulation scheme, that will be the modulation mode that achieves the highest throughput while guaranteeing at the same time the quality requirements. The scheme of this procedure, for a IEEE802.11g terminal is presented in Figure 6.18. In Figures 6.19 through 6.22, maximum throughput and quality for different SNR values are presented, using as ARS switching thresholds those obtained from the procedure presented in Figure 6.18, for a maximum end-to-end delay of 150 ms. SNR-lookup tables for voice and videoconference 802.11g users with different quality of experience requirements are presented in Tables 6.5, 6.6, and 6.7. This values will be used in simulation processes to characterize the performance of the automatic rate selection algorithm for these types of conversational services users.

Several conclusions can be derived from the results presented. Minimum values of SNR that can be permitted, or, in other words, SNR values considered as outage will be higher if quality requirements want to be fulfilled. The more strict the requirement is, the higher the value of SNR considered as outage. Apart from that, stricter requirements in terms of perceived quality will produce a more profound decrease in terms of capacity. Finally, it can be derived that fulfilling multimedia requirements in terms of MOS will imply a higher cost in terms of capacity than fulfilling audio quality requirements.

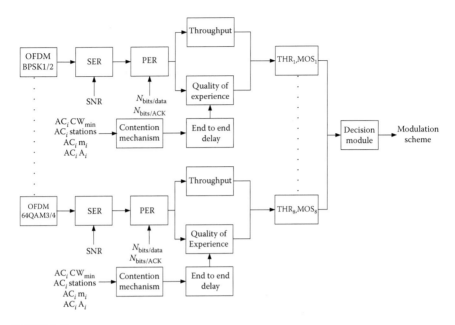

FIGURE 6.18
Procedure to design and deploy a WLAN oriented to deal with real time voice and conversational video services for IEEE802.11g equipment.

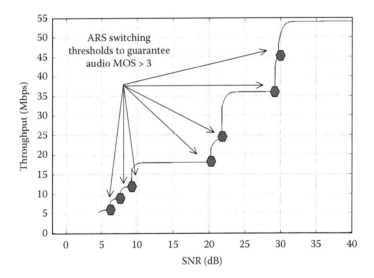

FIGURE 6.19
Correspondence between throughput and SNR using ARS switching values to guarantee audio MOS > 3.

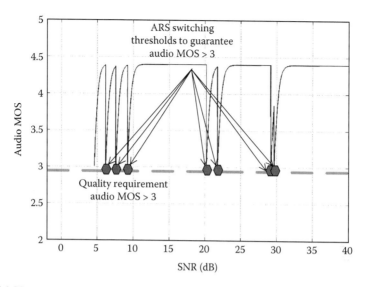

FIGURE 6.20

Correspondence between audio MOS values and SNR using ARS switching values to guarantee MOS > 3.

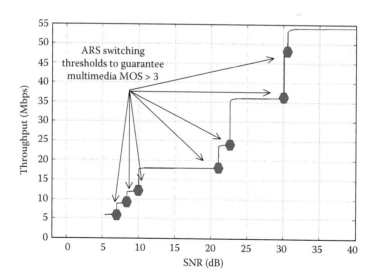

FIGURE 6.21

Correspondence between throughput and SNR using ARS switching values to guarantee multimedia MOS > 3.

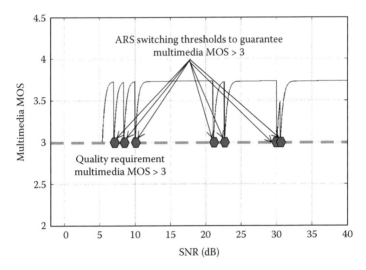

FIGURE 6.22

Correspondence between multimedia MOS values and SNR using ARS switching values to guarantee multimedia MOS > 3.

TABLE 6.5

Estimation of Minimum Values of SNR that Fulfill Audio and Multimedia MOS Requirements for Codec G711PLC

Standard	Modulation	G711PLC			
		Audio MOS 4.39	Audio MOS 4	Audio MOS 3.6	Audio MOS 3
IEEE802.11g	64QAM3/4 to 64QAM2/3	31.7	30.2	29.9	29.6
IEEE802.11g	64QAM2/3 to 16QAM3/4	31.2	29.8	29.5	29.2
IEEE802.11g	16QAM3/4 to 16QAM1/2	23.8	22.3	22.0	21.8
IEEE802.11g	16QAM1/2 to QPSK3/4	22.1	20.7	20.5	20.2
IEEE802.11g	QPSK3/4 to QPSK1/2	11.1	9.7	9.5	9.2
IEEE802.11g	QPSK1/2 to BPSK3/4	9.4	8.1	7.9	7.6
IEEE802.11g	BPSK3/4 to BPSK1/2	8.0	6.7	6.4	6.2
IEEE802.11g	No coverage	6.4	5.1	4.8	4.6
IEEE802.11b	CCK2 to CCK1	9.2	8.0	7.8	7.6
IEEE802.11b	No coverage	6.0	4.9	4.6	4.4
IEEE802.11	DQPSK to DBPSK	7.1	5.7	5.5	5.2
IEEE802.11	No coverage	4.1	2.7	2.4	2.2

TABLE 6.6

Estimation of Minimum Values of SNR that Fulfill Audio and Multimedia MOS Requirements for Codecs G729 and G723

		G729			G723	
Standard	Modulation	Audio MOS 4	Audio MOS 3.6	Audio MOS 3	Audio MOS 3.6	Audio MOS 3
IEEE802.11g	64QAM3/4 to 64QAM2/3	31.0	30.3	30.0	30.5	30.1
IEEE802.11g	64QAM2/3 to 16QAM3/4	30.5	29.9	29.5	30.1	29.6
IEEE802.11g	16QAM3/4 to 16QAM1/2	23.0	22.4	22.1	22.7	22.2
IEEE802.11g	16QAM1/2 to QPSK3/4	21.4	20.9	20.5	21.1	20.6
IEEE802.11g	QPSK3/4 to QPSK1/2	10.4	9.8	9.5	10.0	9.6
IEEE802.11g	QPSK1/2 to BPSK3/4	8.7	8.2	7.9	8.4	8.0
IEEE802.11g	BPSK3/4 to BPSK1/2	7.4	6.8	6.5	7.0	6.6
IEEE802.11g	No coverage	5.7	5.2	4.9	5.4	5.0
IEEE802.11b	CCK2 to CCK1	8.6	8.1	7.8	8.3	8.0
IEEE802.11b	No coverage	5.4	4.9	4.6	5.1	4.8
IEEE802.11	DQPSK to DBPSK	6.4	5.8	5.5	6.0	5.6
IEEE802.11	No coverage	3.4	2.8	2.5	3.0	2.6

TABLE 6.7

Estimation of Minimum Values of SNR That Fulfill Audio and Multimedia MOS Requirements for Audio Codec G711PLC and Video Codec MPEG4

		MPEG4 QQVGA 2.1 in.		
		512 kbps—20 Frames/s		
Standard	Modulation	MM MOS 3.86	MM MOS 3.6	MM MOS 3
IEEE802.11g	64QAM3/4 to 64QAM2/3	32.28	30.99	30.49
IEEE802.11g	64QAM2/3 to 16QAM3/4	31.80	30.53	30.02
IEEE802.11g	16QAM3/4 to 16QAM1/2	24.35	23.09	22.59
IEEE802.11g	16QAM1/2 to QPSK3/4	22.71	21.48	21.00
IEEE802.11g	QPSK3/4 to QPSK1/2	11.60	10.42	9.96
IEEE802.11g	QPSK1/2 to BPSK3/4	9.95	8.80	8.35
IEEE802.11g	BPSK3/4 to BPSK1/2	8.59	7.41	6.95
IEEE802.11g	No coverage	6.94	5.79	5.34
IEEE802.11b	CCK2 to CCK1	9.65	8.62	8.23
IEEE802.11b	No coverage	6.53	5.47	5.07
IEEE802.11	DQPSK to DBPSK	7.66	6.45	5.98
IEEE802.11	No coverage	4.65	3.44	2.97

It was inevitable, because, as presented before, multimedia MOS values are a combination of speech and video values, and therefore, the maximum packet loss rate to obtain a determined multimedia MOS will be lower that the maximum packet loss rate to obtain the same value of MOS referring only to speech.

6.6 Conclusions and Future Work

The concept of user-perceived QoS in audio and video conversational services has been analyzed, and a simple method to predict the impact on speech and conversational video performance of packet loss rate has been proposed. WLAN IEEE802.11 basic planning principles to optimize network performance in terms of capacity has been taken as reference and optimal SNR values to commute between different modulation schemes to maximize overall throughput in the WLANs have been discussed. A new procedure to design and deploy a WLAN oriented to guarantee perceived quality in real time voice and conversational video services has been proposed, evaluating the effect of WLAN planning oriented to maximize capacity on the quality of service of conversational services and the effect of the variability of the indoor channel in perceived quality for audio and conversational video services. Finally, the proposed analysis method has been applied to a real scenario by simulating deployments for different design strategies. The simulations confirm the theoretical foundations presented, showing the degradation of the network performance in terms of perceived quality that will imply using a strategy to optimize network capacity in the deployment, and the trade between capacity, outage points, and quality that will cause planning to guarantee a certain subjective audio or multimedia quality threshold. Subsequent research steps will be the application of these concepts to the development of algorithms to be included in processes of automatic planning and dimensioning WLANs. The inclusion of these concepts will allow taking into account quality of service concepts in the design of WLAN IEEE802.11 networks conceived to provide voice and conversational video services.

Acknowledgments

The authors are grateful for the support of the Spanish Ministry of Education and Science within the framework of the TEC2005-07010-C02-01/TCM project. The authors are also thankful for the support of CELTIC Project EasyWireless II.

Appendix 6.A: Default Values in VoIP and Multimedia QoE Assessment Procedures

TABLE 6.A.1

Default Values for Different Parameters Used in Voice Quality Assessment

Parameter	Acronym	Units	Typical Value
Send loudness rating	SLR	dB	+8
Receive loudness rating	RLR	dB	+2
Sidetone masking rating	STMR	dB	15
Listener sidetone rating	LSTR	dB	18
D-value of telephone, send side	Ds	—	3
D-value of telephone receive side	Dr	—	3
Talker echo loudness rating	TELR	dB	65
Weighted echo path loss	WEPL	dB	110
Mean one-way delay of the echo path	T	ms	0
Round trip delay in a 4-wire loop	Tr	ms	0
Absolute delay in echo-free connections	Ta	ms	0
Number of quantization distortion units	qdu	—	1
Circuit noise referred to 0 dBr-point	Nc	dBm0p	−70
Noise floor at the receive side	Nfor	dBmp	−64
Room noise at the send side	Ps	dB(A)	35
Room noise at the receive side	Pr	dB(A)	35
Advantage factor	A	—	0

TABLE 6.A.2

Coefficients for Video and Multimedia Quality Assessment

Coefficients	MPEG4 QQVGA 4.2 in.	MPEG4 QVGA 2.1 in.	Coefficients	MPEG4 QQVGA 4.2 in.	MPEG4 QVGA 2.1 in.
v_1	1.431	7.160	m_1	-4.457×10^{-1}	-6.966×10^{-1}
v_2	2.228×10^{-2}	2.215×10^{-2}	m_2	-6.638×10^{-1}	-8.127×10^{-1}
v_3	3.759	3.461	m_3	4.042×10^{-1}	4.562×10^{-1}
v_4	184.1	111.9	m_4	2.321	3.003
v_5	1.161	2.091	m_5	-3.255×10^{-1}	-1.638×10^{-1}
v_6	1.446	1.382	m_6	3.309×10^{-1}	3.626×10^{-1}
v_7	3.881×10^{-4}	5.881×10^{-4}	m_7	1.494×10^{-1}	1.291×10^{-1}
v_8	2.116	0.8401	m_8	5.457×10^{-1}	5.456×10^{-1}

(continued)

TABLE 6.A.2 (continued)

Coefficients for Video and Multimedia Quality Assessment

Coefficients	MPEG4 QQVGA 4.2 in.	MPEG4 QVGA 2.1 in.	Coefficients	MPEG4 QQVGA 4.2 in.	MPEG4 QVGA 2.1 in.
v_9	467.4	113.9	m_9	-3.235×10^{-4}	-1.251×10^{-4}
v_{10}	2.736	6.047	m_{10}	3.915	3.763
v_{11}	15.28	46.87	m_{11}	-1.377×10^{-3}	-1.065×10^{-3}
v_{12}	4.170	10.87	m_{12}	0.000	1.465×10^{-2}
			m_{13}	-1.095×10^{-3}	-1.002×10^{-3}
			m_{14}	0.000	0.000

Appendix 6.B: Packet Error Rate Assessment in IEEE802.11

TABLE 6.B.1

Parameters for Different OFDM Modulations Supported by IEEE802.11g

Modulation	Code Rate	N_{BPSC}	N_{CBPS}	N_{DBPS}	$V_{simbOFDM}$ (Ksimb/s)	Throughput (Mbps)
BPSK	1/2	1	48	24	250	6
BPSK	3/4	1	48	36	250	9
QPSK	1/2	2	96	48	250	12
QPSK	3/4	2	96	72	250	18
16-QAM	1/2	4	192	96	250	24
16-QAM	3/4	4	192	144	250	36
64-QAM	2/3	6	288	192	250	48
64-QAM	3/4	6	288	216	250	54

TABLE 6.B.2

BER and SER Values for Different Modulation Schemes Supported by IEEE802.11g

Modulation	Max Cap (Mbps)	Error Probability
DSSS DBPSK	1	$BER_{1Mbps} = Q\left(\sqrt{11 * \dfrac{S}{N}}\right)$
DSSS DQPSK	2	$BER_{2Mbps} = Q\left(\sqrt{5.5 * \dfrac{S}{N}}\right)$

TABLE 6.B.2 (continued)

BER and SER Values for Different Modulation Schemes Supported by IEEE802.11g

Modulation	Max Cap (Mbps)	Error Probability
HR-DSSS CCK	5.5	$BER_{5.5Mbps} \leq \frac{8}{15} * 14 * \left[Q\left(\sqrt{8 * \frac{S}{N}} \right) + Q\left(\sqrt{16 * \frac{S}{N}} \right) \right]$
HR-DSSS CCK	11	$BER_{11Mbps} \leq \frac{128}{255} * \left[24 * Q\left(\sqrt{4 * \frac{S}{N}} \right) + 16 * Q\left(\sqrt{6 * \frac{S}{N}} \right) + 174 * Q\left(\sqrt{8 * \frac{S}{N}} \right) \right.$ $\left. + 16 * Q\left(\sqrt{10 * \frac{S}{N}} \right) + 24 * Q\left(\sqrt{12 * \frac{S}{N}} \right) + Q\left(\sqrt{16 * \frac{S}{N}} \right) \right]$
OFDM BPSK	6/9	$SER_{BPSK} = Q\left(\sqrt{2 * \frac{48}{52} * \frac{16.6}{R} * \frac{S}{N}} \right)$
OFDM QPSK	12/18	$SER_{QPSK} = 2 * Q\left(\sqrt{2 * \frac{48}{52} * \frac{16.6}{R} * \frac{S}{N}} \right) * \left[\left(1 - \frac{1}{2} * Q\left(\sqrt{2 * \frac{48}{52} * \frac{16.6}{R} * \frac{S}{N}} \right) \right) \right]$
OFDM 16QAM	24/36	$SER_{16-QAM} = 1 - \left[1 - \frac{3}{2} * Q\left(\sqrt{\frac{3}{15} \frac{48}{52} \frac{16.6}{R} \frac{S}{N}} \right) \right]^2$
OFDM 64QAM	48/54	$SER_{64-QAM} = 1 - \left[1 - \frac{7}{4} * Q\left(\sqrt{\frac{3}{63} \frac{48}{52} \frac{16.6}{R} \frac{S}{N}} \right) \right]^2$

References

1. 3GPP, Generic access to the A/Gb interface, Stage 2, Release 6, Technical Specification 43.3 18, May 2006.
2. FMCA, Convergence services over Wi-Fi GAN (UMA), FMCA Product Requirement Definitions, Release 2.0, May 8, 2006.
3. 3GPP, Voice call continuity between CS and IMS Study, Release 7, Technical Specification 23.806, December 2005.
4. FMCA, Convergence services using SIP over Wi-Fi, FMCA Product Requirement Definitions, Release 2.0, May 8, 2006.
5. IEEE Std 802.11g-2003, Amendment to IEEE Std 802.11-1999, Amendment 4: Further higher data rate extension in the 2.4 GHz Band.
6. P. Wertz, M. Sauter, F. A. Landstorfer, G. Wolfle, and R. Hoppe, Automatic optimization algorithms for the planning of wireless local area networks, in *Proceedings of the IEEE Vehicular Technology Conference (VTC'04-Fall)*, Los Angeles, CA, 2004, pp. 3010–3014.
7. S. Bosio, A. Capone, and M. Cesana, Radio planning on wireless local area networks, *IEEE Transactions on Networking*, 15(6), 1414–1427, December 2007.

8. K. Jaffres-Runser, J. Gorce, and S. Ubéda, QoS constrained wireless LAN optimization within a mulitobjective framework, *IEEE Wireless Communications*, 13(6), 26–33, December 2006.

9. Y. Lee, K. Kim, and Y. Choi, Optimization of AP placement and channel assignment in wireless LANs, in *Proceedings of IEEE Conference on Local Computer Networks (LCN'02)*, Tampa, FL, 2002, pp. 831–836.

10. A. Gondran, O. Baala, A. Caminada, and H. Mabed, Joint optimization of access point placement and frequency assignment in WLAN, in *Proceedings of the Third IEEE/IFIP International Conference in Central Asia on Internet (ICI 2007)*, Tashkent, Uzbekistan, September 26–28, 2007.

11. J. Liang, K. Runser, J. Gorce, and F. Valois, Indoor WLAN planning with a QoS constraint based on a Markovian performance evaluation model, in *IEEE Second International Conference on Wireless and Mobile Computing, Networking and Communications (WiMob)*, Montreal, Canada, June 2006.

12. T. Vanhatupa, M. Hannikainen, and T. Hamalainen, Genetic algorithm to optimize node placement and configuration for WLAN planning, in *Fourth International Symposium on Wireless Communication Systems 2007 (ISWCS 2007)*, Trondheim, Norway, October 17–19, 2007.

13. C. Prommak, J. Kabara, and D. Tipper, Demand-based network planning for large scale WLANs, in *Proceedings of the First International Conference on Broadband Networks (BROADNETS'04)*, San Jose, CA, 2004.

14. E. Amaldi, A. Capone, M. Cesana, and F. Malucelli, Optimizing WLAN radio coverage, in *Proceedings of the IEEE International Conference on Communications (ICC'04)*, Paris, France, 2004, pp. 180–184.

15. M. Kamenetsky and M. Unbehaun, Coverage planning for outdoor wireless LAN systems, in *Proceedings of the International Zurich Seminar on Broadband Communications*, Zurich, Switzerland, 2002, pp. 49-1–49-6.

16. S. Kouhbor, J. Ugon, A. Kruger, A. Rubinov, and P. Branch, Optimal placement of access point in WLAN based on a new algorithm, in *Proceedings of the International Conference on Mobile Business (ICMB'05)*, Sydney, Australia, 2005, pp. 592–598.

17. IEEE Std 802.11e-2005, Amendment to IEEE Std 802.11-1999, Amendment 8: Medium access control (MAC) Quality of service enhancements.

18. L. Cai, Y. Xiao, X. Shen, and J. Mark, VoIP over WLAN: Voice capacity, admission control, QoS, and MAC, *International Journal of Communication Systems*, 19(4), 491–508, May 2006.

19. A. Saliba, M. Beresford, M. Ivanovich, and P. Fitzpatrick, User-perceived quality of service in wireless data networks, *Personal and Ubiquitous Computing*, 9(6), 413–422, November 2005.

20. D. Skyrianoglou, N. Passas, and A. Salkintzis, Support of IP QoS over wireless LANs, in *59th IEEE Vehicular Technology Conference 2004 (VTC-Spring. 2004)*, vol. 5, Milan, Italy, May 17–19, 2004, pp. 2993–2997.

21. K. Diwakar and S. Iyer, Supporting real time applications with better QoS guarantees in 802.11, in *First International Symposium on Wireless Communication Systems, 2004*, September 20–22, 2004, pp. 373–377.

22. International Telecommunication Union, Objective quality measurement of telephone band (300–3400 Hz) speech codecs, ITU-T Recommendation P.861, International Telecommunication Union, Geneva, Switzerland, February 1998.

23. M. P. Hollier, M. O. Hawksford, and D. R. Guard, Algorithms for assessing the subjectivity of perceptually weighted audible errors, *Journal of Audio Engineering Society*, 43, 1041–1045, December 1995.

24. S. Voran, Objective estimation of perceived speech quality—Part I: Development of the measuring normalizing block technique, *IEEE Transactions on Speech and Audio Processing*, 7, 371–382, July 1999.

25. W. Yang, Enhanced modified bark spectral distortion (EMBSD): An objective speech quality measure based on audible distortion and cognition model, PhD dissertation, Temple University, Philadelphia, PA, May 1999.

26. International Telecommunication Union, Perceptual evaluation of speech quality (PESQ), An objective method for end-to-end speech quality assessment of narrowband telephone networks and speech codecs, ITU-T Recommendation P.862, International Telecommunication Union, Geneva, Switzerland, February 2001.

27. A. Markopolou, F. Tobagi, and M. Karam, Assessing the quality of voice communications over Internet backbones, *IEEE/ACM Transactions on Networking (TON)*, 11(5), pp 747–760, October 2003.

28. A. Clark, Modelling the effects of burst packet loss and recency on subjective voice quality, in *Proceedings of the IP Telephony Workshop*, New York, March 2001.

29. S. Choudhury, N. Shetty, and J. Gibson, MOSx and Voice outage rate in wireless communications, in *Proceedings of the Fortieth Asilomar Conference on Signals, Systems and Computers, 2006 (ACSSC '06)*, Pacific Grove, CA, October–November 2006, pp. 1303–1307.

30. M. Narbutt and M. Davis, An assessment of the audio codec performance in voice over WLAN (VoWLAN) systems, in *Proceedings of the Second Annual International Conference on Mobile and Ubiquitous Systems: Networking and Services, 2005 (MobiQuitous 2005)*, San Diego, CA, July 17–21, 2005, pp. 461–467.

31. International Telecommunication Union, The E-model, a computational model for use in transmission planning, ITU-T Recommendation G.107, International Telecommunication Union, Geneva, Switzerland, 2003.

32. International Telecommunication Union, Methodology for subjective assessment of the quality of television pictures, ITU-R Recommendation BT.500–11, International Telecommunication Union, Geneva, Switzerland, 2002.

33. International Telecommunication Union, Subjective video quality assessment methods for multimedia applications, ITU-R Recommendation P.910, International Telecommunication Union, Geneva, Switzerland, 1999.

34. S. Wolf and M. H. Pinson, In-service performance metrics for MPEG-2 video systems, in *Proceedings of the Measurement Techniques of the Digital Age Technical Seminar*, Montreux, Switzerland, November 1998, pp. 1–10.

35. S. Wolf and M. H. Pinson, Spatial-temporal distortion metrics for in-service quality monitoring of any digital video system, in *Proceedings of the SPIE on Multimedia Systems and Applications II*, vol. 3845, Boston, MA, September 1999, pp. 266–277.

36. P. Pace and E. Viterbo, Fast and accurate video PQoS estimation over wireless networks, *EURASIP Journal on Advances in Signal Processing*, 2008, doi: 10.1155/2008/548741.

37. K. Yamagishi and T. Hayashi, Opinion model for estimating video quality of videophone services, in *Proceedings of the Global Telecommunications Conference 2006 (GLOBECOM '06)*, San Francisco, CA, November 2006, IEEE.

38. S. Winkler and C. Faller, Perceived audiovisual quality of low-bitrate multimedia content, multimedia, *IEEE Transactions on Multimedia*, 8(5), 973–980, October 2006.
39. M. Ries, O. Nemethova, and M. Rupp, Video quality estimation for mobile H.264/AVC video streaming, *Journal of Communications*, 3(1), 41–50, 2008.
40. International Telecommunication Union, Opinion model for video-telephony applications, ITU-T Recommendation G.1070, International Telecommunication Union, Geneva, Switzerland, 2007.
41. M. Molina-Garcia, A. Fernandez-Duran, and J. I. Alonso, WLAN optimization based on the perceived conversational quality using subjective metrics, in *Proceedings of the European Wireless Technology Conference 2009 (EuWIT 2009)*, Rome, Italy, September 28–29, 2009, pp. 53–56.
42. International Telecommunication Union, One way transmission time, ITU-T Recommendation G.114, International Telecommunication Union, Geneva, Switzerland, 2003.
43. International Telecommunication Union, Transmission impairments due to speech processing, ITU-T Recommendation G.113, International Telecommunication Union, Geneva, Switzerland, 2001.
44. I. Haratcherev, K. Langendoen, R. Langendijk, and H. Sips, Hybrid rate control for IEEE 802.11, in *Proceedings of the Second International Workshop on Mobility Management & Wireless Access Protocols, International Conference on Mobile Computing and Networking*, Philadelphia, PA, 2004, pp. 10–18.
45. J. Pavon and S. Choi, Link adaptation strategy for IEEE 802.11 WLAN via received signal strength measurement, in *IEEE International Conference on Communications 2003 (ICC '03)*, vol. 2, Anchorage, AK, 2003, pp. 1108–1113.
46. D. Qiao, S. Choi, and K. Shin, Goodput analysis and link adaptation for IEEE 802.11a wireless LANs, *IEEE Transactions on Mobile Computing*, 1(4), 278–292, October-December 2002.
47. J. G. Proakis, *Digital Communications*, 3rd edn. McGraw-Hill, New York, 1995.
48. A. Hwang, Observations of network traffic patterns at an end network, Master thesis, Harvard University, Cambridge, MA, April 1998.
49. A. Banchs and L. Vollero, A delay model for IEEE 802.11e EDCA, *IEEE Communications Letters*, 9(6), 508–510, June 2005.

7

Convergence and Interworking of Heterogeneous Wireless Access Networks

Peyman TalebiFard and Victor C. M. Leung

CONTENTS

7.1 Introduction

Service providers are in the process of transforming from legacy packet and circuit-switched networks to converged Internet Protocol (IP) networks and consolidating all network services and business units on a single IP infrastructure. The future users of communication systems will require the use of data rates around 100 Mbps in their homes while all the services and applications require high bandwidth. The next generations of heterogeneous wireless networks are expected to interact with each other and be capable of interworking with IP-based infrastructures. The requirements for Next Generation Networks (NGNs) lead to an architectural evolution that requires a converged infrastructure where users across multiple domains can be served through a single unified domain. Convergence is at the core of IP-based NGNs [1]. The aim of IP convergence is to build a single network infrastructure that is cost effective, scalable, reliable, and secure. The aim of standardization has been to enable a mix and match of services bundled to offer innovative services to the end users. Service enabler in this context is the approach to eliminate the vertical silo structure and to transform into a horizontal layered architecture. In a vertical silo approach, all the service components and architectures are designed for a specific service. Examples are security features, charging function, management, and policy enforcement modules. In this approach, the vertical elements are interrelated and tightly coupled and services can only be offered over a specific access network. In a vertical silo–based approach, the following issues exist. Maintenance and upgrade of individual components is difficult and it will take a long time for the system to be replaced, if needed. Other systems cannot take advantage of and reuse the data repository, functional elements, and policy designs that are deployed by other systems. Optimized designs and enhanced service performance cannot be integrated to other systems. This makes the systems isolated, hence difficult to integrate and this increases the capital and operational expenditures of new system deployment. Although the consolidation of IP core networks is advantageous, the business units and providers must have the flexibility to securely manage their own IP networks.

This chapter is organized as follows. In Section 7.2, the concept of service enablers, convergence, and different aspects of it are discussed. Section 7.3 emphasizes on the significance of common charging infrastructure to enable the interworking of heterogeneous networks. Location-based services (LBS) are defined in Section 7.4. OMA service enablers and components for LBS are also reviewed. Finally, Section 7.5 concludes the chapter.

7.2 Convergence

In the following, we describe different aspects of convergence such as device and technology convergence, service and application convergence, as well as network convergence [2].

7.2.1 Device and Access Technology Convergence

The users of wireless technologies prefer to use one device to access various services whether from different providers or the same provider. It enables the service providers to increase revenues and customer loyalty by offering an enhanced service quality at a lower cost. From the manufacturing point of view, it leads to lower cost of manufacturing a device that allows the integration of different technologies on a single device at a cheaper price. From the technological perspective, the capability to connect to any access technology (fixed and mobile) is a driving force of this integration. Furthermore, it leads to the demand for seamless handover among heterogeneous access technologies. The convergence of mobile devices and access technologies refers to wireless devices that can support multiple functions within different radio access technologies such as Global System for Mobile (GSM) Communications, Code Division Multiple Access (CDMA), or Wireless Fidelity (WiFi). These devices are also referred to as multimodal devices; examples are Personal Digital Assistants (PDAs), smart phones, and laptops.

7.2.2 Service and Application Convergence

Service convergence refers to the integration and interworking of various communication services such as text messaging, voice, multimedia services, and LBS. IP Multimedia Subsystem (IMS) is an all-IP core network that enables convergence through common signaling framework for control of multimedia services. Convergence of applications can be supported through the common Service Delivery Platform (SDP). The SDP enables application to interwork with each other by establishing a set of Application Programming Interfaces (APIs) and well defined sets of general services that are useful to most of the applications. Through convergence of services and applications, new services can be utilized by applications and new applications can be developed in a timelier manner. One of the key design principles for convergence of NGNs is the Service Oriented Architecture (SOA) [3]. SOA is a service delivery platform that is aimed toward a modular and horizontal design to support application convergence [4]. The SOA design consists of service layer, session control layer, and access interworking and media layer. Service layer is in charge of policy control and policy enforcement. Policy can be viewed in terms of user policy, service policy, and network policy. It also consists of service enablers and SDP that interfaces with the application layer. The session control layer consists of Call Session Control Functions (CSCFs) and is mainly in charge of signaling via the Session Initiation Protocol (SIP) and in charge of initiating and terminating sessions. The access interworking and media layer consists of the IP core network that provides the means for interworking of heterogeneous access networks.

7.2.3 Network Convergence

The evolution toward all-IP core networks is mainly based on the vision of NGNs. NGNs will capture the flexibility and adaptability of different access technologies and services. Various architectures have been proposed based on providing intelligence to the core network. Examples are IMS [5] that is standardized by European Telecommunications Standards Institute (ETSI) and Third Generation Partnership Project (3GPP), Open Mobile Alliance (OMA), Telecoms and Internet Converged Services and Protocols for Advanced Networks (TISPAN), and the 3GPP Multimedia Broadcast Multicast Service (MBMS) [6].

7.2.4 Convergence and the IMS

IMS [5] defined by ETSI and 3GPP, is one of the enablers of convergence of fixed and mobile networks for NGNs and is aimed at enabling subscribers of third generation (3G) cellular networks to access packet-switched multimedia services over IP-based networks. It also enhances user mobility and enables access-agnostic service deployment. Its service-centric framework makes the development of new revenue-generating services possible.

7.2.5 Open Mobile Alliance

The Open Mobile Alliance (OMA) is an industry forum for development of market driven service enablers to facilitate the interoperability of mobile multimedia services. The architecture group in OMA has defined a horizontal architecture in which the service enablers can interwork. Service enablers link directly with clients on user devices to enable applications. For example, a presence server is a service-enabler that facilitates the delivery of presence-based call routing services orchestrated by applications hosted on servers within the application layer. Some service enablers will be common to both landline and wireless applications, e.g., presence, Intelligent Network (IN), Unified Messaging (UM), Multimedia Messaging Service (MMS), connection manager, content download; other enablers will be relevant to only landline or wireless applications, but not both, e.g., Wireless Application Protocol (WAP), Short Messaging Service (SMS), and LBS.

The following are some examples of service enablers that are explained in details in [7]:

- Security enablers
- Presence and list management enablers
- Push-to-talk over cellular enabler
- Device management enabler
- Digital rights management enabler
- Broadcast enabler
- Dynamic content delivery enabler
- Global permissions management enabler
- Categorization-based content screening enabler
- Game services enabler
- Location enabler
- Charging enabler

7.2.5.1 OMA Service Environment

As mentioned earlier, OMA Service Environment (OSE) is a horizontal service architecture that allows interworking of various service enablers. One of the principles of this design is the idea of reuse. The idea of reuse brings the concept that all enablers are equal and should be considered as a network of peers that are able to connect to any resource through their defined interfaces. The applications are therefore not tightly coupled to enablers in the way that enablers should not be application dependent. The applications are capable to reuse any combination of enablers at the southbound layer in the horizontal architecture. However, it is not sufficient to break the vertical silos into a horizontal architecture. There is a need for a control infrastructure to allow bundling of service enablers to deliver specific real time service to the end users. The control infrastructure can take the form of a *policy enforcement module* that enables the service provider to manage and control access to its applications and network resources from a centralized location. OSE is advantageous from the end user perspective and the service provider perspective. From the end user point of view, the OSE facilitates a more consistent Quality of Experience (QoE) by end users, who subscribe to multiple applications via a single service provider compared to the vertically bundled case where the user has to set the preferences and do the provisions. It furthermore, enables roaming capabilities when the user is accessing an application via a different network or uses a different device. From a service provider point view, it facilitates the resource protection when interfacing with other external or non-trusted domains. Examples are third party application providers who are contracted to offer applications that use the service provider's domain to deliver services. The service provider requires control over usage of resources to protect the integrity of its resources from harmful actions. In addition, the service providers should provision the QoS according to the terms of the Service Level Agreement (SLA) contract that it has with the third party application provider. Parlay X [8] is one of the technologies that deal with controlled third party access to service capabilities. The Parlay X module interfaces with external (non-trusted domain) applications and enables them to securely access network resources, service enablers, and user devices. Policy Enforcement (PE) module can cover other needs of the service provider such as triggering the charging events, enforcement of user privacy such as filtering.

7.2.6 Convergence and TISPAN

The view of IMS toward network convergence requires interworking and cooperation of different access networks, policy modules, QoS mapping, databases, and admission control policies. The aim of TISPAN is to address these issues and its architecture which is based on the idea of cooperative subsystems that share common components. The key contribution of TISPAN is the interworking of these subsystems. In this approach it is possible to add more subsystems and ensure that network resources are allocated commonly to all subsystems. One of these subsystems is the IMS core. Other relevant subsystems are Network Attachment Subsystem (NASS), which is in charge of network connectivity by providing address allocation, authorization, authentication functions, and location management. Another subsystem is the Resource and Admission Control Subsystem (RACS) that is responsible for QoS control such as resources reservation and allocation, gateway control, and admission control, based on SLA and operator specific policy rules and availability of resources [9].

7.2.7 Convergence and MBMS

Another working item of 3GPP is the MBMS [6], which is aimed at delivery of IP multicast datagram to mobile User Equipments (UEs) with a specified QoS. The key element in MBMS is the use of IP to identify the particular distance of the bearer service for managing the multicast services. Other key elements of the MBMS reference architecture are Gateway General Packet Radio Service (GPRS) Support Node (GGSN) that is in charge of reserving resources for UMTS Terrestrial Radio Access Network (UTRAN), Serving Gateway Support Node (SGSN) capable of storing user specific MBMS context. Broadcast Multicast Service Center (BM-SC) is another element of MBMS that is responsible for control of session initiation and termination and management of bearer resources. It provides functionalities for both data path and control/signaling path. When considering NGN convergence, the key functionalities should be abstracted in a reusable manner. Examples of such functionalities are membership, authorization and security, service discovery, QoS provisioning, and user profile management. For membership and authorization functions, IMS uses the Home Subscriber Server (HSS) as its main database and can support the role of membership functions in BM-SC. The security function defined by 3GPP only applies to MBMS technologies. One may consider the development of a security enabler in the subsystem that is interworking with MBMS.

MBMS needs to provide user accounting, billing, security, and QoS for a heterogeneous network environment. Therefore, the integration of IMS and MBMS can enhance the MBMS. The integrated IMS-MBMS architecture should have the following features [9]:

- Scheduling and congestion control mechanism that is adaptive to the feedback from Radio Access Network (RAN)
- Signaling scheme that is compatible with the IMS signaling
- Session management based on RAN selection in a converged environment
- QoS support for multicast and broadcast services

7.3 Charging Enablers and Common Charging Functions

Common billing and charging functions are important as heterogeneous networks and domains are emerging to combine various services, applications, and business units. Subscribers often use various services offered by service providers that can be either through fixed or mobile network communications. Examples are cellular telephony services, Internet services, broadcasting/multicasting, voice, video, and data services. Various mobile devices can be used such as cell phones, PDAs, smart phones, pagers, and Global Positioning Systems (GPSs). OMA, 3GPP, and 3GPP2 have defined various standards and protocols to govern the communications between mobile devices and base stations. They have also defined sets of specifications that interwork with both online charging and offline charging mechanisms [10,11,12].

Online charging that is also known as prepaid charging, refers to charging for services prior to the delivery of services. It involves direct interaction of charging mechanism with the session control functions and real time collection of charging information with the usage of network resources. For this purpose, a Charging Trigger Function (CTF) may be used to collect charging information and assemble the data to be sent to the Online Charging System (OCS) [13]. The OCS will then make a decision to grant the quota. If the quota is expired or is not granted by the OCS, the CTF will enforce the termination of resource allocation to the subscriber [10].

Offline charging or postpaid billing refers to charging for services after service delivery. Similar to online charging, the offline charging is also triggered by the CTF. Charging related data are collected concurrently with the usage of resources by CTF within all network domains but it does not enforce charging related decisions in real time. Collected charging information by CTF will be assembled and sent to a Charging Data Function (CDF) and finally will be delivered to the billing domain via a Charging Gateway Function (CGF) [10]. A more detailed explanation of the aforementioned functions will be provided in Sections 7.3.1 through 7.3.5. In online charging, the charging information is transferred from the network to OCS, whereas in offline charging the charging information is transferred from the network to the billing domain for postpaid billing.

Different domains, subsystems, and services have the charging function elements embedded in a different way from each other. However, the charging requirements shall remain the same across all domains, subsystems, and services. For charging standardization 3GPP has provided a common charging architecture and a set of functional requirements for interworking of heterogeneous domains and systems in release 8 of 3GPP document TS 32.240 [10]. In Sections 7.3.1 through 7.3.2, we review some of the charging functions and requirements.

7.3.1 Offline Charging Functions

The offline common charging architecture consists of the following main elements. CTF, CDF, CGF, and Billing Domain (BD).

7.3.2 Charging Trigger Function in Offline Charging

CTF is the focal point of collecting chargeable event information and responsible for generating charging events based on monitoring the usage of network resources and forwarding this information to the CDF. It consists of two main functional blocks: Accounting Metrics Collections and Accounting Data Forwarding.

Accounting metrics collection is responsible for identifying the user's consumption of network resources and services in real time. It triggers conditions for collection of charging information. It mainly depends on the functions and services that a Network Element (NE) provides and therefore can be considered as the NE-dependent part of CTF.

Accounting data forwarding receives the collected information from accounting metrics collection and determines the occurrence of charging events based on a set of collected metrics. Although collected data received from accounting metrics collection is specific to the type NE, the task of receiving, assembly, and forwarding is a generic task. Therefore, the accounting data forwarding is considered as a NE-independent part of CTF.

7.3.3 Charging Data Function

The CDF is responsible for generating the Charging Data Records (CDRs) with the received charging events from the CTF. Details about the format and contents of CDRs are dependent on the domain, subsystem, and services and are specified in 3GPP document TS 32.250 for CS domains and TS 32.251 for PS domain.

7.3.4 Charging Gateway Function

The CGF acts as a gateway between the 3GPP network and billing domain. It receives the CDRs from CDF and performs the following actions:

- Preprocessing, e.g., validation, consolidation, reformatting of CDRs, error handling, and CDR storage
- Routing and filtering of CDRs
- CDR file management
- Transferring the CDRs to billing domain

7.3.5 Charging Trigger Function in Online Charging

As mentioned earlier, in Section 7.3.2, the CTF is responsible for collecting the charging information and creating the charging events. This is done both for online and offline charging. In order to support the online charging, functional enhancements are required for the accounting data forwarding functional block. The collected information for online charging events are not necessarily identical to the offline charging events and the charging events for online charging should be forwarded to the Online Charging Function (OCF) instead of CDF. The CTF must be able to delay the release of resources until permission is granted by the OCS and it should also be able to keep track of resource availabilities. The CTF must also be able to terminate the network usage if the quota is expired or not granted by the OCS.

7.4 Location-Based Services and Convergence of Technologies

LBS refer to delivery of information and services to a mobile user in the context of user's current geographical location. LBS are heterogeneous technologies that converge in a common framework. These technologies are geographical information systems, the Internet, and communication technologies [14]. It is important to provide an acceptable QoE to the users with service customization and emergency services in terms of subscriber's location, time of day, and so on. This can be partially supported by enhancement of LBS. In addition to the technological challenges of LBS such as display capability, limited battery lifetime, and bandwidth requirement, mobility of LBS brings in additional challenge in the area of subscriber profile management to ensure access to services and applications across all domains and to ensure that the requested information is delivered to the users in real time. The main components of LBS are as follows:

- User equipment or mobile device
- Communication network
- Positioning system
- Application/service provider
- Content provider

In this section, we briefly describe some of the related challenges that exist in LBS.

7.4.1 Context Aware Communication and LBS

Context aware communication refers to the use of any user related contextual information for delivering and provisioning of services that are relevant to the subscriber. It can be used for improving system design and delivery of relevant content for enhanced communication and service delivery. Intelligent networks and context aware communications are other interesting aspects to future mobile subscribers. Service providers often want to offer services to the interested customers but do not have sufficient information on how to create and deploy services. This is an incentive to open their networks to third party service providers. In this manner, they can take advantage of an added value through a variety of context aware services offered by third party service and application providers as they have the potential to reach many customers. Examples of some context information are user location, time, environmental information, web search history, network traffic, and so on [15]. Context awareness is also referred to as the ability to discover and adapt to changes in user context(s). Contextual information can be obtained either by sensing and monitoring devices or by requiring the user to enter data into the system.

Understanding the context is a focal point in the design of LBS. Some of the relevant context information in LBS are location, surrounding environment, time and date, UE technology. It can also be relevant to user characteristics, personal preferences, and behaviors. The choice of context in the design of applications is an important factor as it enhances the ability of service providers to supply the subscribers with information and services that are viewed as having a high level of efficacy [14].

7.4.2 Communication Issues around LBS

One of the aspects of communication in LBS includes the presentation of information and delivery of services to users. The interaction of users with LBS mainly depends on the device interfaces and deployed technologies (e.g., network data transfer rate, screen size, and UE processing power). Another aspect is the interaction of LBS users with the service providers. Furthermore, mobility of the user plays a major role in delivery of LBS.

7.4.3 Multimodal Communication with Context Awareness in LBS

LBS provide mobility of accessing the services and applications using small portable devices with wireless connectivity. Some of the challenges are related to the characteristics of device, which pose various challenges. Examples are size of device and low processing power, bandwidth, display capability, and battery life. Other challenges are related to mobility of LBS where it is necessary to deliver services in real time, according to the requests of users. In choosing the appropriate mode of communication, often, the user's preferences play a role. User preferences may depend on physical conditions, habits, knowledge, and background of users. Mode of communication can also be chosen on a context aware manner with regards to locations, users, environment, and types of devices. Multimodal communication with context awareness is explained in details in [14].

7.4.4 OMA Location Services Enabler

In this section, we review the Mobile Location Service (MLS) and Secure User Plane Location (SUPL) enablers that are defined by OMA [7].

7.4.4.1 OMA Mobile Location Services Enabler

The MLS enabler is responsible for supporting a control plane location service that helps the network locate the phone. The MLS consists of three protocols:

1. *OMA Mobile Location Protocol (MLP)* describes the interaction protocol between a LBS application and the location server. It defines XML content in the form of Document Type Definition (DTD) elements that can be transmitted over various transport protocols, including HTTP.
2. *OMA Privacy Checking Protocol (PCP)* is an application level protocol that emphasizes on privacy settings of a subscriber specific to LBS by accessing the privacy profile of the user. Privacy consideration for MLS is an important service requirement and mandatory for LBS.

3. *OMA Roaming Location Protocol (RLP)* is responsible for inter-location server communication, where a mobile subscriber is roaming in another network and exchange of information need to take place between the home location server and visiting location server. In this case, the location server in the home network is responsible for privacy checking.

7.4.4.2 OMA Secure User Plane Location Enabler

The SUPL enabler is responsible for assisting the device to calculate the location and communicate the location information. Detailed explanation on the development phases of SUPL architecture by OMA is provided in [7]. There are two dedicated entities in the SUPL architecture. The SUPL Location Center (SLC) is responsible for service management function and the SUPL Positioning Center (SPC) is responsible for providing GPS assistant data to the SUPL Enabled Terminal (SET) and in calculating the position of SET. This task is performed by the Positioning Determination function in SPC. The SLC is also in charge of coordinating the operation of SUPL and managing SPC.

7.5 Summary

The future users of communication systems require the use of high data rates at their home since the use of services and applications require a high bandwidth. The industry is in the process of transforming to a converged IP infrastructure and consolidating all network services. The convergence is at the core of NGN that lead to an architectural evolution that requires a converged infrastructure that is cost effective, scalable, reliable, and secure. In this chapter, we have reviewed different aspects of convergence, i.e., device and access technology convergence, service and application convergence, and network convergence. In Section 7.2, we have reviewed perspectives of IMS, OMA, TISPAN, and MBMS on the convergence. Common billing and charging functions are important because of the interworking requirements of heterogeneous access networks, services, and applications. Subscribers often use different services that might be either offered by different providers or the same provider. In Section 7.3, the concepts of offline and online charging is explained and charging infrastructure that is provided by 3GPP is reviewed. LBS is a good example of enabled services that motivates the interworking of heterogeneous technologies. In Section 4, we have introduced the LBS and reviewed communication issues as well as LBS enablers that are defined by OMA.

Acknowledgments

This work is partially supported by a grant from TELUS and the Natural Sciences and Engineering Research Council of Canada under grant CRDPJ 341254-06.

References

1. Third Generation Partnership Project, All-IP core network multimedia domain Rev. A, X.S0013, 3GPP2, Sept. 2005.
2. P. TalebiFard, T. Wong, and V. C. M. Leung, Integration of heterogeneous wireless access networks with IP-based core networks, in: E. Hossain (Ed.), *Heterogeneous Wireless Access Networks*, Chapter 2, Springer, NewYork, 2008, pp. 19–54.
3. Organization for the advancement of structured information standards, http://www.oasis-open.org
4. T. Erl, *Service-Oriented Architecture: Concepts, Technology, and Design*, Prentice Hall PTR, Upper Sadle River, NJ, August 2005.
5. Third Generation Partnership Project, IP multimedia subsystem, 3GPP TS 23.228, Release 8, Sept. 2008.
6. Third Generation Partnership Project, Multimedia broadcast/multicast service (MBMS)—Architecture and functional description, 3GPP TS 23.246, Release 9, June 2009.
7. M. Brenner and M. Unmehopa, *The Open Mobile Alliance*, John Wiley & Sons, Chichester, U.K., 2008.
8. Third Generation Partnership Project, Open service access (OSA): Parlay X Web services, 3GPP TS 29.199–1, Release 6, Dec. 2005.
9. J. Santos, D. Gomes, S. Sargento, R. L. Aguiar, N. Baker, M. Zafar, and A. Ikram, Multicast/broadcast network convergence in next generation mobile networks, *Comput. Netw.*, 52(1), 228–247, Jan. 2008.
10. Third Generation Partnership Project, Charging architecture and principles, 3GPP TS 32.240, Release 8, V8.5.0, Dec. 2008.
11. X. Y. Li and Y. Cai, Converged network common charging controller function, *Bell Lab. Tech. J.*, 13(2), 161–183, Aug. 2008.
12. Charging architecture, OMA-AD-Charging Draft v1.1, OMA, Oct. 2007.
13. Third Generation Partnership Project, Online charging system (OCS): Applications and interfaces, 3GPP TS 32.296, Release 7, V7.0.0, Dec. 2006.
14. A. Brimicombe and C. Li, *Location-Based Services and Geo-Information Engineering*, John Wiley & Sons, Chichester, U.K., 2009.
15. A. Ranganathan, J. Al-Muhtadi, J. Biehl, B. Ziebart, R. Campbell, and B. Bailey, Towards a pervasive computing benchmark, In: *Workshop on Support for Pervasive Computing (PerWare 05)*, Kauai, HI, 2005, pp. 194–198.

8

Application-Controlled and Power-Efficient Personal Area Network Architecture for FMC

S. R. Chaudhry and H. S. Al-Raweshidy

CONTENTS

The wired–wireless integrated network (WWIN) can be categorized as fixed mobile convergence (FMC). FMC means the convergence of the existing wired network and wireless network. Therefore, a mobile device needs the function of connection and control to the FMC infrastructure. This chapter has been divided into two main folds. Section 8.1 provides an application-controlled handover (ACH) development that keeps channel continuity in the wired–wireless synergy personal area network (PAN) environment that consists of 3G (UMTS) + WLAN + WPAN (UWB) and optical fiber network. Section 8.1 presents a handover mechanism that transmits and receives data by using the proposed application selection criteria. It maintains the channel and the seamless transmission from mobile device to the remote optical fiber network, to provide real-time service continuity for multimedia traffic. Section 8.2 consists of a distance-based power control scheme applied on ACH-PAN, for the development and simulation of the proposed technique.

Section 8.1 presents the results that prove that ACH-PAN has a reduction up to 83% in packet drop, 74% reduction in BER, 85% reduction in power consumption, and 100% enhancement in application response time (delay) as compared with the IEEE standard PAN without handover technique. Results from Section 8.2 have proven that the system performance improvements not only in throughput but the average data drop have been decreased 40% when the proposed approach is applied on ACH-PAN. The average power consumption in the entire network is reduced by 10% in the power- and application-controlled handover (P&ACH-PAN).

8.1 Introduction to ACH-PAN

The integration of all modes of electrical communication is a practical fait accompli under the progress and aegis of information theory, the fact remains that there is an undeniable and healthy competition between wired and wireless networks. With the rapid development of communication and network technologies, traditional voice networks and data networks have been gradually merged together into multimedia network that supplies various services ranging from e-mail, web browsing to telephony, visual conference, and video on demand. The horizon of the integrated multimedia network is extending further to incorporate wired, wireless, and cellular networks.

The progress of the network beyond the third or fourth generation (B3G/4G) will largely depend on the close coordination between mobility, resource, and quality of service (QoS) management schemes [1]. The B3G/4G network is expected to serve stationary as well as mobile subscribers under dynamic network conditions and offer any type of service—anytime, anywhere, and anyhow. Here a discussion of a hierarchical, layered, and modular B3G/4G network architecture with wired–wireless integrated network coordination and distributed network functionalities is presented. An application-controlled mechanism for control/signaling for multi-radio is adopted to facilitate enhanced wireless system performance. Reconfigurable architecture for multi-application, multi-homed, and multiservice-supported mobile device is presented to enable seamless mobility across different access technologies including wired and wireless infrastructures.

In an integrated 3G + WLAN + WPAN, as illustrated in Figure 8.1, more data and multimedia applications are carried end-to-end over the current Internet protocol infrastructure. Fundamentally, WWIN will reshape the tele-communications industry. Technological changes in the telecommunication industry have not only led to the creation of new trends in the deployment

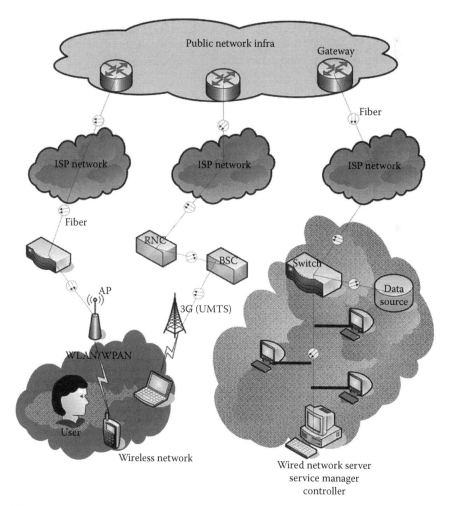

FIGURE 8.1
An integrated wired–wireless PAN formation.

of next generation networks, but also the convergence with information technology sector. In particular, the so-called WWIN has become a reality due to market demand and industrial competition.

FMC enables users to work with existing wired and wireless networks. UMTS has nationwide coverage and it accesses the network by using the cellular network. WLAN and UWB are free of charge communication air interfaces. WLAN, UWB, and UMTS can serve the Internet core network via the transmission control protocol/Internet protocol (TCP/IP) that enables a user to adopt client server communication by accessing individual network.

8.1.1 Related Research Work

In the future Internet where wired and wireless networks are integrated, the traffic volume of wireless users should expand greatly and the QoS of wireless user will be one of the most important technical issues. In the wireless network environment, TCP, one of the most important protocols of TCP/IP, suffers from significant throughput degradation due to the lossy characteristics of a wireless link. Therefore, in order to design the next generation networks, it is necessary to know how much improvement a WWIN brings. The throughput of TCP connection is well known to be dependent on both packet loss and application response delay. Several books and papers have been published in wireless TCP; for example, reference [2] evaluates throughput performance of a single wireless TCP connection experimentally, and reference [3] analyzes the throughput performance mathematically using a fluid flow model. These books and papers only dealt with wireless TCP and did not take care of the interaction of wireless and wired TCP in the FMC environment.

The architecture classification for WWIN is defined as: intracellular, intercellular, and extracellular. In intracellular architectures, each access point or base station (AP/BS) communicates with any mobile node (MN) in its coverage area in a multi-radio mode. On the other hand, in intercellular architectures, each AP/BS communicates with any MN in the coverage area of other APs/BSs in a multi-radio mode. Finally, extracellular architectures enable an AP/BS to communicate with MNs that are not in the coverage area of any AP/BS. Any of these classifications may employ dedicated radio stations. Furthermore, intra/intercellular ad hoc mode architectures can be classified as multi-radio systems, in which MNs act either in single-radio mode or multi-radio mode. Figure 8.2 illustrates this classification with respect to the well-known references such as [4–14] in this research era.

8.1.2 The Problem and Research Challenges

The original work in this section is the development and realization of a handover technique for multiple access technologies, which can provide data transmission and service continuity in the FMC environment. The algorithm uses a cross-layer process to operate on the transport layer between

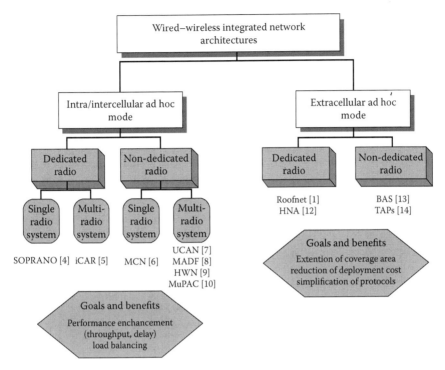

FIGURE 8.2
Hierarchical classification of wired–wireless network architecture.

the wired and wireless network infrastructures. A PDA, mobile phone, and a laptop were used in the wired and wireless network simulation environment. The devices have installed handover algorithms and move freely in the WWIN environment. The handover provides service continuity and data transmission by selecting the appropriate wireless technology, depending on the required data rate for a specific application. The traffic was recorded at the remote server that was located in the wired network and traffic was transported from devices that were located in the wireless network.

Section 8.1 is organized as follows. Section 8.1.3 presents the proposed WWIN architecture and organization of the communication environment for FMC. In Section 8.1.4, the design of ACH has been discussed in detail. Section 8.1.5 demonstrates the network model, scenarios, and layout for the proposed technique. Section 8.1.6 presents the network performance in the terms of results and discussions, leading to precise conclusions in the end.

8.1.3 Proposed WWIN Architecture for PAN

The existing high-speed wired and wireless networks are integrated and shown in Figure 8.3, which reveals the general constitution of FMC. Both wired and wireless networks have separate service platforms, also shown

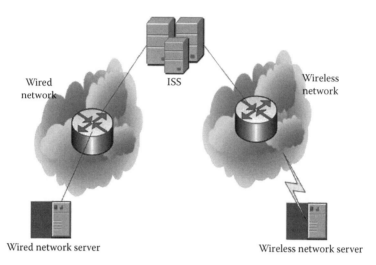

FIGURE 8.3
Proposed FMC architecture for WWIN.

in Figure 8.3. FMC is managed by using integration service server (ISS) [15]. FMC network supports data service in the existing voice/data-based wireless infra, and the existing high-bandwidth wired network infrastructure.

8.1.3.1 Constitution of Communication Environment for FMC

The proposed FMC simulation environment is based on multi-radio such as UMTS, WLAN, UWB, and optical fiber networks. The proposed constitution of communication environment for FMC is that shown in Figure 8.1. This chapter realizes the FMC environment by using TCP/IP and a transport layer between wired and wireless networks not by using ISS. As it can be seen from Figure 8.1, there is no physical ISS for the proposed FMC. ACH uses the transport layer and TCP/IP for each network constituting FMC. So accordingly, there is no physical ISS. In this architecture for ACH, mobile devices select radio access technology according to the application, that is, light FTP data can be accessed using the UMTS cellular networks. For heavy FTP and database applications, such as remote account management and payroll systems for a multinational corporate, a WLAN is a widely used solution. For real-time video calling and videoconferencing, UWB is prominent candidate supporting up to 200 Mbps [16].

Figure 8.1 shows a user carrying mobile devices in the wireless network domain having access to UMTS, WLAN, and UWB air interfaces. The mobile device network protocol stack can communicate cross-layer PHY to APP layer to detect the transmission of data, handover to the suitable air interface, and send it to the remote server. Once the server gets the request, it records real-time data transmitted from user's device.

TABLE 8.1

Classification of Handover

At the point of domain-controlling network

- *Interdomain handover*: The action of handover in different domain networks. It is called roaming.
- *Intradomain handover*: The action of handover in same domain networks.

At the point of an object-controlling handover

- *Controlled handover*: The device checks the signal strength and required data rate and controlling manager decides to handover.

At the point of an access platform

- *Vertical handover*: The action of handover in different platform environment.
- *Horizontal handover*: The action of handover in same platform environment.

At the point of handover service

- *Connectionless*: During handover, the home server keeps the position information and route of device.
- *Switching platform*: Handover depends on the required bandwidth to perform a particular task, that is, for heavy real-time application such as videoconferencing; a handover from WLAN to UWB due to high bit rate requirements.

8.1.3.2 ACH Classification

The definition of ACH is explained as follows. When the mobile device moves in the wireless network domain having access to UMTS, WLAN, and UWB, a certain process is required in order to switch and maintain a network connection channel, such a process is called handover. In the proposed handover technique when the average required data rate for a specific application is greater than the threshold of the achievable data rate, a handover to the suitable radio access occurs. The handover is classified at the point of various communication resources as shown in Table 8.1.

In this chapter, the proposed FMC architecture is an intradomain environment, and vertical handover using an application-controlled mechanism provides the service.

8.1.4 Developments of ACH Algorithm

In this chapter, an ACH that keeps the continuous channel service for wired and wireless networks has been developed. The mobile devices not only sense the signal strength but also determine the application type and initialize the handover process. A soft handover has been considered in the simulation network due to the session handling for connections to more than two radios, for example, connections to UMTS, WLAN, and UWB can be maintained by one device at the same time. It is a trade-off between the one-channel occupancy (hard handover) and the QoS in terms of soft handover's make-before-break (session continuity). This section elaborates the

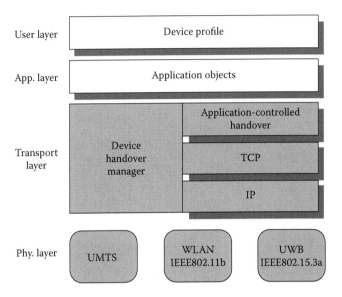

FIGURE 8.4
Application-controlled based network protocol stack.

application-controlled based network protocol stack, the design of network components, and finally, the soft handoff process model for ACH.

8.1.4.1 Application-Controlled Network Protocol

Figure 8.4 depicts the protocol stack for the proposed handover technique. Seamless bridging is natively supported by this protocol that enables both information and process data to be easily exchanged in heterogeneous environment.

The device handover manager adopts the lower layers of the protocol stack (PHY). Despite the relatively low speed of UMTS (up to 384 Kbps), it is able to guarantee deterministic behavior. As long as the load on the network is well below the theoretical available bandwidth, the handover technique guarantees that any low or high data is transmitted to the server within the suitable application response time.

8.1.4.2 Application-Controlled Network Components

The design of components for mobile device is illustrated in Figure 8.5. It consists of (1) the data component that can exchange data with external device, (2) the signal acquisition component that can receive the radio signal strength of the wireless network, and (3) the socket control component that controls the communication resources inside the device and the application control handovers to the wireless network platform according to the demand of quality data transmission requirements.

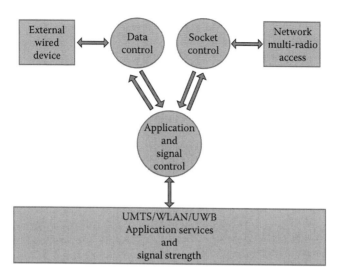

FIGURE 8.5
The components of ACH.

The ACH used a virtual socket for all three multiple access, for example, UMTS, WLAN, and UWB require three sockets [17,18]. In this chapter, a design of three virtual socket for the three multiple access (UMTS, WLAN, and UWB) has been developed.

As shown in Figure 8.6, the signal control component can control all the three signals together between the data control and the socket control component. The signal acquisition component provides the signal strength of AP/BS. The application controller is connected to the application manager for the handover process.

Figure 8.6 shows the implementation of how the components of Figure 8.5 are connected with the TCP. The data control component is connected with an external device and collects data from the external device (server) periodically.

8.1.4.3 ACH Process Model

Soft handover sets up a new channel on condition that it maintains the existing channel, whereas hard handover sets up a new channel after breaking the existing channel. The socket control component controls a virtual UMTS, WLAN, and UWB socket in ACH. A socket control component controls three sockets: UMTS, WLAN, and UWB, respectively. The buffer is connected to hidden UMTS, WLAN, and UWB sockets. A soft handover process design is shown in Figure 8.7.

8.1.5 Network Model

The simulation model has been divided into two major scenarios: (1) multiple radios over fiber network without ACH (base network) and (2) multiple

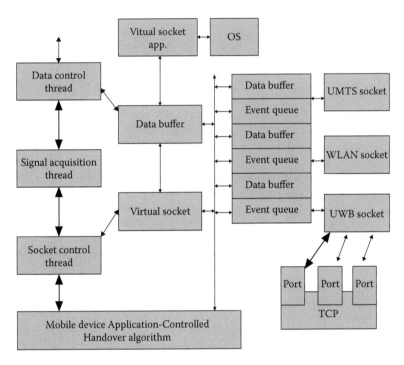

FIGURE 8.6
The design of components of ACH.

radios over fiber network with ACH. Both scenarios are simulated in parallel and compared on the basis of same parameters; the details are given in Sections 8.1.5.1 and 8.1.5.2.

The client subnet in Figure 8.8 represents the wireless client with multi-radio mobile devices capable of WLAN, UWB, and UMTS connectivity. The AP supports both PHY layers of WLAN and UWB. The UWB model used here is 10 m in range with a data rate up to 200 Mbps. This small network was simulated to analyze the scope of the simulation and to achieve reasonable simulation time.

8.1.5.1 Multiple Radios Over Fiber Network (Base Model)

In this scenario, the behavior of three wireless technologies UMTS, WLAN, and UWB were examined within the framework of a deployed FMC to simulate the WWIN without an efficient handover technique. First, the mobile device communicates to the remote server using a simple http application with UMTS infrastructure. The data traffic changed from HTTP to FTP (databases) and then to videoconferencing. The WLAN and UWB were accessed via AP, which was connected to a wired backbone

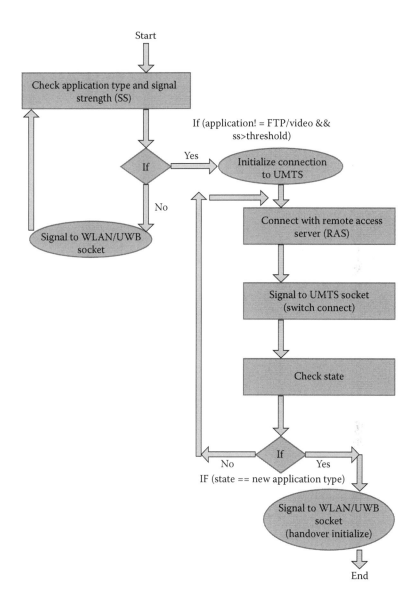

FIGURE 8.7
Handover process design for ACH.

server through a central hub using optical fiber (2 Gbps) simulating a real life office environment [19,20]. An IP gateway (i.e., an enterprise router) connected the wireless network to an IP cloud used here to represent the backbone Internet connectivity. The remote server supports three kinds of traffic FMC such as HTTP, FTP/heavy FTP (databases), and real-time video calling/conferencing.

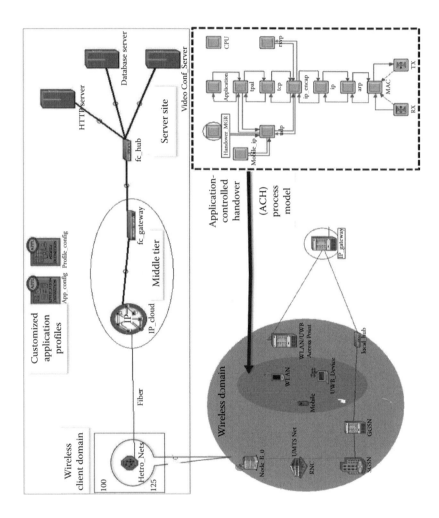

FIGURE 8.8
Network model hierarchical layout.

TABLE 8.2

Network Data Traffic Definition for Scenarios

| UWB/WLAN and | Applications | | |
UMTS-Based WPAN	HTTP	FTP	Real-Time Videoconferencing
No handover	L	H	H
Application-controlled handover	L	H	H

H, heavy; L, light.

TABLE 8.3

Network Simulation Parameters

Number of nodes	3
Simulation area	$500 \times 500\,\text{m}$
Simulation time	60 min
Multimode fiber channel bit rate	2 Gbps

8.1.5.2 Multiple Radios Over Fiber Network with ACH

In this second entire simulation, the setting and configuration were kept the same as for the first scenario. The only functionality addition is that the mobile devices have been upgraded with ACH.

ACH made the data exchange very bandwidth efficient and enhanced system performance as discussed and presented in Section 8.16. The simulation network traffic type and parameters for both scenarios are described in Tables 8.2 and 8.3, respectively.

8.1.6 Performance Evaluation and Discussions

This chapter carried out an imperative performance evaluation for multiple radios over fiber network with a special focus on the performance of the UMTS, WLAN, and UWB when subjected to rich multimedia applications. Real-time traffic was introduced in the network in the form of videoconferencing. Videoconferencing encompasses data, voice, and images and adequately represents the ultimate heavy multimedia traffic. No extensive network management techniques were configured but the ACH technique has been used to identify the effect of multiple wireless radio interfaces on the performance of the optical fiber network.

8.1.6.1 Network Performance Metric

The performance measurements that are relevant to network traffic such as wireless and cellular link utilization in terms of throughput were obtained and discussed. More specifically, the response times of applications (HTTP, database, and videoconferencing) at the mobile devices were also compared to ascertain the user-expected QoS. Additionally, some critical network parameters such as media access delay, end-to-end packet drop, bit error rate, and network power consumption have also been evaluated for both scenarios. A simulation time of 1h was set for both scenarios to achieve feasible results.

8.1.6.1.1 Network Throughput

Comparing the multiple radios over fiber network in both scenarios showed that a high network throughput has been achieved for the application-controlled scenario. Figure 8.9 presents a comparison of average network throughput for same type data traffic run for both scenarios. One of the scenarios used the ACH mechanism, and the result in Figure 8.9 shows that it has a large effect—83% improvement in the throughput of the networks. The application-controlled optical network has reached up to an average of 12,000 Bps as compared with 2,000 Bps by the network without handover technique.

8.1.6.1.2 HTTP Client Server Response Time

There was great improvement in the HTTP response time of the optical network with ACH when subjected to http traffic as shown in Figure 8.10.

The result shows that the optical network with handover can handle higher loads than the base model (without handover) due to the selection of the appropriate wireless air interface. It can be seen clearly from Figure 8.10 that both network models performed equally when the network data traffic was kept simple at the start of simulation time. Whereas, after introducing

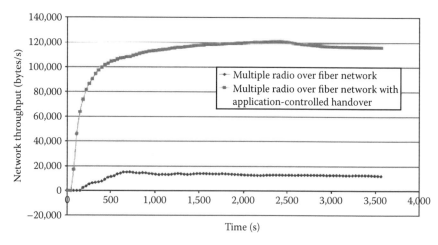

FIGURE 8.9
Network average throughput.

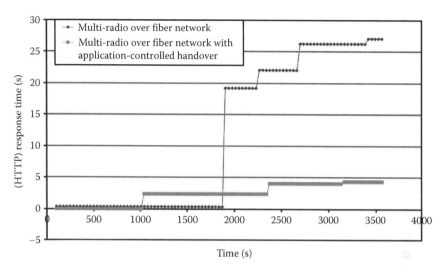

FIGURE 8.10
HTTP average response time.

heavy http browsing, the response time reached 27.5 s for the optical network without handover technique. Whereas the optical network with handover performed excellently comparatively and reached an average response time of up to 4.5 s, that is 23 s less than the base model.

8.1.6.1.3 Database Client Server Response Time

A similar improved performance was seen in database response time (FTP) during the simulation. A database needs higher authentication and security information compared with a simple HTTP application. Therefore, it enhanced the load on the communication technology. After 3000 s of simulation time, the database response time reached up to 1000 s for the base model while using UMTS for database access as shown in Figure 8.11. However, ACH improved network response time by switching to the WLAN. Database response time for controlled handover reached 10 s, which is 100 times less than the base network.

8.1.6.1.4 Videoconferencing Response Time

The optical fiber network without handover used UMTS service for real-time videoconferencing, but heavy multimedia video data transmission overloaded the radio link. The handover technique in the optical network enhanced the response time from the video server. An average improvement of 3 packet/s for video transmission delay can be seen in Figure 8.12.

8.1.6.1.5 Network Multiple Media Access Delay

In previous results, it was seen that ACH improved the network efficiency and enhanced the performance enormously. ACH works on the transport

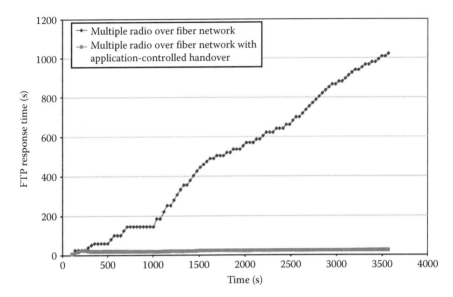

FIGURE 8.11
FTP (database) average response time.

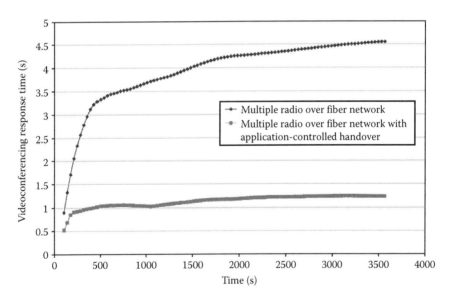

FIGURE 8.12
Videoconferencing average response time.

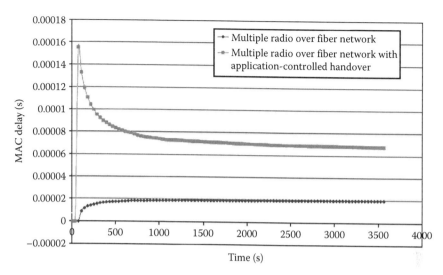

FIGURE 8.13
Network average media access delay.

layer; the algorithm's working architecture is based on the selection of particular wireless technology for specific application type. The handover mechanism increased the medium access delay of the network because of handover between different wireless access technologies, whereas with base network, it resulted in 50 μs less delay as shown in Figure 8.13.

The medium access delay is a function of the handover algorithm and the traffic characteristics. That is why, when this algorithm enhanced the network efficiency, it led to an increase in medium access delay too. Once the network was established, a linear behavior was observed from both network scenarios.

8.1.6.1.6 Network Packet Drop

Figure 8.14 shows the peak-to-peak network average packet drop comparison for both network scenarios. The base network shows very nonlinear behavior because of high variation in transmission. It has dropped up to 18 packets/s and gradually improved to an average of 10 packets/s. On the other hand, ACH behaved well by reducing the packet drop up to 83% comparatively.

8.1.6.1.7 Network BER and Power Consumption

The base network showed very nonlinear behavior for both BER and power transmitted. The glitches in the graphs are because of the variation in data traffic and the high bit rate demand. It can be seen that ACH has reduced the BER and power consumption to 74% and 85% as shown in Figures 8.15 and 8.16, respectively. Proposed scheme has reduced an average of 0.005 BER/packet and 8 nJ power in comparison to the base model.

Minimum power consumption and high bandwidth in any wireless networks are of most importance. The wireless devices mostly have limited

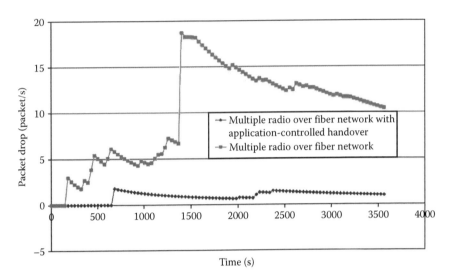

FIGURE 8.14
Network average packet drop.

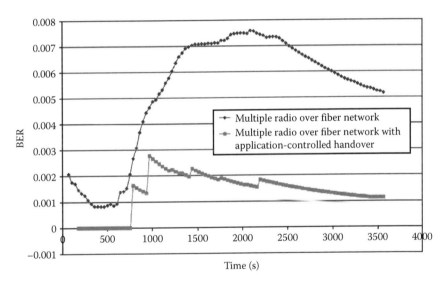

FIGURE 8.15
Network average BER.

battery power, and the selection of appropriate air interface in ACH has proved the significant advantage of the proposed technique. Reduced packet drop and BER by ACH technique not only enhances the bandwidth efficiency but also reduces the number of retransmitted packets, which is directly proportional to the power and complex error recovery techniques.

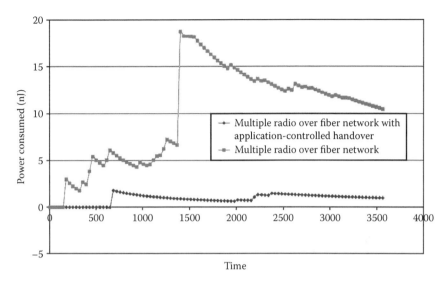

FIGURE 8.16
Network average power consumption.

8.2 Introduction to Power-Controlled ACH-PAN

In this section, the focus is on power-controlled WWIN PAN. WWINs are formed with existing infrastructure and there is a handshake of wireless and wired network to enhance the end-to-end network performance. It enables users to work with existing wired and wireless networks.

8.2.1 The Research Challenge

The original work in this chapter is the enhancement and realization of a efficient power control scheme for ACH [21] using multiple access technologies, which can provide data transmission and service continuity in the FMC environment.

The devices have installed power and ACH algorithms and move freely in the FMC environment. The handover provides service continuity and data transmission by selecting the appropriate wireless technology depending on the required power and data rate for a specific application. The traffic was recorded at the remote server that was located in the wired network and traffic was transported from devices that were located in the wireless network as shown in Figure 8.1.

The Section 8.2 is organized as follows. Section 8.2.2 presents the proposed power scheme for ACH-PAN. Section 8.2.3 discusses the design and implementation of P&ACH-PAN in detail. Section 8.2.4 demonstrates the P&ACH-PAN model, scenario, and layout for the proposed technique. Section 8.2.4.2 expresses the network performance results and discussions leading to the conclusion and summary in the end.

8.2.2 Proposed Power-Controlled Scheme for ACH

Power management in wireless networks has been proposed in [23]. Lower transmission power leads to lower interference level and better end-to-end throughput. Power saving in wireless radios has been studied in order to increase the MNs operating time [24].

As a multiple-hops network, wireless network protocols can manage power consumption for each individual node and the total power distribution in the entire network. As shown in Figure 8.17a, the source data is node (A), while destination is node (B).

In this case, the destination node (B) is still in communication range of the data source node (A) but requires high transmit power to meet the necessary signal quality. Instead, the node (A) can use the multihop feature of ad hoc network and build a route from (A) to (B) via an intermediate node (C), where (C) is located in midway between (A) and (B) as shown in Figure 8.17a. The decision to route data from (A) to (B) via (C) should take into consideration the battery level in both nodes (A) and (C) as well as the link stability.

8.2.2.1 Power Saving Using Multiple-Hops Algorithm

Power level is a scale for the minimum power required by the sender node to reach the next hop. If the next hop can be reached via another node by lower transmission power, then this new path should be adapted. Figure 8.17b illustrates this concept. For node (A), less transmission power is required to

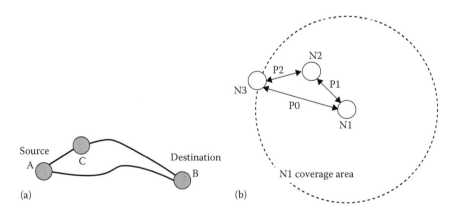

FIGURE 8.17
(a) Wireless networks power management. (b) Direct and in-direct transmission.

reach the destination node (B) via intermediate node (C) instead of sending data directly from (A) to (B). The decision to change the route from direct link between (A) and the destination node (B) to an indirect route via (C) is taken by calculating the battery life of node (C) and the total throughput and efficiency of the new route.

In the source node:

- When (battery/expected use time) = threshold: Then start transmission save mode.
- Check all the next hop neighbor nodes about a possible volunteer (the old destination node should be included in this message).
- If a neighbor node can reach the old destination and source-neighbor required transmission power is less than the original: then a volunteer request message will be sent back to the source node.
- The route will be changed from the old destination to the new volunteered neighbor.

In the neighbor node:

- When a volunteer request message is received.
- If the battery/expected use time) > threshold: then start volunteer process.
- If the old destination is in reasonable reach of transmission: then send volunteer acknowledge message.

8.2.2.2 Transmission and Receiving Power

In Figure 8.18, node (A) can reach node (C) directly using P_{ac} watts. However, node (C) could be reached indirectly via node (B). The second path power is the sum of two parts (P_{ab} and P_{bc}).

The wireless nodes transmit data to a certain range around the data source. The coverage range is a function of the transmission power and the medium coefficient. The following formula calculates the received power [25]:

$$P_{rcv} = \frac{\varsigma}{d^4} \cdot P_{tx} \tag{8.1}$$

where
P_{rcv} is the received power
ς is the log-normal coefficient
d is the distance between the source and the destination
P_{tx} is the transmission power

FIGURE 8.18
Minimum required power for direct and indirect transmission.

Received power values (of transmission signals from source nodes) are assumed to be equal at the destination nodes. Furthermore, the medium coefficient is the same for all data transmissions since it is the same ambience as shown in Equation 8.2.

$$K = \frac{P_{rcv}}{\varsigma} \qquad (8.2)$$

where K is a constant value.

$$d_{ac} = d_{ab} + d_{bc} \qquad (8.3)$$

where
d_{ac} is the distance between "A" and "C"
d_{ab} is the distance between "A" and "B"
d_{bc} is the distance between "B" and "C"

Based on these two assumptions, the transmission power of the two paths could be calculated as the following:
The direct path power:

$$P_{ac\_direct} = K(d_{ac})4 \qquad (8.4)$$

Based on Equation 8.3, d_{ac} could be replaced as

$$P_{ac\_direct} = K\{d_{ab} + d_{bc}\}4 \qquad (8.5)$$

$$P_{ac\_direct} = K\left\{(d_{ab})4 + (d_{bc})4 + 4(d_{ab})3d_{bc} + 6(d_{ab})2(d_{bc})2 + 4(d_{ab})(d_{bc})3\right\} \qquad (8.6)$$

The indirect path power:

$$P_{ab} = K(d_{ab})4 \quad \text{and} \quad P_{bc} = K(d_{bc})4 \qquad (8.7)$$

From Equation 8.7, the required power for the indirect path between "A" and "C" is

$$P_{ac\_indirect} = P_{ab} + P_{bc} \qquad (8.8)$$

$$P_{\text{ac\_indirect}} = K(d_{ab})4 + K(d_{bc})4 \tag{8.9}$$

Comparing Equation 8.6 with Equation 8.9, the direct transmission power is higher by

$$P_{\text{ac\_direct}} - P_{\text{ac\_indirect}} = K\left\{4(d_{ab})3d_{bc} + 6(d_{ab})2(d_{bc})2 + 4(d_{ab})(d_{bc})3\right\} \tag{8.10}$$

Equation 8.10 is an evident that the multiple-hops path requires less power than the direct path.

8.2.3 Implementation of Power-Controlled ACH-PAN

In this chapter, the proposed algorithm for ACH, which is shown in Figure 8.19, is implemented and carried out corresponding to simulation using the OPNET model. The running algorithm by each node calculates the set of parameters toward all possible destinations at periodic time intervals.

In each node, the adjacent neighbor nodes are listed in a table stored with the minimum required power to reach. Source routing is used at the nodes according to a QoS set of requirements, which specifies the optimization function to be used for the selection of the paths. When a data packet is generated at the source node, the node applies the function to the cost vectors (check destination node, battery check, and power threshold for transmission) of the nondominated paths to select the optimal path. If the source node battery power level is over threshold, the packet is transmitted to the destination in peer-to-peer mode. If the battery power is under threshold value, the algorithm looks for a lower cost node with minimum required power in neighborhood and forward it the data packet.

Using this method, less power consumption will be drained from the battery, which leads to longer operation time for the battery-based nodes.

8.2.4 Network Performance Evaluation

In this chapter, authors have simulated and compared basic model with and without ACH in comparison with P&ACH. The considered statistics include end-to-end network delay, data drop, and average power consumption for the data sent/received.

8.2.4.1 Network Setup

The wireless network consists of five MNs that are equipped with multiple radios such as WiMAX, WiFi, UWB, and UMTS as shown in Figure 8.1. The nodes are randomly located and destinations are connected at remote site with public access network. The proposed scheme is an enhancement for ACH model. The initial transmission power of the nodes is assumed to be 3 mW. The transmission range depends on the mobility of the device and it

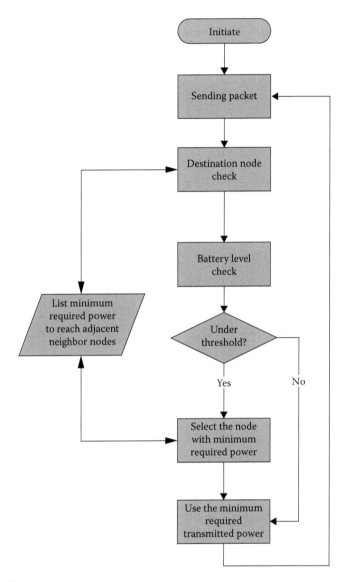

FIGURE 8.19
The power-controlled multiple-access protocol structure.

is varying from node to node (distributed between 5 and 15 m in the experiments). The topologies and simulation parameters for the network setup scenario are presented in Section 8.2.4.2.

8.2.4.2 Simulation Configuration Parameters

The simulation setup is based on five nodes running multiple applications such as FTP, HTTP, and video up to 100 Mbps. The mobility considered here is average person mobility of 3 m/s. Total simulation time is kept to 60 min to gain achievable results about the network. The power varies in the proposed scheme from 3 mW, whereas basic standard MAC uses fixed power for variable distance.

8.2.4.2.1 Network Delay

The first statistic (end-to-end network delay) indicates the comparison between base model, ACH model, and P&ACH model. The network general average delay has been increased about 50% when the proposed power control scheme is applied on ACH as shown in Figure 8.20.

The ACH performed excellent in comparison to base model and P&ACH as multihop communication led to increased delay. This result is expected when the proposed approach is applied due to increased throughput in P&ACH scheme.

8.2.4.2.2 Network Data Drop

Figure 8.21 proves that when the new approach is applied, the data drop has been reduced due to the extra process of power determining and multihop approach, which has taken place at MAC and PHY cross layer. This is the part of the trade-off between the data drop and the delay.

8.2.4.2.3 Network Power Consumption

As shown in Figure 8.22, the average power consumption is reduced by about ten times compared with the base network model. The P&ACH reduced power consumption 5% for ACH as shown in Figure 8.22. The average power consumption due to data transmission is reduced when the proposed approach is applied. The advantage of applying the proposed approach in this work is that the interference is reduced and the battery power consumption is minimized as the results have proven.

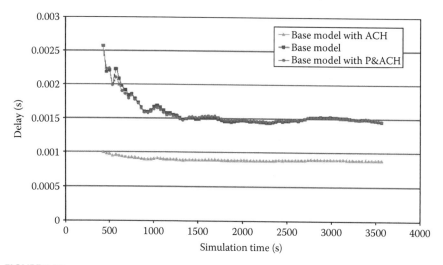

FIGURE 8.20
End-to-end average network delay.

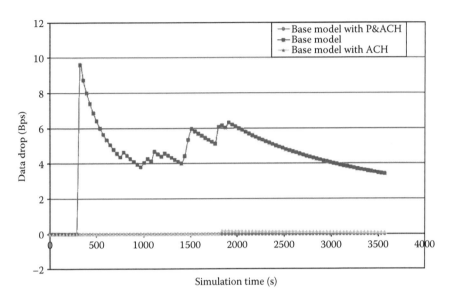

FIGURE 8.21
Network average data drop.

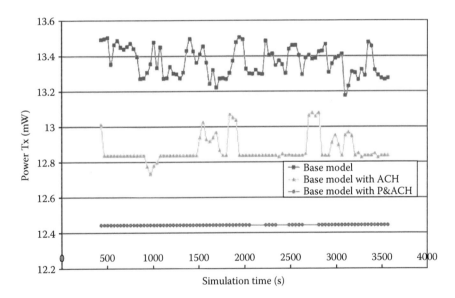

FIGURE 8.22
Network average power consumption.

8.3 Conclusion

To the authors' knowledge, this development of a novel ACH technique for FMC has been achieved for the first time. In this chapter, MAC, PHY, and transport cross layer enhancement (ACH) for the selection of multiple radios according to application type has been successfully developed in Section 8.1, which concludes that an optical network with an application-controlled technique would act as a better convergence platform for future broadband network communication systems. In comparison to the PAN model without ACH, the handover algorithm not only enhanced the network overall performance but also reduced packet drop by up to 83%, BER by up to 74%, response time by up to 100%, and power consumption by up to 85%, which is an enormous achievement for any wired–wireless converged platform.

In Section 8.2, applying the proposed power-controlled handover leads to significant reduction in power consumption and network data drop in modified wired–wireless converged PAN. However, the network delay is increased due to the proposed handover. This is the trade-off between the power consumption and the network delay.

References

1. B. Evans, M. Werner, E. Lutz, M. Bousquet, G. E. Corazza, G. Maral, and R. Rumeau, Integration of satellite and terrestrial systems in future multimedia communications, *IEEE Wireless Communications*, 12(5), 72–80, Oct. 2005.
2. H. M. Chaskar, T. V. Lakshman, and U. Madhow, TCP over Wireless with link level control: Analysis and design methedology, *IEEE Transactions on Networking*, 7(5), 605–615, Oct. 1999.
3. IEEE Std 802.1Q-2005, IEEE standard for local and metropolitan area networks virtual bridged local area networks, 2006.
4. A. N. Zadeh, B. Jabbari, R. Pickholtz, and B. Vojcic, Self-organizing packet radio Ad Hoc networks with overlay (SOPRANO), *IEEE Communications Magazine*, 40(6), 149–157, June 2002.
5. H. Wu, C. Qiao, S. De, and O. Tonguz, Integrated cellular and ad hoc relaying systems: iCAR, *IEEE Journal on Selected Areas in Communications*, 19(10), 2105–2115, Oct. 2001.
6. R. Ananthapadmanabha, B. S. Manoj, and C. Siva Ram Murthy, Multi-hop cellular networks: The architecture and routing protocols, *Proceedings of the 12th IEEE International Symposium*, vol. 2, San Diego, CA, Sept./Oct. 2001, pp. 78–82.
7. H. Luo, R. Ramjeey, P. Sinhaz, L. E. Liy, and S. Lu, UCAN: A unified cellular and adhoc network architecture, *Proceedings of the Ninth ACM MobiCom*, San Diego, CA, Sept. 2003, pp. 353–367.

8. X. Wu, S. H. Gary Chan, and B. Mukherjee, MADF: A novel approach to add an ad-hoc overlay on a fixed cellular infrastructure, *Proceedings of the IEEE WCNC*, vol. 2, Chicago, IL, Sept. 2000, pp. 549–554.
9. H.-Y. Hsieh and R. Sivakumar, Performance comparison of cellular and multi-hop wireless networks: A quantitative study, *Proceedings of the ACM Sigmetrics*, Cambridge, MA, June 2001, pp. 113–122.
10. K. J. Kumar, B. S. Manoj, and C. Siva Ram Murthy, MuPAC: Multi-power architecture for cellular networks, *Proceedings of IEEE PIMRC*, vol. 4, San Diego, CA, Sept. 2002, pp. 1670–1674.
11. J. Bicket, D. Aguayo, S. Biswas, and R. Morris, Architecture and evaluation of an unplanned 802.11b mesh network, *Proceedings of the ACM MobiCom*, Cologne, Germany, Sept. 2005, pp. 31–42.
12. R. Zoican and D. Galatchi, Mobility in hybrid networks architectures, *Proceedings of the IEEE TELSIKS*, Sept. 2005, pp. 273–276.
13. T. Li, C. K. Mien, J. L. S. Arn, and W. Seah, Mobile Internet access in BAS, *Proceedings of the 24th IEEE ICDCSW*, Lisboa, Portugal, Mar. 2004, pp. 736–741.
14. R. Karrer, A. Sabharwal, and E. Knightly, Enabling large scale wireless broadband: The case for TAPs, *Proceedings of the HotNets*, Cambridge, MA, 2003.
15. J. Chen, Mixed mode wireless networks: Framework and power control issues, PhD thesis, Computer Science Department, Hong Kong University of Science and Technology, Hong Kong, May 2004.
16. M. P. Wylie-Green, P. A. Ranta, and J. Salokannel, Multi-band OFDM UWB solution for IEEE 802.15.3a WPANs, *Proceedings of the IEEE/Sarnoff Symposium on Advance in Wired and Wireless Communication*, Princeton, NJ, Apr. 2005, pp. 102–105.
17. S. R. Chaudhry and H. S. Al-Raweshidy, An application controlled mechanism for heterogeneous Wired Wireless Integrated Network, *Proceedings of the Wireless Personal Multimedia Communications IEEE (IST 07)*, Budapest, Hungary, July 2007.
18. H.-D. Cho, J.-K. Park, K. Lim, J. Kim, and W. Kim, The mobile controlled handover method for fixed mobile convergence between WLAN, CDMA and LAN, *Proceedings of the Seventh International Conference Advanced Communication Technology 2005 (ICACT 2005)*, vol. 1, Phoenix Park, Republic of Korea, Feb. 2005, pp. 559–563.
19. N. Khashjori and H. S. Al-Raweshidy, Macrodiversity evaluation in WCDMA with radio over fibre access network, *Proceedings of the Fifth IEEE International Conference on Mobile and Wireless Communications Networks (MWCN2003)*, Singapore.
20. H. Al-Raweshidy and S. Komaki (eds.), *Radio Over Fiber Technologies for Mobile Communication Networks (Universal Personal Communications Library)*, Artech House, Norwood, MA, 2002.
21. S. R. Chaudhry and H. S. Al-Raweshidy, Application-controlled handover for heterogeneous multiple radios over fibre networks, *IET Communications*, 2(10): 1239–1250, Nov. 2008.

22. E.-S. Jung and N. H. Vaidya, An energy E_cient MAC protocol for wireless LANs. in *INFOCOM 2002*, New York, June 2002.
23. E. S. Jung and N. H. Vaidya, A power control MAC protocol for ad hoc networks, *Wireless Networks*, 11(1), 55–66, 2005.
24. A. Muqattash and M. M. Krunz, A distributed transmission power control protocol for mobile ad hoc networks, *IEEE Transactions on Mobile Computing*, 3(2), 113–128, 2004.

9

Mobility Management Protocols Design for IPv6-Based Wireless and Mobile Networks

Li Jun Zhang, Liyan Zhang, Laurent Marchand, and Samuel Pierre

CONTENTS

9.1 Introduction

Advancement in wireless technologies and mobile computing enables mobile users to benefit from disparate wireless networks such as wireless personal area networks (WPANs), wireless local area networks (WLANs), wireless metropolitan area networks (WMANs), wireless wide area networks (WWANs) that use mobile telecommunication cellular network technologies such as Worldwide Interoperability for Microwave Access (WIMAX), Universal Mobile Telecommunications System (UMTS), General Packet Radio Service (GPRS), code division multiple access 2000 (CDMA2000), Global System for Mobile communications (GSM), Cellular Digital Packet Data (CDPD), Mobitex, High-Speed Downlink Packet Access (HSDPA), or third generation (3G) to transfer data. On the other hand, mobile nodes (MNs) are equipped with multimode radio interfaces so that they can perform roaming among these different access technologies. Under the circumstances, mobility management (MM) becomes one of the major challenges in fixed and mobile-converged next-generation telecommunications network.

In next-generation telecommunications network, mobile users perform two types of movements: intrasystem roaming and intersystem roaming. *Intrasystem roaming* refers to a mobile terminal's movement between the location areas within a telecommunication system (Furht and Ilyas 2003, 364), and its MM solutions are based on similar network interfaces and protocols. On the other hand, *intersystem roaming* refers to the movements between different systems (Furht and Ilyas 2003, 364) that exploit diverse technologies, protocols, backbones, or service providers. Generally, MM comprises two components: *location management* and *handover management* (Quintero et al. 2004, 1509). Based on intrasystem or intersystem roaming, the corresponding location management and handoff management can be further classified into intrasystem and intersystem location management and handoff management (Akyildiz et al. 2004, 17).

Location management contains a set of functions executed to discover the current point-of-attachment/access point (AP) of a mobile station for call or data delivery (Boudriga et al. 2008, 3756). This process enables wireless network to discover the current location of a mobile terminal and deliver data to it (Taheri and Zomaya 2007, 714–716). It can be further split into two steps: location update and call (or data) delivery. The former requires MNs to provide the network with their location information, while the latter indicates that the network is queried for the location information of mobile users in order to deliver calls (or data) to them. The challenges of designing protocols for intersystem location management include (Akyildiz et al. 2004, 17)

- Reducing signaling loads and latencies that pertain to service delivery.
- How to guarantee on demand quality of service (QoS) in different wireless networks/systems.
- How to select the most suitable network for mobile users to perform location registrations (or updates) when the service areas of heterogeneous wireless networks fully overlap. That is, the concerning mobile user is not only always connected, but also connected through the best available access technology at all times (Gustafsson and Jonsson 2003, 49).
- Where and how to store the mobile user's updated location information when the service areas of heterogeneous wireless networks fully overlap.
- Determining mobile users' exact location within a specific time constraint when the service areas of heterogeneous wireless networks fully overlap.

Typically, *handover management* aims to maintain network connectivity as mobile users change their attachment points to the network. Obviously, handoff protocols need to preserve connectivity as mobile users move about, while simultaneously curtailing disturbance, or disruption from ongoing call (or data sessions) transfers. Therefore, minimal handoff disruption and session continuity are the primary goals of handoff management (Dimopoulou et al. 2005, 14). Generally, handoffs require certain key features such as low latency, minimal packet loss, minimal jitter for multimedia streaming, high reliability, and sustainable scalability to large networks. The challenges encountered while designing handoff management protocols (Akyildiz et al. 2004, 17) are described as follows:

- Minimizing signaling overheads and power consumption related to handoff management.
- Making efficient use of network resources during handoff.
- Improving network scalability, reliability, and robustness.
- Guaranteeing on demand QoS during handoff: (1) reducing intra- and intersystem handoff latency that is composed of signaling message processing time, resources allocation and route setup delay, format transformation time, etc., (2) alleviating user-perceptible service degradation, (3) decreasing handoff failure to a near-zero level, and (4) mitigating packet loss rate to a near-zero level to achieve seamless mobility.

This chapter elaborates MM protocols design for IPv6-based wireless and mobile networks. First, we elaborate the MM protocols such as mobility support in IPv6 (MIPv6), MM process in hierarchical mobile IPv6 (HMIPv6), mobile IPv6 fast handovers (FMIPv6), and proxy mobile IPv6 (PMIPv6). And then, new schemes to support seamless MM are proposed for next-generation all-IP-based wireless networks. The novel approaches, also called seamless mobile IPv6 (SMIPv6), consist of pre-configuring bidirectional tunnels among adjacent access routers (ARs) before actual handover (Zhang and Marchand 2006a) and using such tunnels to guarantee seamless communication while a mobile user changes its connectivity from one network to another (Zhang and Marchand 2006b). Thereafter, we present mathematical models to evaluate the performance of the above-mentioned MM protocols. From the analytical results, we find that the proposed SMIPv6 yields better performance than MIPv6 and its extensions like HMIPv6, FMIPv6, and F-HMIPv6. Research trend is also elaborated for MM in next-generation fixed and mobile-converged networks. Finally, we conclude this chapter with a discussion of future works. We hope that understanding the MM protocols cannot only help researchers to design more advanced solutions in this field but also provide a training toolkit for worldwide telecom operators.

9.2 Mobility Management Protocols

In new-generation fixed and mobile-converged networks, MNs are able to freely change their APs while communicating with one or more correspondent nodes (CNs). Accordingly, MM becomes a critical issue to track mobile users' current location and to efficiently deliver packets to them. Several typical MM protocols are designed within the Internet Engineering Task Force (IETF) working groups, such as mobile IP, which include IP mobility support for IPv4 (MIPv4) and MIPv6, and the extensions of HMIPv6, FMIPv6 and its variants, PMIPv6, etc. The mobile IP protocol allows location-independent routing of IP datagrams on the Internet, and provides an efficient, scalable mechanism for roaming within the Internet. Using such protocol, MNs may change their point-of-attachment to the Internet without changing their home IP address. This allows them to maintain transport and higher-layer connections while roaming.

Generally, IP mobility includes macro- and micromobilities. *Macromobility* designates mobility over a large area, that is, a node moves between the gateways that belong to different access network (Baumann et al. 2007). Usually, this refers to situations where MNs move between IP domains (Manner and Kojo 2004). Typically, MIPv4 (Perkins 2008) and MIPv6 (Johnson et al. 2004) are best suited for macroMM. Nevertheless, this chapter only addresses MIPv6 for the next-generation all-IP-based fixed and mobile-converged networks. *Micromobility* refers to mobility over a small area, that is, within an IP domain. Usually, a node moves between the gateways belonging to the same access network. Typically, micromobility protocols confine movements related signaling within a local domain without propagating to the mobile user' home network and those hosting their communicating peers. As a result, location updates to a distant home agent (HA) and CNs are eliminated as long as MNs remain inside their local domain. Hence, micromobility protocols provide better performance than macromobility solutions for intradomain roaming. Micromobility protocols are proposed to limit the signaling messages within a domain (He et al. 2003). This section presents macromobility protocol, such as MIPv6, and micromobility protocols, such as HMIPv6, FMIPv6 and its variants, and PMIPv6.

9.2.1 Mobile Internet Protocol Version 6

MIPv6 is a protocol developed as a subset of Internet Protocol version 6 (IPv6) (Deering and Hinden 1998) to support mobile connections. Typically, this protocol was proposed to solve the problem of global mobility in the new era of Internet by handling routing of IPv6 packets to MNs that have moved away from their home network (Habaebi 2006, 611). And it was designed for future all-IP-based wireless networks (Johnson et al. 2004). It enables MNs to remain connected while moving around within the Internet topology. An MN is always identified by its home address (HoA), regardless of its current location on the Internet. While away from its home network, the MN is associated with a care-of-address (CoA).

A router on the MN's home link, which is called HA, manages a binding between the MN's HoA and its CoA in order to route correctly the packets to the MN's current location. To do so, packets destined for the MN are intercepted by the HA and tunneled to the MN's current CoA. This leads to the unavoidable triangular routing problem. Furthermore, the HA also becomes a bottlenecks in the network. In order to solve the problem of triangular routing, MIPv6 protocol defines route optimization (RO) procedure that enables an MN to register its current location with a CN. As a result, the CN can directly send packets to the MN without passing though the HA.

The MM procedure in protocol of MIPv6 consists of movement detection, new IP address (or CoA) configuration, duplicate address detection (DAD), and binding registration (or update) with the HA and/or all CN(s) with RO mode. Movement detection aims to determine whether the mobile has moved to a new network or not. Such detection is important since, in situations where the mobile has moved, the addresses configured within its previous network become invalid, requiring additional configuration to establish or maintain upper layer connectivity. Generally, this is accomplished by assessing the reachability of the current (or default) router, checking the validity of the configured addresses, and finding a new available router on the network. To accelerate movement detection, nodes can multicast *router solicitation* (RS) messages requesting immediate *router advertisements* (RAs) from ARs, rather than waiting for the subsequent periodic advertisements to arrive (Narten et al. 2007).

Based on the advertisement message it received from neighboring ARs, the MN then configures a CoA on the new link. The address can be configured in two different manners: either using a stateless way, by combining the MN's respective interface identifier with the network prefix found in the advertisement message (Thomson et al. 2007) or via a stateful way during which a server assigns an address by executing the dynamic host configuration protocol (Droms et al. 2003).

When the new IP address is added to the specific interface, detection on duplicate address is performed in order to preserve the uniqueness of the new address. Hence, the mobile multicasts *neighbor solicitation* message to its neighbors. And then the mobile has to wait for responses from other nodes during at least 1 s. If no responses are received before the time limit expires, the new IPv6 address is considered available or valid (Narten et al. 2007). To alleviate the lengthy delay of more than 1 s, optimistic duplication address detection procedure (Moore 2006) is preferred for fast address configuration.

Home registration involves the exchange of signaling messages (*binding update* (BU) and *binding acknowledgment* (BA)) between an MN and its specified HA to enable the creation and maintenance of a binding between the mobile's HoA and its CoA. Accordingly, packets destined for the mobile will be intercepted by the HA and tunneled to its current location, identified by the CoA. Such registration is authorized by using IPsec protocols (Arkko et al. 2004).

Following by a successful home registration, the RO mode may be configured at a node. In this context, the MN executes correspondent registration via sending out a BU message to one or more CN with the purpose of creating the same binding between the HoA and the CoA. However, as empowering nodes with the ability to redirect packets from one IP address to another raises security concerns, *return routability* is performed prior to the correspondent registration. In addition, network administrator can apply ingress filtering in a proper way to prevent the source IP address spoofing, which is used by the malicious nodes to launch some sort of attacks (Malekpour et al. 2005).

The return routability procedure consists of HoA test and CoA test. During the HoA test, an MN sends a *home test init* message to a CN via its HA. The correspondent responds with a *home test* message containing a secret home keygen token. Addressed to the mobile's HoA, the *home test* message is encapsulated or tunneled to the mobile's current CoA by the HA. While CoA test is in progress, the mobile sends a *care-of-test init* message to a CN, which returns a *care-of-test* message that contains a secret care-of-keygen token. The *care-of-test init* and *care-of-test* messages are directly routed to the mobile by its correspondent, without going through the HA.

The MN needs both the home keygen token and the care-of-keygen token to validate the ensuing correspondent registration. The former token proves the mobile to be the legitimate owner of its HoA, while the latter token confirms that the mobile is currently present at the new IP address, CoA. The execution of return routability also enables the CN to be reasonably confident that the mobile can be reached at both IP addresses. Afterward, the mobile sends directly a BU message to its correspondent. As a result, the correspondent can send packets directly to the mobile's CoA without involving the mobile's home network. Figure 9.1 illustrates the MM process in MIPv6.

Even though MIPv6 is a protocol designed for the future all-IP mobile Internet access (Haseeb and Ismail 2007, 849), it has some limitations due to long delays and signaling load during handover operation (Vivaldi et al. 2003), and therefore several improvements such as HMIPv6 and FMIPv6 have been proposed by the IETF working groups, as well as academic and industrial researchers.

The protocol HMIPv6 employs mobility anchor points (MAPs) to reduce signaling overhead and delays related to intradomain location update procedures (Soliman et al. 2008), while FMIPv6 enables a mobile to exploit various link layer (L2) handoff event notifications (or triggers) and to configure a new care-of-address (NCoA) on the link of the new access router (NAR) before disconnecting with its current associated AR, which is also called previous access router (PAR). In addition, FMIPv6 utilizes a bidirectional tunnel between the PAR and NAR to reduce packet losses during handoff (Koodli 2008).

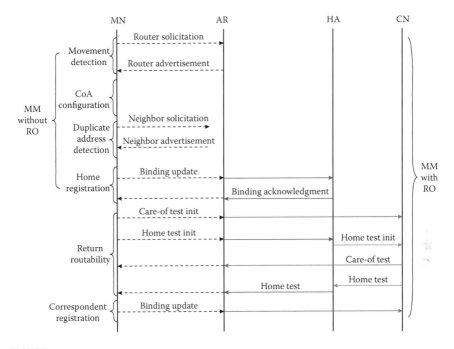

FIGURE 9.1
Mobility management (MM) process in mobile IPv6.

Researchers also design an improvement of FMIPv6, called enhanced forwarding (EFWD) scheme, to enable MNs to directly control bidirectional tunnel to redirect data from the PAR to NAR. Accordingly, packet losses during handover are prevented. Moreover, the EFWD scheme reduces handover delay by expediting the mobile's movement detection and the NAR discovery, as well as by eliminating time to acquire an NCoA. Its major advantage is the removal of link layer pre-triggers required by FMIPv6 (Gwon and Yegin 2004).

As both FMIPv6 and HMIPv6 are designed in their own ways to improve MIPv6 performance in terms of handover delay and signaling overhead, it is necessary to amalgamate these two schemes. In this context, a new protocol, named fast handover for hierarchical mobile IPv6 (F-HMIPv6), is proposed that enables MNs to exchange handoff-related signaling messages with local MAPs, and establish bidirectional tunnel between the MAP and NAR, instead of between the PAR and NAR (Jung et al. 2005a, 798–799).

The protocol MIPv6 and its extensions like HMIPv6, FMIPv6, EFWD, and F-HMIPv6 are host-based MM protocols, which require certain IPv6 client functionalities at MNs. That is, MNs must be capable to manage mobility by themselves. However, as certain mobiles have not such functionalities, PMIPv6 is designed to enable network-based MM for MNs without their participation in mobility-related signaling activities (Gundavelli et al. 2008). Sections 9.2.2 through 9.2.4 elaborate MM process in protocols HMIPv6, FMIPv6, and PMIPv6.

9.2.2 Hierarchical Mobile IPv6

Even though MIPv6 provides appropriate macromobility management features, certain pitfalls such as high signaling overhead, long handover latency, and unacceptable packet loss rate pertaining to MM procedures remain. Hence, improvements are required to enhance the performance of MIPv6. In this context, HMIPv6 was designed to reduce signaling overhead and location update delays in MIPv6 using hierarchical mobility agents called MAPs. Such entities control the local movements of mobiles roaming in their domain (Soliman et al. 2008).

When an MN enters a new MAP domain, it receives RA messages from nearby ARs. Based on the received information, the mobile configures two IP addresses: a regional care-of-address (RCoA) on a specific MAP subnet and an on-link local care-of-address (LCoA). After performing DAD for the LCoA, the mobile sends a *local BU* message to the selected MAP in order to establish a binding between the RCoA and LCoA. Upon receiving this message, the MAP performs DAD for the mobile's RCoA and returns a BA message to the MN.

Once the MN moves within the local MAP domain, it only needs to inform the MAP of its new LCoA without further updating the binding at its HA and all CNs. However, if the MN performs interdomain movements, that is, moving from one MAP domain to another, the MN then needs to register new binding at the new MAP. After receiving the BA from the MAP, the MN sends a BU message to its HA to establish a binding between its HoA and new RCoA. With RO, the MN also needs to execute the return routability process and correspondent registration in order to create (or update) the identical binding for every CN. Packets addressed to the mobile are intercepted by the MAP and tunneled to the mobile' current LCoA.

On the whole, the protocol HMIPv6 confines BU messages to a local MAP for intradomain movements, rather than propagating back to the HA and all CNs, thus improving the MIPv6 performance. However, this approach still needs further improvements to support real-time applications as it is mainly concerned with location update delays pertaining to intradomain roaming, which does not improve the latencies caused by movement detection, CoA configuration, and verifications (Jung et al. 2005a, 798).

9.2.3 Fast Handovers for Mobile IPv6

FMIPv6 enables MNs to rapidly detect their movements and formulate a prospective CoA while still connected to their current subnets (Koodli 2008). It also provides MNs with the opportunity to use available link-layer event notifications (or triggers) to accelerate network layer handovers. Consequently, delays due to network prefix discovery and the generation of new IP addresses are completely eliminated during handoff. Moreover, a bidirectional tunnel is setup between the PAR and NAR to avoid packet drops during handover. In addition, the PAR maintains a binding between MNs' PCoA and NCoA. Hence, packets addressed to a mobile are intercepted by the PAR and tunneled to the mobile's NCoA. Such an approach avoids sending BU messages to MN's HA and all CNs in the event of RO.

FMIPv6 supports two operation modes: predictive and reactive modes. The former implies that MNs receive *fast binding acknowledgment* (FBack) messages from their previous links. The latter indicates that MNs do not receive FBack messages from their attached PARs.

After discovering one or more nearby APs, an MN sends a *router solicitation for proxy advertisement* message to its default router prior to handover, the PAR, to resolve AP's identifiers (AP-IDs) to subnet router information. In response to this message, the PAR sends a *proxy router advertisement* to the MN to indicate whether it holds information regarding the NAR or not. Upon receiving the proxy router advertisement, the MN formulates a prospective CoA and sends a *fast binding update* (FBU) message to the PAR. The PAR then sends the NAR a *handover initiate* message to set up a bidirectional tunnel. In return, the NAR validates the proposed NCoA via duplication address detection. If the NCoA is invalid, the NAR allocates a new IP address for the MN and sends the PAR a *handover acknowledgment*. The PAR then binds the mobile's PCoA to the NCoA, and forwards an FBack message to the MN. Afterward, the PAR intercepts packets destined for the MN's PCoA and tunnels them to its new location. The NAR then buffers these packets.

Upon receiving the FBack, the MN disconnects from its current default router and initiates link-layer switching procedures. Once attached to the new link, the MN sends an *unsolicited neighbor advertisement* (UNA) message to the NAR (Koodli 2008), so that both arriving and buffered packets can be forwarded immediately to the MN.

Predictive MM enables MNs to receive FBack on their previous link. Furthermore, receiving FBack from the PAR means that packet tunneling is already in progress when MNs handover to the NAR. If MNs do not receive the FBack message on the previous link, they must follow the reactive MM procedures.

Performing reactive MM can occur for two reasons: either the mobile did not send the FBU message, or it left the link after sending the FBU, which was possibly lost, but before receiving the FBack from the PAR. Without receiving an FBack, the MN is not sure either the PAR successfully processed the FBU or that the tunnels are ready to use after handover to the NAR. Consequently, the mobile sends an UNA message to the NAR immediately after attaching to the new link and launches a duplication address detection process for its new configured CoA. Upon receiving the UNA, the NAR may detect the invalidity of the NCoA. In this case, the NAR assigns an alternate IP address for the MN, and sends it an RA with a *neighbor advertisement acknowledgment* option containing the new IP address.

Once the MN confirms the validity of its NCoA either via DAD or provided by the NAR, it sends a *neighbor advertisement* (NA) to the all nodes multicast address to join the multicast group. The MN then sends an FBU message to the PAR via the NAR.

Upon receiving the FBU, the PAR binds the mobile's PCoA to NCoA. If necessary, *handover initiate* and *handover acknowledgment* messages are exchanged to establish a bidirectional tunnel between the PAR and NAR. Accordingly, packets destined to the mobile's PCoA are intercepted by the PAR and tunneled to the NCoA via the NAR. Moreover, the FBack message can also be piggybacked with these packets.

Basically, FMIPv6 addresses the following issues: how can mobile users send packets as soon as they detect a new subnet link? How can they receive packets as soon as the NAR detects their attachment? As the FMIPv6 protocol utilizes pre-handover triggers, its performance, in terms of number of lost packets, depends dramatically on the pre-handover trigger time, thus becoming unreliable when the pre-handover trigger is delivered too closely to the actual link switch (Kempf et al. 2003).

In summary, FMIPv6 tries to leverage information from the link-layer technology to either predict or rapidly respond to a handover event, it provides IP connectivity in advance of actual IP mobility registration with the HA or CN, and this allows real-time services to be reestablished without waiting for the completion of such mobile IP registration (McCann 2005). Furthermore, this protocol can be implemented in different networks, such as IEEE-802.11 wireless local area networks (McCann 2005, Ryu et al. 2005), IEEE 802.16e networks (Jang et al. 2008), and 3G CDMA networks (Yokota and Dommety 2008).

9.2.4 Proxy Mobile IPv6

MIPv6 requires certain IPv6 client functionalities for MNs, that is, MNs must be capable to manage mobility by themselves. However, as certain MNs lack such functionalities, PMIPv6 is designed to enable network-based MM for MNs without their participation in mobility-related signaling activities (Gundavelli et al. 2008). PMIPv6 extends MIPv6 signaling messages and reuses the functionality of HA to support mobility for MNs without host involvement. In the network, new mobility entities are introduced to track the movements of MNs, initiate mobility signaling on behalf of mobiles, and set up the routing state required.

When an MN enters a PMIPv6 domain and attaches to a new access link, the mobile access gateway (MAG) on that access link authenticates and authorizes the MN before providing PMIPv6 services. While access is authenticated or network attachment events are notified, the MAG acquires the MN's identifier and uses it to access the MN's policy store and retrieves the address of the local mobility anchor (LMA) that serves as the MN's HA.

The MAG also configures a proxy-CoA for the MN, and then sends a *proxy binding update* (PBU) message to the LMA. This message is used for establishing a binding between the MN's home network prefix assigned to a given interface of an MN and its current care-of-address (proxy-CoA) (Gundavelli et al. 2008). The LMA then updates its binding cache entry (BCE) for that MN and replies with a *proxy binding acknowledgment* (PBA) message, which contains the MN's home network prefix assigned by the LMA. Upon receiving the PBA, the MAG establishes a bidirectional tunnel between its proxy-CoA and the LMA address.

Then, the MAG periodically sends RA messages to the MN on the access link advertising the MN's home network prefix as the hosted on-link prefix. The MN can also send an RS to request for an immediate RA from the MAG. In return, the MAG replies with an advertisement equipped with the MN's home network prefix. Hence, the MN can configure an HoA using either stateful or stateless mode depending on the modes permitted on that access link. As a result, the MN only uses its HoA for all its communication and the proxy-CoA is invisible to the MN. Packets sent to the MN are intercepted by the LMA, tunneled to the MAG that removes the tunnel header before forwarding them to the MN. Likewise, packets sent by the MN are intercepted by the MAG and tunneled to the LMA that then removes the outer header and forwards them to the CN. Figure 9.2 illustrates the MM process in PMIPv6.

When an MN handover from a previous MAG (P-MAG) to new MAG (N-MAG), the MN detached event is detected by the P-MAG that sends a PBU to the LMA in order to update the MN's binding and routing states. After receiving the PBU, the LMA waits for a certain *MinDelayBeforeBCEDelete* amount of time before deleting the MN's entry from its BCE. Then, the LMA sends a PBA to the P-MAG.

When the MN attaches to the new link, the N-MAG obtains its identifier either via access authentication or notified network event. Then, the N-MAG accesses the MN's profile and obtains the LMA address. The N-MAG also configures a new proxy-CoA for the MN on one of its interfaces. Subsequently, it sends a PBU to the LMA that assigns a home network prefix to the MN and sends a PBA to the N-MAG that contains this home network prefix. Upon receiving the PBA, the N-MAG sets up a bidirectional tunnel with the LMA before configuring a data path for the MN's traffic. Finally, it emulates the MN's home link by sending *RA* messages to the MN.

9.3 Proposed Seamless Mobility Management Protocol: SMIPv6

Since none of the above-mentioned protocols provides a perfect solution for seamless MM, hence, we propose new schemes to support seamless MM, which is called SMIPv6, for IPv6-based wireless and mobile networks.

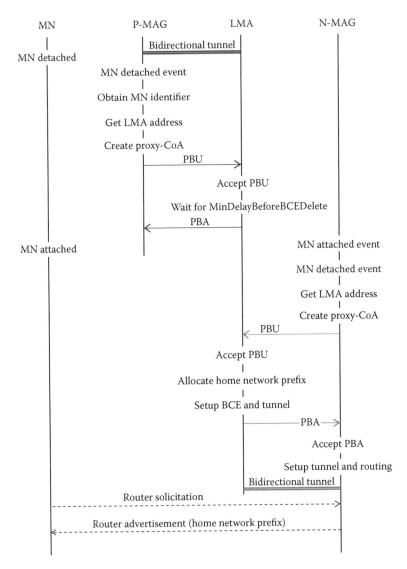

FIGURE 9.2
MM process in protocol of PMIPv6.

The basic idea is to pre-configure *bidirectional secure tunnels* (BSTs) among adjacent AR before actual handover. The QoS-related parameters (such as delay jitter and packet loss) and security aspects (like authentication method and tunneling keys) are specified for each unidirectional tunnel in a context of contracted *service level agreement* (SLA), which is made among telecommunication operators. These SLAs enable service operators to provide services to mobile users from other operators on condition that the involved two parties have made an SLA, within which the mobile users are given the opportunity to exploit pre-configured BSTs for their ongoing multimedia sessions during handoff. In addition, with the assistance of preestablished bidirectional tunnels, mobile users can utilize their previous valid IP addresses in a new visiting network or domain; this minimizes the service interruption during handover. Additionally, new routing policies are added to ARs to enable the delivery of packets to mobile users that have a topologically invalid address within the attached network.

The proposed MM process in SMIPv6 comprises two stages: making bilateral SLAs between neighboring radio access networks prior to actual handover and using such SLAs during handoff. The first stage consists of establishing bidirectional tunnels among adjacent ARs (Zhang and Marchand 2006a) through SLA. Such SLAs enable radio access networks to establish business and security relationship with their neighborhood. Consequently, communication services are offered to a list of subscribers from other mobile operators. Within an SLA, the conditions of bidirectional tunnel are specified according to the traffic type. For example, QoS related parameters, traffic classification aspects, QoS mapping algorithm, and shared keying materials are elaborated in detail. As a result, bidirectional tunnels are established prior to actual handovers. So signaling for setup, such tunnels are eliminated completely during fast handovers. The second stage consists of utilizing the pre-configured BSTs during handover (Zhang and Marchand 2006b). In doing so, new routing policies are added to ARs. During handoff, the NAR may receive a type of data packets from the PAR, as shown in Table 9.1.

Upon receiving such a packet, the NAR usually decapsulates it and verifies its ultimate destination: MN's previous care-of-address (PCoA). According to the normal IP routing policy, the NAR forwards the packet to the PAR, because the destined node MN is supposed to be located in the PAR's subnet. This leads to loop routing problem presented in FMIPv6. To resolve the problem, our proposed SMIPv6 force the NAR to buffer such packets and wait for the attachment of the MN.

During handoff, the NAR may receive another type of data packets from an MN, shown in Table 9.2. Generally, such kind of packets is dropped by the NAR due to ingress filtering. However, using SMIPv6, the NAR encapsulates such packets and forwards them to the PAR, which then decapsulates and sends them to the CN after successful decryption and authentication.

Since SMIPv6 empowers MNs to use their valid PCoAs, MN-related context information can be kept intact at the PAR. Hence, delays caused by the context transfer process are removed completely during handover.

TABLE 9.1

An Example of a Packet Sent by the PAR and Received by the NAR

Source Address 1	Destination Address 1	Source Address 2	Destination Address 2	Data
PAR	NAR	CN	MN's PCoA	Data

TABLE 9.2

An Example of a Packet Sent by an MN and Received by the NAR

Source Address	Destination Address	Encrypted Data
MN's PCoA	CN	Encrypted data using pre-shared key between MN and PAR

In Addition, the MN can resume or initiate its communication on the new link using its valid PCoA with the help of pre-configured tunnels within the SLA. Compared with the bidirectional edge tunnel handover for IPv6 (BETH) (Kempf et al. 2001), SMIPv6 does not require *handover request* and *handover reply* messages exchanged to set up bidirectional tunnels during handover, neither does it exploit link-layer pre-triggers to facilitate IP layer handoff. Given that both FMIPv6 and BETH protocols utilize pre-handover triggers, their performance, in terms of the number of lost packets and handoff latency, depends greatly on the pre-handoff trigger time, thus becoming unreliable when the pre-handoff trigger is delivered too closely to the actual link switch (Kempf et al. 2003, Gwon and Yegin 2004).

We assume that an MN roams from an AR to another in the IPv6-based wireless networks. And it acquires a valid CoA within the range of the first AR, called the PAR. And such CoA is named PCoA. In addition, the MN establishes a security association (SA) with the PAR, and configures a shared secret key.

In the overlapping zone of the PAR and its neighboring ARs, the MN receives one or more beacons from nearby APs. Such beacons contain the AP-ID. In case of horizontal handover (handoff within the same access technology), the MN may select the most suitable AP according to the received signal strength. In case of vertical handoff (handover among different access technologies), the MN may select the most appropriate AP using the score function presented in McNair and Zhu (2004, 11–13). Thereafter, the MN sends a *seamless binding update* message to the PAR before breaking its connection between them. Such message contains the selected new AP's identifier (NAP-ID) and a session token. The token is generated by the MN and is used to avoid replay attack.

The PAR is supposed to have pre-knowledge about its neighboring ARs, such as their associated AP-ID, the AR's IP address, etc. Therefore, upon receiving the *seamless binding update*, the PAR maps the NAP-ID to the associated AR's IP address and starts intercepting packets addressed to the MN. The PAR caches one copy of these packets, at the same time, it begins tunneling them to the NAR. The session token is then inserted into the first tunneled packet. In a meanwhile, the PAR adds an entry to its *forwarding tunnels list*. Such list is utilized to track the state of tunnels. Note that packets buffered by the PAR will be forwarded to the MN in case of ping pong and erroneous movements. The former implies that MNs move between the same two ARs rapidly while the latter connotes that MNs think that they enter into a NAR network, but they are actually either moving to a different AR or abort their movements by returning to the PAR.

Upon receipt of the tunneled packets from the PAR, the NAR decapsulates and buffers them. At the same time, the NAR adds an entry into its *forwarding tunnels list*. The session token is extracted from the first tunneled packet and added by the NAR into its *token list*, which is indexed by MN's IP address. At the meantime, the NAR creates a host route entry for the MN's PCoA and allocates a unique NCoA for the MN. Here we advocate that each AR manages a private address pool and guarantees the uniqueness of each individual address from the pool. By this means, the DAD process is removed from the overall handover process, thus expediting handoff process and reducing handover latency.

Once attached to the new link, the MN sends a *seamless neighbor advertisement* (SNA) message to the NAR immediately. Besides all the fields in the message format of UNA defined in FMIPv6 (Koodli 2008), the message SNA contains the same session token sent by the MN to PAR and the PAR's IP address. The IP source address of the SNA is the MN's PCoA, and its IP destination address is typically the all-nodes multicast address. In addition, the source link layer address (LLA) is the MN's LLA, while the destination LLA is the NAP's LLA. Here it is assumed that the NAP's LLA equals to the NAP's ID, which the MN acquires at the moment of preparing the handover.

The NAR then checks its forwarding tunnels list and the buffer for packets addressed to the MN. With the assistance of the token list, the NAR verifies whether the received session token from the MN is the same as the one from the PAR. In case of equality, the NAR forwards packets and the assigned NCoA to the MN. Otherwise, the NAR extracts the PAR's IP address from the received SNA and generates an FBU message on behalf of the MN, and sends the FBU to the PAR. Such message contains the MN's MAC address and PCoA. The PAR then verifies the MN's identities and returns an FBack to the NAR. At the same time, the PAR adds an entry into its *forwarding tunnels list* and *reverse tunnels list*. Upon receiving the FBack, the NAR forwards packets and the NCoA to the MN. Consequently, the MN becomes reachable on the new link under both CoAs: PCoA and NCoA. Figure 9.3 illustrates the predictive MM procedure for SMIPv6.

In case where the MN initiates a new communication session with CNs using its PCoA, it uses the pre-shared key with the PAR to encrypt the data packets. Note that instead of using the pre-shared key, the encapsulating security payload header may also be used to provide confidentiality, data origin authentication, connectionless integrity, an antireplay service, and limited traffic flow confidentiality (Kent and Atkinson 1998). And then, these packets are intercepted by the NAR. Before executing ingress filtering,

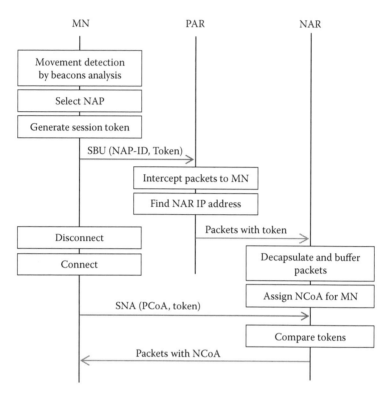

FIGURE 9.3
Predictive MM process in proposed SMIPv6.

the NAR extracts the subnet prefix information from the PCoA, and checks whether there is a preestablished SLA with the PAR using its *contract database*. In case where the MN is on the list of subscribers specified in the SLA, the NAR begins tunneling the MN's packets to the PAR, and adds an entry into the *reverse tunnels list* for further tunnel maintenance and billing issues. Otherwise, the NAR discards the intercepted packets.

Upon receipt of the tunneled packets from the NAR, the PAR decapsulates the packets and performs ingress filtering for the MN's PCoA. And then, PAR decrypts packets using the pre-shared key with the MN. Afterward, the PAR forwards the decrypted packets to the CN and adds an entry into its reverse tunnels list. Figure 9.4 illustrates the reactive MM procedure for SMIPv6.

Note that in FMIPv6, even though the MN is IP-capable on the new link, it cannot use NCoA directly with a CN before the CN first establishes a BCE for the MN's NCoA. However, our proposed fast and seamless scheme bypasses this problem by allowing MNs to utilize its valid PCoA immediately after connected to the new link. Hence, SMIPv6 provides not only expedited forwarding packets to MNs but also accelerated sending packets to their

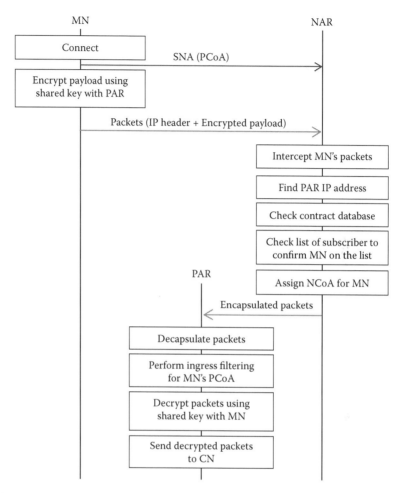

FIGURE 9.4
Reactive MM process for SMIPv6.

correspondents, thus optimizing handover performance in terms of signaling overhead, handoff latency, packet drop rate, etc. In addition, SMIPV6 is independent of the architecture. For example, if bidirectional tunnels are pre-configured between adjacent MAPs, SMIPv6 is also applicable to reduce handoff delay and packet losses.

Once terminating its ongoing session using the PCoA on the NAR's link, the MN can follow the legacy MIPv6 or HMIPv6 registration procedures. To facilitate tunnel maintenance, the MN sends a *tunnel bye* message to the NAR, which then releases the reserved bandwidth for the MN, and forwards the same message to the PAR. As a consequence, entries in *forwarding tunnels list* and *reverse tunnels list* are removed or refreshed (Zhang and Marchand 2006b). However, SMIPv6 requiring bidirectional tunnel remains active until

MNs complete their BU procedures with its correspondents, similar idea as that of the FMIPv6 protocol.

Typically, a session is identified by a group of information such as session ID, source address, destination address, source port number, destination port number, etc. When moving from one network to another, an MN loses its network connectivity and becomes unreachable because its previous source address is invalid in the visiting network. Under the circumstances, the MN has to acquire an NCoA and register the NCoA with its HA and all active CNs. Prior to successful registration, the MN cannot receive and send packets in the foreign network, thus the ongoing session is disrupted during handoff. In case where the MN executes multimedia applications such as video-streaming, it cannot tolerate the degraded quality of the session. SMIPv6 resolves such problem by allowing MNs to utilize their previous IP addresses on the new link without experiencing unacceptable QoS, thus guarantee seamless roaming with ongoing sessions.

9.4 Performance Analysis

Generally, performance evaluation of MM schemes is based on simulation and test bed approaches (Perez-Costa and Hartenstein 2002, 191–200, Perez-Costa et al. 2003, 5–6, Gwon et al. 2004). Nevertheless, network scenarios for simulations vary greatly, and performance comparison of the abovementioned handoff protocols is rarely viable. Hence, this paper proposes analytical models (based on the random-walk and the fluid-flow mobility models) to analyze the handoff performance.

We adopt the IPv6-based wireless cellular networks to evaluate the handover performance of the above-mentioned protocols. We assume that mobile service areas are partitioned into cells of equal size. Each cell is surrounded by rings of cells (Akyildiz and Wang 2002, 182). And each domain is composed of n rings of the same size. We name the inmost cell "0" central cell. Cells labeled "1" constitute the first ring around cell "0," and so on. Each ring is labeled in accordance with the distance to the cell "0." To simplify the analysis, it is assumed that each cell is managed by one AR. Figure 9.5 illustrates an example of an MAP domain with three rings.

9.4.1 Mobility Models

The literature documents two common mobility models: the fluid-flow model (Markoulidakis et al. 1998, 390–394) and random-walk model (Ho and Akyildiz 1995, 413–425) (Pack and Choi 2003, 2004). The fluid-flow model is most appropriate for users with high mobility, few speed, and direction changes, while the random-walk model is best suited for

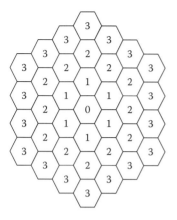

FIGURE 9.5
Network topology for a MAP domain
with three rings.

pedestrian movements where mobility is confined to a limited geographical area, such as a residential neighborhood or a commercial building
(Akyildiz and Wang 2002, 182). Our investigation considers both models.
In the random-walk mobility model, the MN's subsequent position is determined by adding a random variable with an arbitrary distribution to its
previous position (Zhang and Pierre 2008a, 38). Using the fluid-flow model,
the MN's directional movement is distributed uniformly in the range of $(0, 2\pi)$
(Wan and Lin 1999, 250–252).

9.4.2 Cost Functions

We assume that HMIPv6, F-HMIPv6, FMIPv6, and SMIPv6 support RO
and only a pair of messages (neighbor solicitation and NA) exchanged for
the DAD process. Both MN and CN processing costs are ignored during
analysis. The signaling overhead functions for MIPv6 with tunnel mode,
MIPv6 with RO mode, intradomain HMIPv6, interdomain HMIPv6, predictive FMIPv6 (P-FMIPv6), reactive FMIPv6 (R-FMIPv6), intradomain
F-HMIPv6 (Jung et al. 2005b), and interdomain F-HMIPv6 are given in
paper by Zhang and Pierre (2008a,b, 38–40). The signaling overhead functions for predictive SMIPv6 (P-SMIPv6) and reactive SMIPv6 (R-SMIPv6)
are listed as follows:

$$S_{P-SMIPv6} = 2\kappa \tag{9.1}$$

$$S_{R-SMIPv6} = \kappa \tag{9.2}$$

where κ represents the unit transmission cost in a wireless link (Zhang
2008, 88).

Using the random-walk model, MM-related signaling cost function for MIPv6 with tunnel mode, MIPv6 with RO mode, P-FMIPv6, R-FMIPv6, and F-HMIPv6 are given in paper by Zhang and Pierre (2008a, 38–39). The MM-related signaling cost functions for P-SMIPv6 and R-SMIPv6 are expressed as follows:

$$C^1_{\text{P-SMIPv6}} = S_{\text{P-SMIPv6}} * (1-q)/E(T) \tag{9.3}$$

$$C^1_{\text{R-SMIPv6}} = S_{\text{R-SMIPv6}} * (1-q)/E(T) \tag{9.4}$$

where
  q denotes the probability that an MN remains in its current cell
  $E(T)$ the average cell residence time (Zhang 2008, 89)

Using the fluid-flow model, MM-related signaling cost function for MIPv6 with tunnel mode, MIPv6 with RO mode, P-FMIPv6, R-FMIPv6, and F-HMIPv6 are given in the Ph.D. dissertation by Zhang (2008, 89–90). The MM-related signaling cost functions for P-SMIPv6 and R-SMIPv6 are expressed as follows:

$$C^1_{\text{P-SMIPv6}} = R_c * S_{\text{P-SMIPv6}} * (1-q) \tag{9.5}$$

$$C^1_{\text{R-SMIPv6}} = R_c * S_{\text{R-SMIPv6}} * (1-q) \tag{9.6}$$

where
  R_c denotes cell crossing rate
  q denotes the probability that an MN remains in its current cell (Zhang 2008, 90)

9.4.3 Numerical Results

Figure 9.6 shows the relationship between MM-related signaling costs and cell residence time for $q = 0.2$, using the random-walk model. This figure shows dynamic mobiles eager to move to another cell. We observe that longer cell residence time yields lower signaling costs. This is to be expected given that fewer handoffs are required as MNs remain longer in their current cells. Additionally, our proposed SMIPv6 schemes deliver better performance than other handover schemes, while MIPv6 with RO mode requires the most handoff signaling cost. The mean costs are 32.80 for MIPv6 with RO, 25.77 for F-HMIPv6, 19.92 for HMIPv6, 6.56 for MIPv6 with tunnel mode, 4.69 for P-FMIPv6, 3.28 for R-FMIPv6, 0.94 for P-SMIPv6, and 0.47 for R-SMIPv6 (Zhang 2008, 96).

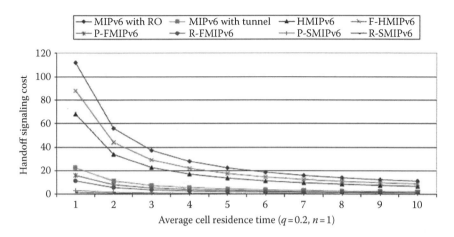

FIGURE 9.6
Handoff signaling costs versus cell residence time.

Figure 9.7 shows the relationship between MM-related signaling costs and user velocity for an MAP domain with one ring, using the fluid-flow model. Here the probability that the MN remains at its current cell is set to 0.2. Handoff related signaling costs increase linearly as MNs' average velocity augments. Since MNs with a higher mean velocity are more likely to cross a cell and a domain, resulting in higher signaling costs. This figure shows that MIPv6 with RO engenders the most exorbitant cost, its signaling costs rising to 113.12, on average. In comparison, F-HMIPv6 climbs to 28.74, 22.62 for MIPv6 with tunnel mode, and 16.16 for P-FMIPv6, 15.85 for HMIPv6, 11.31 for R-FMIPv6, 3.23 for P-SMIPv6, and 1.62 for R-SMIPv6 (Zhang 2008, 97).

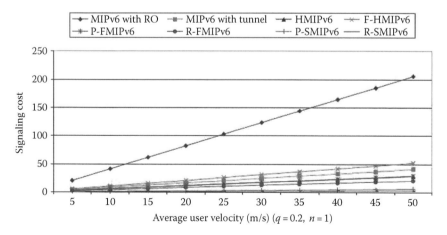

FIGURE 9.7
Signaling costs versus user velocity.

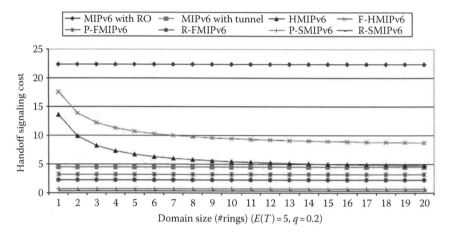

FIGURE 9.8
Handoff signaling cost versus domain size.

Figure 9.8 shows handoff related signaling costs compared with domain size for $q = 0.2$ and $E(T) = 5s$ under the random-walk model. As domain size increases, MM-related signaling costs are largely reduced for both HMIPv6 and F-HMIPv6, although the increasing of domain size does not affect the performance of MIPv6, FMIPv6, and SMIPv6. The average signaling cost for MIPv6 with RO mode is 22.40; 10.22 for F-HMIPv6, 6.22 for HMIPv6, 4.48 for MIPv6 with tunnel mode, 3.20 for P-FMIPv6, 2.24 for R-FMIPv6, 0.64 for P-SMIPv6, and 0.32 for R-SMIPv6. In addition, our proposed SMIPv6 schemes yield better performance than other approaches (Zhang 2008, 99).

9.5 Research Trend

Besides from standardization within IETF, researchers try to find IP multimedia subsystem-based solution for value-added service provisioning and handover between fixed and mobile operators in next-generation wireless and mobile networks. Within the third Generation Partnership Project (3GPP) and the 3GPP2, research activities are well underway to provide handoff capabilities between disparate technologies and different domains. This 3GPP activity falls under the projects undertaken by System Architecture Evolution and long-term evolution.

Researchers in the IEEE 802.21 working group are putting their efforts on the media-independent handover project, while those in the IEEE 802.11u attempt to design new intertechnology handover methods. The IEEE 802.11r task group is currently attempting to enhance handoff performance within

IEEE 802.11. Additionally, a number of R&D activities are ongoing by academia, research institutes, and industries simultaneously. So far, numerous research projects have focused on certain facets pertaining to MM, such as security, QoS provisioning, and routing. Nevertheless, it is essential to develop an end-to-end MM solution that covers the entire network span from radio access to packet delivery with guaranteed QoS. In the future, mobile users carrying an integrated terminal may access to a wide range of applications provided by multiple wireless networks (Hui and Yeung 2003, 54).

It is expected that the next-generation all-IP wireless networks integrate a large number of heterogeneous wireless technologies (Zhang et al. 2003, 102) toward universal seamless access and omnipresent computing through seamless mobility, which enables users to access sought services anywhere, at any time. One of the main research challenges for seamless mobility consists of providing available and reliable intrasystem and intersystem handoff solutions (Nasser et al. 2006, 96). This implies the design of new and efficient MM schemes to optimize QoS provisioning while supporting flawless mobility.

Furthermore, as the fourth generation (4G) wireless networks tend to revolve around universal mobile access and ubiquitous computing via seamless mobility, another major challenge for seamless mobility remains in the design of effective and efficient vertical handoff protocols for MNs' roaming among different types of networks. As traditional operations for handoff detection policies, decision metrics and radio link transfers cannot adapt to dynamic handoff criteria, nor deliver context-aware services or ensure network interoperability, it is obvious that new techniques are required to manage intersystem mobility (McNair and Zhu 2004, 8–9).

9.6 Conclusion

This chapter presents some basic concepts and background related to MM protocols design in IPv6-based wireless and mobile networks. The protocol MIPv6 made significant contributions to enable MNs access to the Internet while roaming around. However, the MIPv6 MM process leads to impaired handoff performance in terms of signaling load, handoff delays, and packet losses. Hence, improvements are crucial to enhance network performance. In this context, HMIPv6, FMIPv6, F-HMIPv6, and PMIPv6 were designed for localized MM. All of these improvements have their own merits and shortcomings. In this context, we proposed SMIPv6, which consists of pre-configuring BSTs among ARs before actual handoff, and utilization of such tunnels to accelerate MM during handover. Furthermore, performance evaluation is conducted through analytical models. Of which results show that our proposal delivers better performance than other existing protocols such as MIPv6, HMIPv6, FMIPv6, and F-HMIPv6. Additionally, this chapter also describes the research trends for MM in next-generation networks; several standardization activities are underway to create intelligent MM schemes that enable mobile users to benefit from its cost-effectiveness, enhanced features, location-independence, and seamless mobility across different systems via multimode radio interfaces.

References

Akyildiz, I. F. and W. Wang. 2002. A Dynamic location management scheme for next-generation multitier PCS systems. *IEEE Transactions on Wireless Communications* 1(1): 178–189.

Akyildiz, I. F., J. Xie, and S. Mohanty. 2004. A Survey of mobility management in next-generation All-IP-based wireless systems. *IEEE Wireless Communications* 11(4): 16–28.

Arkko, J., V. Devarapalli, and F. Dupont. 2004. Using IPsec to protect mobile IPv6 signaling between mobile nodes and home agents. IETF RFC 3776. http://www.ietf.org/rfc/rfc3776.txt (accessed June 10, 2009).

Baumann, R., O. Bondareva, S. Heimlicher, and M. May. 2007. A protocol for macro mobility and multihoming notification in wireless mesh networks. *Paper Presented at 21st International Conference on Advanced Information Networking and Applications Workshops (AINAW'07)*, Niagara Falls, Canada, May 21–23, 2007.

Boudriga, N., M. S. Obaidat, and F. Zarai. 2008. Intelligent network functionalities in wireless 4G networks: Integration scheme and simulation analysis. *Computer Communications* 31(16): 3752–59.

Deering, S. E. and R. M. Hinden. 1998. Internet protocol, version 6 (IPv6) specification. IETF RFC 2460. http://www.ietf.org/rfc/rfc2460.txt (accessed June 10, 2009).

Dimopoulou, L., G. Leoleis, and I. S. Venieris. 2005. Fast handover support in a WLAN environment: Challenges and perspectives. *IEEE Network* 19(3): 14–20.

Droms, R., J. Bound, B. Volz, T. Lemon, C. E. Perkins, and M. Carney. 2003. Dynamic host configuration protocol for IPv6 (DHCPv6). IETF RFC 3315. http://tools.ietf.org/rfc/rfc3315.txt (accessed June 10, 2009).

Furht, B. and M. Ilyas. 2003. *Wireless Internet Handbook: Technologies, Standards, and Applications*. Boca Raton, FL: CRC Press.

Gundavelli, S., K. Leung, V. Devarapalli, K. Chowdhury, and B. Patil. 2008. Proxy mobile IPv6. IETF RFC 5213. http://tools.ietf.org/rfc/rfc5213.txt (accessed June 11, 2009).

Gustafsson, E. and A. Jonsson. 2003. Always best connected. *IEEE Wireless Communications* 10(1): 49–55.

Gwon, Y. and A. Yegin. 2004. Enhanced forwarding from the previous care-of address (EFWD) for fast handovers in mobile IPv6. *Paper Presented at the IEEE Wireless Communications and Networking Conference (WCNC 2004)*, Atlanta, GA, March 21–25, 2004.

Gwon, Y., J. Kempf, and A. Yegin. 2004. Scalability and robustness analysis of mobile IPv6, fast mobile IPv6, hierarchical mobile IPv6, and hybrid IPv6 mobility protocols using a large-scale Simulation. *Paper Presented at the 2004 IEEE International Conference on Communications (ICC 2004)*, Paris, France, June 20–24, 2004.

Habaebi, M. H. 2006. Macro/micro-mobility fast handover in hierarchical mobile IPv6. *Computer Communications* 29(5): 611–17.

Haseeb, S. and A. F. Ismail. 2007. Handoff latency analysis of mobile IPv6 protocol variations. *Computer Communications* 30(4): 849–55.

He, Xiaoning, D. Funato, and T. Kawahara. 2003. A dynamic micromobility domain construction scheme. *Paper Presented at 14th IEEE International Symposium on Personal, Indoor and Mobile Radio Communications (PIMRC2003)*, Beijing, China, September 7–10, 2003.

Ho, J. S. M. and I. F. Akyildiz. 1995. Mobile user location update and paging under delay constraints. *Wireless Networks* 1(4): 413–425.

Hui, S. Y. and K. H. Yeung. 2003. Challenges in the migration to 4G mobile systems. *IEEE Communications Magazine* 41(12): 54–59.

Jang, H., J. Jee, Y.-H. Han, S. D. Park, and J. Cha. 2008. Mobile IPv6 fast handovers over IEEE 802.16e networks. IETF RFC 5270. http://www.ietf.org/rfc/rfc5270.txt (accessed June 17, 2009).

Johnson, D. B., C. E. Perkins, and J. Arkko. 2004. Mobility support in IPv6. IETF RFC 3775. http://www.ietf.org/rfc/rfc3775.txt (accessed June 9, 2009).

Jung, H. Y., E. A. Kim, J. W. Yi, and H. H. Lee. 2005a. A scheme for supporting fast handover in hierarchical mobile IPv6 networks. *ETRI Journal* 27(6): 798–801.

Jung, H. Y., H. Soliman, S. J. Koh, and J. Y. Lee. 2005b. Fast handover for hierarchical MIPv6 (F-HMIPv6). IETF Draft. http://www.join.uni-muenster.de/Dokumente/drafts/draft-jung-mobopts-fhmipv6-00.txt (accessed June 25, 2009).

Kempf, J., P. Calhoun, G. Dommety, S. Thalanany, A. Singh, P. J. McCann, and T. Hiller. 2001. Bidirectional edge tunnel handover for IPv6. IETF Draft. http://www.join.uni-muenster.de/Dokumente/drafts/draft-kempf-beth-ipv6-02.txt (accessed June 23, 2009).

Kempf, J., J. Wood, and G. Fu. 2003. Fast mobile IPv6 handover packet loss performance: Measurements for emulated real time traffic. *Paper Presented at the 2003 IEEE Wireless Communications and Networking Conference (WCNC 2003)*, New Orleans, LA, March 16–20, 2003.

Kent, S. and R. Atkinson. 1998. IP encapsulating security payload (ESP). IETF RFC 2406. http://www.ietf.org/rfc/rfc2406.txt (accessed June 23, 2009).

Koodli, R. 2008. Mobile IPv6 fast handovers. IETF RFC 5268. http://www.ietf.org/rfc/rfc5268.txt (accessed June 10, 2009).

Malekpour, A., D. Tavangarian, and R. Daher. 2005. Optimizing and reducing the delay latency of mobile IPv6 location management. *Paper Presented at the Fifth International Workshop on Innovative Internet Community Systems*, Paris, France, June 20–22, 2005.

Manner, J. and M. Kojo. 2004. Mobility related terminology. IETF RFC 3753. http://tools.ietf.org/rfc/rfc3753.txt (accessed June 9, 2009).

Markoulidakis, J. G., G. L. Lyberopoulos, and M. E. Anagnostou. 1998. Traffic model for third generation cellular mobile telecommunication systems. *Wireless Networks* 4(5): 389–400.

McCann, P. 2005. Mobile IPv6 fast handovers for 802.11 networks. IETF RFC 4260. http://www.ietf.org/rfc/rfc4260.txt (accessed June 17, 2009).

McNair, J. and F. Zhu. 2004. Vertical handoffs in fourth-generation multinetwork environments. *IEEE Wireless Communications* 11(3): 8–15.

Moore, N. S. 2006. Optimistic duplicate address detection (DAD) for IPv6. IETF RFC 4429. http://www.rfc-editor.org/rfc/rfc4429.txt (accessed June 10, 2009).

Narten, T., E. Nordmark, W. A. Simpson, and H. Soliman. 2007. Neighbor discovery for IP version 6 (IPv6). IETF RFC 4861. http://tools.ietf.org/rfc/rfc4861.txt (accessed June 10, 2009).

Nasser, N., A. Hasswa, and H. Hassanein. 2006. Handoffs in fourth generation heterogeneous networks. *IEEE Communications Magazine* 44(10): 96–103.

Pack, S. and Y. Choi. 2003. Performance analysis of hierarchical mobile IPv6 in IP-based cellular networks. *Paper Presented in the 14th IEEE International Symposium on Personal, Indoor and Mobile Radio Communications (PIMRC 2003)*, Beijing, China, September 7–10, 2003.

Pack, S. and Y. Choi. 2004. A study on performance of hierarchical mobile IPv6 in IP-based cellular networks. *IEICE Transactions on Communications* E87-B(3): 462–469.

Perez-Costa, X. and H. Hartenstein. 2002. A simulation study on the performance of mobile IPv6 in a WLAN-based cellular network. *Computer Networks* 40(1): 191–204.

Perez-Costa, X., M. Torrent-Moreno, and H. Hartenstein. 2003. A performance comparison of mobile IPv6, hierarchical mobile IPv6, fast handovers for mobile IPv6 and their combination. *Mobile Computing and Communications Review* 7(4): 5–19.

Perkins, C. E. 2008. IP Mobility Support for IPv4, Revised. IETF draft, draft-ietf-mip4-rfc3344bis-07. http://tools.ietf.org/id/draft-ietf-mip4-rfc3344bis-07.txt (accessed June 9, 2009).

Quintero, A., O. Garcia, and S. Pierre. 2004. An alternative strategy for location update and paging in mobile networks. *Computer Communications* 27(15): 1509–23.

Ryu, S., Y. Lim, S. Ahn, and Y. Mun. 2005. Enhanced fast handover for mobile IPv6 based on IEEE 802.11 network. *Paper Presented at the International Conference on Computational Science and Its Applications (ICCSA 2005)*, Singapore, May 9–12, 2005.

Soliman, H., C. Castelluccia, K. ElMalki, and L. Bellier. 2008. Hierarchical mobile IPv6 (HMIPv6) mobility management. IETF RFC 5380. http://www.ietf.org/rfc/rfc5380.txt (accessed June 10, 2009).

Taheri, J. and A. Y. Zomaya. 2007. A simulated annealing approach for mobile location management. *Computer Communications* 30(4): 714–30.

Thomson, S., T. Narten, and T. Jinmei. 2007. IPv6 stateless address autoconfiguration. IETF RFC 4862. http://tools.ietf.org/rfc/rfc4862.txt (accessed June 10, 2009).

Vivaldi, I., M. H. Habaebi, B. M. Ali, and V. Prakesh. 2003. Fast handover algorithm for hierarchical mobile IPv6 macro-mobility management. *Paper Presented at the Ninth Asia-Pacific Conference on Communications (APCC 2003)*, Penang, Malaysia, September 21–24, 2003.

Wan, G. and E. Lin. 1999. Cost reduction in location management using semi-realtime movement information. *Wireless Networks* 5(4): 245–56.

Yokota, H. and G. Dommety. 2008. Mobile IPv6 fast handovers for 3G CDMA networks. IETF RFC 5271. http://www.ietf.org/rfc/rfc5271.txt (accessed June 17, 2009).

Zhang, L. J. 2008. Fast and Seamless Mobility Management in IPv6-based Next-Generation Wireless Networks. PhD dissertation, Ecole Polytechnique de Montreal, University of Montreal, Montreal, Canada.

Zhang, L. J. and L. Marchand. 2006a Tunnel establishment. U.S. Patent Application 11,410,205, filed April 25, 2006.

Zhang, L. J. and L. Marchand. 2006b. Handover enabler. U.S. Patent Application 11,410,206, filed April 25, 2006.

Zhang, L. J. and S. Pierre. 2008a. Evaluating the performance of fast handover for hierarchical MIPv6 in cellular networks. *Journal of Networks* 3(6): 36–43.

Zhang, L. J. and S. Pierre. 2008b. Performance analysis of fast handover for hierarchical MIPv6 in cellular networks. *Paper Presented at the IEEE 67*th *Vehicular Technology Conference (VTC2008-Spring)*, Marina Bay, Singapore, May 11–14, 2008.

Zhang, Q., C. Guo, Z. Guo, and W. Zhu. 2003. Efficient mobility management for vertical handoff between WWAN and WLAN. *IEEE Communications Magazine* 41(11): 102–108.

10

SIP-Based Mobility Management and Multihoming in Heterogeneous Wireless Networks

**Chai Kiat Yeo, Bu Sung Lee, Teck Meng Lim,
Dang Duc Nguyen, and Yang Xia**

CONTENTS

10.1 Introduction

The advance in wireless technologies and its ubiquitous deployment, accel-erated with the proliferation of more sophisticated mobile devices and applications, has triggered a new trend in computer communication: the convergence of fixed-line and wireless networking, as well as the conver-gence of telecommunications and networking. Worldwide, new netbook and notebook sales for 2008 were recorded at 146 million units, with the first half of 2009 already surpassing 162 million units according to Display Search [1]. Global mobile handset production reached 1.2 billion units in 2008 [2]. The International Telecommunication Union (ITU), in its press release in September 2008, estimated that worldwide, mobile cellular subscribers will reach the 4 billion mark by the end of 2008 [3]. In March 2009, *Infonetics Research* reported in its fourth quarter 2008 edition of Mobile Broadband Cards, Routers, Services, and Subscribers Report, that the number of mobile broadband subscribers worldwide, in 2008 had registered a 125% increase over 2007, reaching 210.5 million, and was expected to exceed 1 billion by 2013 [4]. This can be attributed to the increase in broadband speeds with the deployment of technologies including High-Speed Packet Access (HSPA+), Long Term Evolution (LTE) and WiMax as well as their relatively lower cost compared to 3G roaming charges.

Allot Communications, in its inaugural Global Mobile Broadband Traffic Report for the second quarter of 2009 [5] showed a significant 30% increase in worldwide mobile data bandwidth usage in Q2 2009. As testament to the Fixed Mobile Convergence (FMC) phenomenon, the report also concluded that subscribers do not differentiate between fixed and mobile data networks and seemed to expect the same service from the Internet, irrespective of their access method. The report was based on data collected from more than 150 million subscribers of leading mobile operators worldwide. The International Telecommunication Union (ITU) also announced in May 2009, its consent to permit two new standards on FMC where operators will be able to pro-vide both services with a single phone, switching between networks on an ad hoc basis. The recommendations are ITU-T Y.2018 [6] and ITU-T Y.2808 [7]. The former describes a mobility management and control framework, and architecture for the Next Generation Network which includes support to handle mobile location management, handover decision, and control.

The latter describes the principles, service, and network capabilities as well as architectures for the IP Multimedia Subsystem (IMS)-based FMC where the same services are provided to user terminals regardless of the use of fixed or mobile access network technologies, and the changes in the point of attachment as the terminal traverses different access network technologies is transparent to the services.

Despite the huge increase in mobile broadband subscribers, Wi-Fi hotspots have continued to remain relevant in part due to its low cost if not free international data roaming, better indoor coverage, and the ubiquity of Wi-Fi built-in support on mobile phones and notebooks. Network aggregator iPass allows subscribers to access about 100,000 hotspots covering 92 countries for a flat rate while Qwest has partnered with AT&T to offer its high-speed Internet subscribers free access to 17,000 hotspots in May 2009. Similarly, Verizon Communications is getting ready to partner with Boingo Wireless while Tele2 Sweden has signed a deal with The Cloud, a European network covering 7000 locations in 12 countries, to offer its customers Internet access via 700 Wi-Fi hotspots, including some in McDonald's restaurants, in the Scandinavian countries. For the first quarter of 2009, AT&T announced that it had 10.5 million Wi-Fi connections, which was triple that of the first quarter of 2008.

The prevalence of multihomed devices, the ubiquity of wireless network access either via Wi-Fi or mobile broadband, and the promise of higher speed of delivery with such wireless and mobile networks have fuelled the phenomenal increase in the use of mobile wireless networks. The increased efficiency of mobile wireless networks also increases user expectations. Users will definitely wish to take full advantage of the technology to maximize their benefits and the Quality of Service (QoS) when using the Internet especially for data streaming, such as video streaming or VoIP services. However, different access technologies and complicated network protocols prevent applications from effectively exploiting them. In most cases, applications need to rely on either a network-based solution or a terminal-based solution due to the following reasons:

- Applications are unable to access the interfaces or to handle the different interfaces all at once.
- A complicated protocol needs a sophisticated solution, such as how to address the issue of maintaining QoS while using a multihomed solution.
- A virtual interface is needed every time to keep the application alive and maintain the seamless communication while interfaces are being sought for, switched over to, or they simply fail as the user roams around and latency is incurred in the handover. Hence, Care-of-Addresses (CoAs) and Home addresses (HoAs) mapping are complicated.

- The supporting Transmission Control Protocol (TCP) further complicates the problem. As TCP provides congestion control, it is crucial to notify the transport layer at both ends to adjust the rate of data transfer when the mobile node moves to a new network. If the mobile node's sending rate is too fast for the new network link, substantial packet loss will result if the transport layer does not reinitialize its congestion control state for the new network path. In addition, a slow start will occur during handover, and if the receiver side senses a link degradation and requests a reduction of the window size of the TCP flow, it will impact the performance, especially for real-time applications.

There is thus, an urgent need for a mobility and a multihoming framework that can not only support user mobility across the different access networks and but also abstract the network layer from the user applications, as well as effect a seamless handover across such heterogeneous networks. In addition, such a framework should be able to exploit the multihoming feature of the mobile device to both support a seamless handoff and to improve QoS. Such, is the proposed framework that will be discussed in this chapter.

10.2 Personal Mobility Support by SIP

Protocols such as Session Initiation Protocol (SIP) [8] and H.323 support personal mobility in terms of allowing the user to establish a connection from different terminal devices as he roams. For SIP, a single user is identified by a logical SIP Uniform Resource Identifier (URI).* The user can establish a connection using his SIP URI from different terminal devices as he roams around. However, once the connection is established, should the user move into another access network that results in a change in the point of attachment which in this case is the Internet Protocol (IP) address, the connection will be broken. The mobility support provided by SIP is therefore confined to one network once a session has been established. SIP on its own is adequate to support seamless mobility as users roam across different networks. For the current scenario, where there is an abundance of narrow range and often nonoverlapped IEEE 802.11 (WLANs/Wi-Fi) together with wide area networks offered by mobile broadband networks such as the Universal Mobile Telecommunication System (UMTS) and/or IEEE 802.16 WiMax Wireless

---

* An SIP URI is an email-like address comprising two components, namely, a user component and a server component, separated by "@." An example is client1@SIPserver. The user component contains the user name to identify the user while the server component refers to the domain name of the SIP server or its IP address.

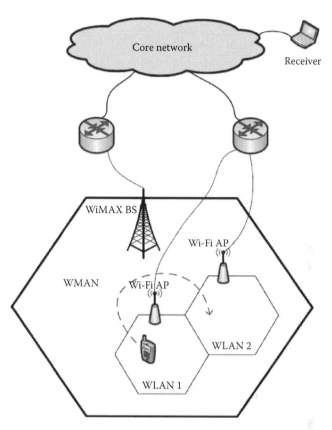

FIGURE 10.1
Mobility in a heterogeneous wireless network environment.

Metropolitan Area Networks (WMAN), the mobility management scheme must be able to seamlessly support the mobile user and the handoffs from one access technology to another as he leaves one Wi-Fi hotspot to enter the WiMAX/UMTS coverage and then enters another Wi-Fi hotspot. Such a scenario is shown in Figure 10.1. Given the cheaper rate of Wi-Fi access, intuitively, it will always be the preferred access technology.

10.3 Related Work

Mobile IP (MIP) [9] is one of the most popular mobility frameworks proposed. In MIP, each Mobile Node (MN) is associated with a Home Agent (HA) in its home network which is responsible for MN's packet redirection. When MN moves away from its home network to a foreign network, it obtains a

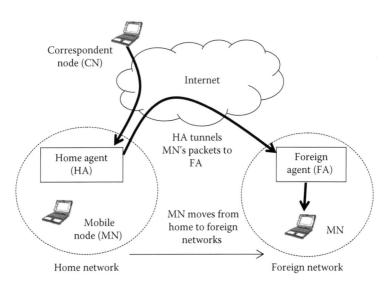

FIGURE 10.2
Architecture of a mobile IP.

Care-of-Address (CoA) from the Foreign Agent (FA) and sends a Binding Update (BU) to its HA in its home network. Upon receiving the BU, MN's data packets are encapsulated in an IP-in-IP tunnel by HA sent to MN's current CoA. This indirect routing of traffic is commonly known as triangular routing and is shown in Figure 10.2. MIP has since been extended to Mobile IPv6 (MIPv6) [10] and Mobile IPv4 (MIPv4) [11] to support IPv6 and IPv4 protocols, respectively. MIPv4 suffers from a lack of security constructs for authorization, authentication, and accounting, as well as for source routing. The IP packet from an MN to a CN will be generated with a Home Address Option in the Destination Option extension header of IPv6. This is used in an IP packet sent by an MN while it is away from the HA to inform the CN of MN's home address. To eradicate the triangular routing problem, Route Optimization (RO) has been proposed in MIPv6. In RO mode, MN will send BU to both its HA and the communicating correspondent nodes (CNs) so that data traffic can be sent directly from CN to MN's CoA without having to be routed to the MN's HA, thus effectively reducing network overhead and delay. When sending an IP packet from the CN to the MN, CN uses a Type 2 Routing Header, a new routing header variant introduced by MIPv6 that allows the IP packet to be routed directly from the CN to the MN's CoA. The newer version of MIPv6 even includes authentication header processing to provide validation of the mobile nodes.

Hierarchical MIPv6 [12] extends to MIPv6 to enable micro-mobility support. It differentiates between local mobility (within the same network domain) and global mobility (across different domains). Local handovers are managed locally and transparently to MN's CNs in other network domains, thus improving handover latency and isolating signal overhead to the local domain. An adaptive route optimization algorithm is incorporated by Pack et al. in [13] into the Hierarchical MIPv6 to improve data throughput based on the measured session-to-mobility ratio (SMR) of MN.

Mobility support using SIP has been discussed by Schulzrinne and Wedkund [14]. The former's proposal only works for SIP applications and only supports User Datagram Protocol (UDP) connections. For TCP, the authors propose the use of MIP for SIP applications [15]. This scheme incurs double the signaling cost as BUs have to be sent to both the HA as well as the SIP Redirect server when MN moves into a foreign network. When RO is used, it will send both BUs and SIP invite request messages to all MN's CNs.

In [16], MIPv6 is extended to provide multihoming support and flow distribution over heterogeneous links. To enable multihoming support, two new flags are incorporated into the BU message, namely the M and the S flags. When a BU with M flag set reaches the HA, HA will know that the CoA contained in this BU should be appended to the current CoA list without removing the previous CoAs. The S flag is to indicate the default CoA for the MN. The modified BU thus allows the HA to keep track of the multiple CoAs and the respective tunnels to the different CoAs.

SHIM6 [17,18] is a well-known proposal to address the multihoming issue for the IPv6 network. A number of mobility proposals for multi-homed nodes, Refs. [19,20] have employed it to address the multihoming issue. SHIM6 attempts to decouple the node identifier from the node locator. It uses a constant IP address as an identifier for connections of the upper layers. Multiple IP addresses are used as the node locator for routing packets. By separating the node identifier and the node locator, the change in IP address is made transparent to the upper layers. Reference [21] also attempts to separate the node identifier and the locator using a different approach by redefining the IPv6 header. Instead of using the 128 bit source and the destination IPv6 address fields, the source and the destination identifier and the locator are used in the header. The disadvantage of this scheme is that it requires a modification of the existing IPv6 protocol architecture.

10.4 Overview of Terminal Mobility Support Protocol

The proposed multihomed mobility management framework is an extension of the basic mobility management framework proposed by Lim, Yeo et al. in [22,23] called the Terminal Mobility Support Protocol (TMSP). By providing terminal mobility support, TMSP provides the similar functionality as set out in ITU-T Y.2808 [7] to abstract the change in IP addresses when the user roams across different networks from the user applications. Such IP changes are rendered transparent to user applications so that the established connection is maintained seamlessly across the different networks. This is accomplished by installing the TMSP software on MN and CN which among other functions, implements an IP-to-IP address mapping. TMSP uses the pervasiveness of the SIP services to provide an URI-to-IP address mapping to locate the MNs. Thus, users can use any mobile device and start roaming using a URI.

Compared to MIP, TMSP provides a new engineering alternative for route optimization without the need for new infrastructure support such as HA in MIP to support mobility. It does not require the assignment of a permanent IP address to each MN unlike the home address in MIP. It uses a single IP address at any time for each network interface. There is no triangular routing since no tunnels are required, as packets are routed directly between MN and its CN. TMSP does not incur extra IP header extension on each packet and it can support IPSec without the need to rerun the Internet key exchange (IKE). Analytical and testbed results show that TMSP is much more efficient than MIP in terms of hop count and protocol overhead. Experiments conducted on the TMSP testbed also show that mobility over both TCP and UDP can be supported seamlessly. The trade-off for TMSP's host-of-benefits is that both the MN and CN have to be modified to incorporate the TMSP software.

TMSP essentially comprises three modules— a User Agent (UA) and 2 kernel modules, namely, an IP-to-IP address mapping module (AMM) called after the IPSec function and an IP header restore module (HRM) called before the IPSec function. In TMSP, each MN is registered with an SIP server using an SIP URI as a unique identifier. The SIP redirect server/ SIP registrar (SIP-RS) is responsible for the URI-to-CoA resolution. When MN enters a foreign network, it will acquire a new IP address as its CoA. MN will use the SIP registration to inform SIP-RS of MN's new CoA as well as re-invite its CN using the SIP invite request to update the CN of MN's new CoA. The purpose of the UA is to manage the SIP messages. AMM intercepts outgoing and incoming IP and IPSec packets at the network layer to perform an IP-to-IP address mapping function. For outgoing packets arriving from the transport layer, the original source and the destination IP addresses in the packet header have to be swapped with the respective CoAs to ensure that the traffic can be routed directly between the mobile MN and/or CN. Conversely, for incoming packets, the same source and destination headers have to be swapped back to the original ones set during the connection establishment before forwarding the packets to the application. This, therefore, ensures that the IP changes are transparent to the application. Tables have to be maintained at CN and MN to record the original source and the destination IP addresses, port numbers used, as well as the respective CoAs so that AMM knows the mapping between the original IP addresses and the CoAs. For IPSec packets, AMM appends/removes a session identifier (SID) onto/from the packets' AH after/before they are processed by the IPSec function. HRM restores the IP addresses in the IP header of an IPSec packet to those used at the instant when the connection was first established, before the IP packet is passed into the IPSec module of the kernel. HRM thus, removes the need for establishing an IKE negotiation due to the change in the destination IP address of the data packet as a result of mobility. As further details on the multihomed TMSP will be provided in the following sections, a detailed description of TMSP (non-multihomed) will not be repeated here. Readers can refer to [23] for a full documentation of TMSP.

The underlying principles of TMSP have been extended into a new network mobility management scheme to provide IP mobility support for the seamless mobility of mobile routers (MR) and mobile network nodes (MNN). The new scheme is called Terminal-Assisted Route-Optimized NEMO Management (TNEMO) [24,25]. It enjoys similar benefits as TMSP in terms of optimized route between terminal nodes, no tunneling for packet redirection nor dynamic-size IP headers for route recording, no infrastructure support needed nor the need for source routing, the latter of which results in frequent packet fragmentation. TNEMO uses the URI and the SIP server to map the URI to the IP address to support terminal mobility. Similar to TMSP, TNEMO does not require the binding of a permanent IP address to a terminal node or an MR, and each terminal node only requires one IP address at any time. TNEMO makes use of an identifier and a locator pair to provide IP mobility where the former is the IP address used during connection establishment and the locator being the current IP address of the terminal node. As the locator (CoA) of destination nodes is always updated, TNEMO always replaces the identifier (destination IP address) of outgoing packets with the locator of the destination node. The converse is true for incoming packets. In a nested MANET formed by MRs, a Local Prefix Routing Protocol (LPRP) is used to provide optimized routing without the need for hierarchical IP continuity. MRs replace IP addresses using LPRP to route the packets directly between the terminal nodes without any need for packet redirection. In general, TNEMO's AMM works as a virtual layer between terminal nodes to provide IP mobility. Its IP-Swap Extension Module (SEM) creates an in-band storage for each packet to preserve the IP addresses for restoration before AMM. Quantitative comparisons with the Network Mobility Basic Support Protocol (NMBSP) [26] and the RRH (Reverse Routing Header) [27] shows that TNEMO performs better in terms of average Packet Delivery Overhead and IP packet overhead. RRH is an IPv6 header which is appended onto the IP header of each packet. RRH records the route path out of the nested MANET that can thus be used to route packets back to the network. RRH increases the size of each packet, which leads to packet fragmentation. The tradeoff in TNEMO is that the TNEMO software needs to be installed on the MRs and MNNs.

10.5 SIP-Based Multihomed Mobility Management (SM3)

Figure 10.3 shows an example of a typical operating scenario of a multihomed MN in a heterogeneous wireless network environment. The multihomed terminal can simultaneously, if it so wishes, be connected to three different access networks, shown in the figure as Wi-Fi, WiMAX, and UMTS. Each MN is identified by a unique SIP URI, which is in turn associated with the multiple CoAs of the MN. The SIP URI-to-CoA resolution is managed by an SIP server as shown in Figure 10.3. Both horizontal and vertical handovers are supported. When a CN wants to communicate with an MN, it can query the SIP server using MN's SIP URI. The SIP server will reply with a list of CoAs of the MN. CN then chooses one or more CoAs from the list for packet distribution. When MN detects that one of its interfaces is about to move to a different network, it will send a BU to the CN, informing it of the impending change in CoA. CN will then adjust its distribution policy and redirect the packets to other available CoAs.

10.5.1 System Architecture of SM3

Figure 10.4 shows the system architecture of SM3. It comprises four modules, namely, the SIP User Agent (UA), the Virtual Interface, the Header Restore

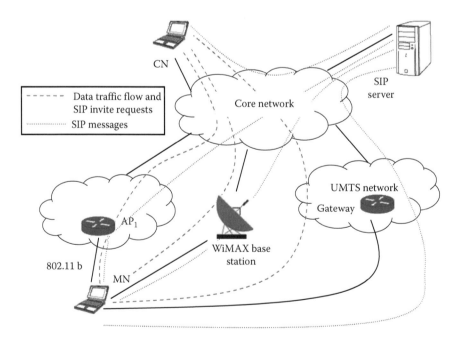

FIGURE 10.3
An example of a multihomed MN in a heterogeneous wireless network environment under SM3.

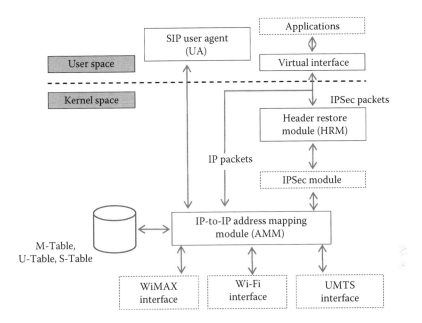

FIGURE 10.4
System architecture of SIP-based multihomed mobility management (SM3).

Module (HRM) and the IP-to-IP Address Mapping Module (AMM). UA and the Virtual Interface sit in the user space while HRM and AMM are kernel modules. UA handles and manages all the SIP messages while the Virtual Interface is meant to hide the heterogeneity of the physical interfaces from the applications. It is the contact point for all the applications. HRM is used primarily for IPSec packets as mentioned in Section 10.4 above while AMM takes care of the IP address swapping.

Outgoing data packets from the upper layers and the incoming data packets from the link layer are intercepted by the kernel modules. As shown in Figure 10.4, outgoing IP packets are directed to AMM to perform the necessary IP address swap while the IPSec packets are processed by HRM before being passed to the IPSec kernel function. HRM restores the IP addresses in the IP header of an IPSec packet to those used at the instant when the connection was first established, before the IP packet is passed into the IPSec module of the kernel. HRM thus removes the need for establishing an IKE negotiation due to the change in the destination IP address of the data packet as a result of mobility. AMM then appends a Session Identifier (SID) onto the packets' AH after the IPSec function. For incoming packets, AMM does the necessary IP address swap and also removes the SID from the packets' AH before sending them to the IPSec function. For IP packets, AMM only needs to do the necessary IP address swap before releasing them to the upper layers.

SM3 keeps states for each data connection in MN and the corresponding CN. Three linked state tables are maintained as illustrated in Figure 10.4.

They are the Mapping Table (M-Table), the URI Table (U-Table), and SID Table (S-Table). The M-Table maintains an entry of each network port that comprises a Reference Source IP Address (RSA), a Reference Destination IP Address (RDA), a Session ID (SID) of CN, a local port number, a remote port number, an index to the S-Table and an index to the U-Table. Each entry corresponds to one connection and is assigned a unique index that is generated from the local port and the remote port. The U-Table maintains data for each CN. Each entry comprises a URI and a CoA. The S-Table keeps a mapping between each network port and a SID for each network port at MN. RSA and RDA refer to the IP addresses of MN and CN at the instant when a data connection is established. SID is created for each IPSec data connection. As the network port is encrypted and encapsulated in an IPSec header for each IPSec packet, AMM uses SID as the reference key to locate the network port to perform the IP-to-IP address mapping on the IPSec header.

10.5.2 Registration with SIP Server

Both MN and CN must be registered with their respective SIP servers identified by the server component of their SIP URI provided by the user MN and CN by issuing an SIP register message. The signaling flow for the registration process among the various entities is shown in Figure 10.5. MN will publish its SIP address and the list of CoAs using a REGISTER message with

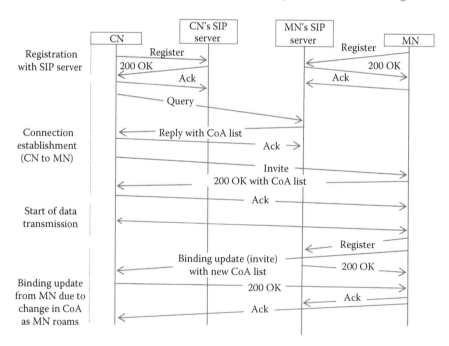

FIGURE 10.5
SIP signaling flow of SM3.

the SIP server. The same applies to CN. It is to be noted that CN and MN do not need to be registered with the same SIP server. All SIP messages are handled by the SIP UA.

10.5.3 Connection Establishment

With reference to the connection establishment signaling flow in Figure 10.5, when CN wants to transfer data to MN, CN can obtain a list of MN's CoAs from the SIP server via the QUERY message. It will then attempt to set up a session with MN using the INVITE message. When MN receives the INVITE message, it will accept and reply with 200 OK messages as well as a list of CoAs. These steps ensure a SIP URI to a CoA resolution. After successful establishment of the session, both MN and CN will record each other's IP address as RDA and RSA, respectively, as well as the network port used for data transfer. This pair of IP addresses is the address used by the transport layer connection and they will be processed by the AMM during data transmission. The network ports are used to identify data flows. Data transmission then commences.

10.5.4 Multihome and Mobility Support—CoA Update

SM3 supports horizontal and vertical handovers both proactively and reactively. When MN senses that one of its wireless interfaces is experiencing lower signal strength, and it is picking up stronger signals from a new network, it will proactively send a BU to the SIP server and any communicating CNs to update its CoA list. See signaling flow in Figure 10.5, for BU. BU will also include information on the traffic condition for each CoA. CN can then re-distribute the load based on the new CoA list. Upon receiving the re-invitation from MN, CN will update the MN's list of CoAs in its U-Table to map the MN's URI to its list of CoAs. The U-Table is used by AMM to swap the destination IP address so that data packets bound for MN are delivered to its current CoA on the correct interface. A proactive handover is an example of the "make-before-break" handover whereby a new connection is established before the existing one is taken down. This is always preferred as it ensures little, if any, data loss. The reactive handover, on the other hand, takes place when the MN is suddenly dropped out of the network. In such an instance, there will be data loss as this a case of the "break before make" handover when MN takes time to switch over to another network.

10.5.5 Virtual Interface

With reference to the system architecture shown in Figure 10.4, a virtual interface is created to decouple the node identifier from the node locator. A constant IP address from an unused private prefix is arbitrarily assigned to the virtual interface as the permanent address. This permanent address is used by applications and can be perceived as the node identifier that is only valid within the node itself. The CoAs of the interfaces are the node locators for routing packets. The virtual interface renders any changes in CoAs to be transparent to the applications hence maintaining ongoing session connections even when one physical interface in use is down while SM3 searches and attempts to handover to another interface.

10.5.6 IP-to-IP Address Mapping Module

To illustrate SM3's support for IP mobility, Figure 10.6 chronicles the path of two outgoing IP packets from CN's application. Each IP packet uses a different wireless interface. The following description will focus on the packet using interface *src_CoA_x* and *dst_CoA_m* but the description applies similarly to the other packet using another interface. The packet is generated by a connection between CN and MN when both are connected to the Internet via wireless access points. This connection is established earlier and at that point in time, CN and MN were holding IP addresses *CoA ref_src* and *ref_dst*, respectively. At the instant when this packet is generated, CN has acquired

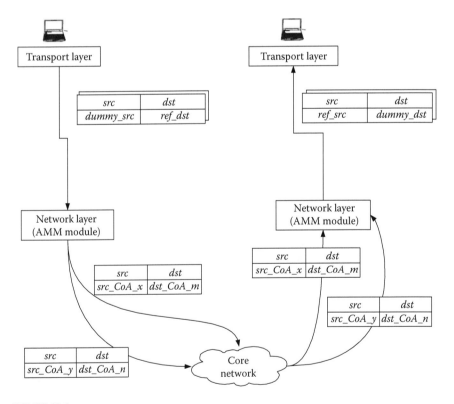

FIGURE 10.6
Illustration of IP-to-IP address mapping in SM3.

a new CoA, *src_CoA*_x while MN has acquired a new CoA, *dst_CoA_m*. Both CN and MN have also informed each other about their new CoAs using the SIP invite request procedure described in Section 10.5.4. Both MN's and CN's U-Tables are updated with their respective correspondent node's new CoA. As the U-Table maintains the list of CNs with whom a node has established connections against the respective CNs' current CoAs, MN's U-Table will be updated with CN's CoA value of *src_CoA_x* and vice versa. The packet arrives via the virtual interface and is directed to AMM. It bypasses HRM and the kernel IPSec function. *dummy_src* and *dummy_dst* are the permanent IP addresses of the virtual interfaces in CN and MN, respectively. Both the M-Tables of CN and MN are also updated. Hence, for this connection, in CN's M-Table, RSA is *dummy_src*, RDA is *ref_dst* and MN's CoA is *dst_CoA_m* while in MN's M-Table, RSA is *ref_src* and RDA is *dummy_dst* and CN's CoA is *src_CoA_x*. As shown in Figure 10.6, right up to the transport layer, the source and destination headers are as per those seen by CN's application. Hence, CN's application is unaware that MN has changed its CoA from *ref_dst* to *dst_CoA_m* and it continues to use the RDA value, *ref_dst*, as parameters for the network layer to generate IP packets towards MN. In fact, CN's application is not even aware that CN has changed its CoA from *ref_src* to *src_CoA_x*. Likewise for MN's application. Such is the ability of SM3 to support seamless multihomed mobility without disrupting the ongoing applications of the users.

AMM in the network layer replaces the destination IP address of the IP packet by MN's CoA value of *dst_CoA_m* based on its entry in the M-Table. The M-Table is referenced by the network port number for IP packets. AMM also replaces the source IP address to CN's current CoA of *src_CoA_x*. This is necessary to prevent network ingress filtering [28].

By rewriting the addresses in the IP header of the packet with the CoA of an interface, packets are thus marked to be transmitted via the corresponding physical interface to the current IP addresses. Figure 10.6's illustration of CN and MN, each using two different interfaces, serves to underscore the success of SM3's multihomed mobility support.

When packets arrive at the destination node, MN, AMM will perform the reverse operations. It will check MN's U-Table for CN's CoAs that match the port number as well as the CoA of the arriving packet to determine that the packet is sent by a CN in MN's CN list. If there is a match, AMM will check MN's M-Table based on the port number to determine the RDA to be used to replace the destination IP address of the packet. In this case, the RDA entry is *dummy_dst* and so it will replace *dst_CoA_m* as the destination IP address. Likewise, AMM will also replace the source IP address with RSA of value *ref_src* from MN's M-Table entry. In this way, MN's application only sees the source and destination IP addresses at the point of connection establishment, totally unaware of the IP address changes as CN and MN roam. In this way, seamless mobility support is provided and no triangular routing is required.

For incoming IPSec packets, AMM will reference the M-Table via the temporarily stored network port number as it cannot decrypt the port number directly from the IPSec packet. AMM has to perform an additional function of pasting the MN's SID from the S-Table into the reserved bits of the AH in the IPSec Header. At the destination, the incoming IP addresses of the IPSec packets will be replaced by the RSA and RDA of the M-Table as per IP packets except that SID is used as the reference for the M-Table as opposed to the port number for IP packets. Thereafter, the IPSec packet is decrypted at the IPSec module. There is no further IP-to-IP address mapping on the decrypted packet as HRM at CN has already restored the IP header of this packet to the state when the connection is first established.

The routing process kicks in after the IP header is modified at the network layer. The usual routing process is changed due to the modification in the IP header. An in-kernel re-route service (IKRS) is incorporated to help maintain the routing information of each packet so that it can be retrieved easily at the network layer. Routing information is stored in different tables and assigned to different physical interfaces. Therefore, IKRS performs a routing table lookup to retrieve the corresponding entry so as to route the packet to the correct physical interface.

10.5.7 Header Restoration Module

For IPSec packets, HRM will replace the source IP address in the IPSec packet's header with the RSA. This restoration of the IP header is needed so that the packet can pass the security check by the IPSec module. It allows the IPSec module to use the share key that is established at the instant of connection establishment. HRM also grabs the network port number used by the data connection and temporarily stores it to facilitate AMM's subsequent processing of the packet. This temporary storage is necessary as the IPSec module will encrypt the packet as a payload and encapsulate it with a IPSec header. AMM will not be able to retrieve the port number from the encrypted packet later on in the processing. AMM will use this port number to reference the M-Table for the outgoing IPSec packets.

10.5.8 SM3 for TCP Communication

The preceding discussion of SM3's architecture, modules and operations are all applicable to TCP. However, TCP does present more complications in the design and implementation of the SM3 and it is worthwhile to highlight some of them in this section.

TCP poses more problems than UDP as there are issues of control overhead and re-establishment or the "slow start" leading to the reduction in window size. The TCP packet is sequenced and sent in order in a normal channel. For multiple links, each packet is redirected to multiple network paths, each with an unknown number of hops. This may increase the end-to-end delay of a certain packet. If the delay differences between the links are significant, the overall average delay will be substantial and the link quality will be degraded. Out-of-order incoming packets could spawn a series of redundant retransmission packets, causing the slowest link to be further jammed, further compounding the out-of-order problem.

For UDP packets, In-Kernel Packet Re-ordering (IKPR) is proposed to address this out-of-sequence packet problem in which the packet is marked and buffered at the kernel level. More details are given in Section 10.5.9. For the TCP packet, there is no out-of-sequence packet problem, as the AMM that modifies the IP addresses, lies in the network layer. The transport layer only processes the packet after it has been modified at the network layer where the checksum and QoS controls are handled here by the kernel functions.

There is a problem with the TCP packet when the window size of each packet is reduced during handover as the receiver will perceive that the connection is interrupted due to network congestion. It will therefore send a message to reduce the window size of the subsequent packets. This is called the slow start and this problem could be solved by a buffer at the kernel level.

Recall the virtual interface in the SM3 architecture shown in Figure 10.4, where an arbitrarily assigned IP address from an unused private prefix is assigned to the virtual interface as the permanent address of the local machine. For the UDP connection, since there is no relation between the two communicating parties, the connection could be broken at any time, and hence, the term "connectionless." For UDP, each connection could be uniquely represented with a remote port and the destination IP address. However, for the TCP connection, each connection is maintained with a session controlled by the transport layer. The TCP process creates a sub-connection with a random port to the other party without informing the AMM in the network layer. To overcome this problem, we hash both the local port and the remote port to generate a unique index for each active connection in SM3. This unique ID is used to match incoming packets, if they belong to an existing connection, else a new connection is established.

As a peer node may have multiple CoAs, the M-Table is used to direct packets to the respective connections and to look up the relevant CoAs to replace the IP addresses in the packet headers. As the entire process of changing IP addresses to correspond to the different interfaces is transparent to the upper layers, it enables ongoing application sessions to survive when multiple interfaces in use are brought down.

10.5.9 QoS Enhancement to SM3

Although the focus here is on seamless SIP-based multihomed mobility management (SM3), it is worth noting that SM3 has been further enhanced with the QoS management incorporated. This work is reported in [29] for UDP. Two modules need to be added to the architecture of the SM3 in Figure 10.4. A Traffic Distribution Module (TDM) is added at the kernel space and a QoS Support Module is added to the user space. Data packets from the virtual interface will be routed to TDM, first to manage the flow distribution of the outgoing traffic based on the QoS policy. The application can specify the QoS requirements. The network traffic can be distributed across multiple links at a flow level or a packet level depending on the QoS policy. Both the QoS policy and the flow distribution can be adjusted by application through the QoS Support Module. The QoS-enhanced SM3 can detect and select optimal paths to satisfy the set QoS requirements such as low latency. It allows different algorithms to be used for path delay evaluation such as sending periodic probe packets to sense end-to-end delays for the different possible paths. With reference to Figure 10.5, when MN receives the INVITE message from N to establish a connection, it will accept and reply with a 200 OK message as well as a list of CoAs. With QoS management, this list of CoAs can be a preferred list determined by the QoS policy. The preferred list can be a subset of MN's interfaces that have better signal strength or lighter traffic load that are metrics used to evaluate an optimal path.

Moreover, the above implies that the list of CoAs used in different sessions may be different and this makes it possible to balance the traffic load among the multiple links. Supporting traffic distribution at the packet level is important, especially when the bandwidth requirement is too high to be fulfilled by a single link. By aggregating multiple links, an effective throughput is increased enabling bandwidth intensive applications to be supported even in a low bandwidth wireless network environment.

To minimize out-of-sequence packets in packet level distribution across multiple links, an In-Kernel Packet Re-ordering (IKPR) service is introduced to SM3. IKPR makes use of a small buffer in the kernel space to re-order the incoming packets without violating the delay constraint that proves to be especially useful for UDP applications that are sensitive to the out-of-sequence problem as vindicated in the experiments carried out on the SM3 testbed and reported in [25].

10.5.10 Incorporation of MIH Framework into QoS-Enhanced SM3

Andi et al. have proposed in [30,31] to incorporate the Media Independent Handover (MIH) protocol into SM3. A virtual device agent is introduced within the framework that performs tasks relating to session management, as well as the execution of policy. The results show that the framework ensures seamless communication during handover by utilizing the link layer event upon which a make-before-break is implemented. In addition, both communicating parties do not experience any disruption due to the IP address transparency provided by the agent. An adaptive load-balancing policy is implemented for a delay-sensitive application [31]. To render the MIH-based QoS-Enhanced SM3 more aware of the prevailing access networks conditions, a set of representative link-layer parameters, extracted from the user device's MACs, is used to classify the current access networks into more crisp states; uncongested, moderately congested, and highly congested. These states are also closely related to the throughput offered by the access networks. The device is equipped with prior knowledge that helps it to estimate the prevailing network condition and identify the network state. Load balancing schemes are then applied adaptively across the different interfaces. Among the several load-balancing schemes implemented, are the Round Robin, the Adaptive Weighted Round Robin, the Adaptive Distribution, and the Hybrid scheme. The Hybrid scheme results in the best performance in terms of throughput, end-to-end delay, inter-packet delay variation, and packet drop. Results show that exploiting the terminal's awareness could greatly improve the user's perceived quality over and above the seamless mobility that he enjoys [29].

10.6 Performance Evaluation

10.6.1 Performance Analysis

The performance analysis of SM3 is similar to that of TMSP [23] using signaling cost, transmission cost, and traffic overhead as the evaluation metrics. The signaling cost comprises movement cost and session maintenance cost. When MN moves from one network to another, a series of message exchanges is needed to keep MN's SIP server/HA up-to-date of MN's current CoA. The movement cost, C_{move}, is thus defined as the total number of hops traversed by these messages. The session maintenance cost, $C_{maintain}$, records the total number of hops traversed by messages needed to maintain existing sessions between CNs and MN. Such messages are usually binding update messages to inform the CNs of MN's new CoA. The transmission cost, C_{tx}, measures the total number of hops in the data path between CN and MN. The traffic overhead provides an insight into the amount of protocol overhead introduced into the network under CBR traffic.

A comparison is made between SM3 and MIPv6 on the one hand, as well as with an SIP-based protocol proposed by Wedkund and Schulzrinne [15] on the other. The latter is an integration of SIP and MIPv6 for both TCP and UDP (SIPMIPv6). For simplicity of comparison, the SIP servers are assumed to be co-located. $H_{A\text{-}B}$ is used to denote the number of hops between the network domain A and the network domain B.

10.6.1.1 Average Signaling Cost

For SM3, when MN moves into a foreign network, it acquires a new IP address via DHCP (whose signaling cost is proportional to $H_{MN\text{-}DHCP}$) and starts a registration procedure using a Register-and-OK message pair between itself and the SIP server (whose signaling cost is proportional to $H_{MN\text{-}SS}$) as per Figure 10.5.

MIPv6 adopts a similar process except that a binding update procedure using BU-and-Acknowledgement message pair takes place between the MN and its HA (whose signaling cost is proportional to $H_{\text{MN-HA}}$). For SIPMIPv6, when MN moves into a foreign network, MN has to perform the registration procedure as per SM3 with the SIP server (SS) and the binding update procedure with its HA as per MIPv6. The C_{move} for SM3, MIPv6, and SIPMIPv6 are thus given in Equations 10.1 through 10.3, respectively.

$$CSM3_{\text{move}} = 4H_{\text{MN-DHCP}} + 2H_{\text{MN-SS}} \qquad (10.1)$$

$$CMIPv6_{\text{move}} = 4H_{\text{MN-DHCP}} + 2H_{\text{MN-HA}} \qquad (10.2)$$

$$CSIPMIPv6_{\text{move}} = 4H_{\text{MN-DHCP}} + 2H_{\text{MN-SS}} + 2H_{\text{MN-HA}} \qquad (10.3)$$

To maintain the current sessions as MN roams into a new foreign network, it has to inform its CNs of its new CoA. In SM3, MN informs its CNs using a binding update procedure that consists of a SIP Invite-and-OK message pair. This signaling cost is proportional to $H_{\text{MN-CN}}$. In MIPv6, MN does not need to inform its CNs of its new IP address, and hence its signaling cost to maintain the existing connection, $CMIPv6_{\text{maintain}}$, is 0. SIPMIPv6 will incur the same signaling cost as SM3 due to its SIP based protocol. The C_{maintain} for SM3, MIPv6, and SIPMIPv6 are shown in Equations 10.4 through 10.6, respectively, where n is the number of CN.

$$CSM3_{\text{maintain}} = 2nH_{\text{MN-CN}} \qquad (10.4)$$

$$CMIPv6_{\text{maintain}} = 0 \qquad (10.5)$$

$$CSIPMIPv6_{\text{maintain}} = 2nH_{\text{MN-CN}} \qquad (10.6)$$

Consider the active handover probability α that refers to the probability that an MN visits a foreign network with active connections. When MN visits a foreign network, it will always incur C_{move}. However, if it does not have active connections, no C_{maintain} will be incurred since there is no CN for MN to update MN's new CoA. Hence, the average signaling cost,

$$C_{\text{signaling}} = C_{\text{move}} + \alpha C_{\text{maintain}} \qquad (10.7)$$

We set $H_{\text{MN-DHCP}}$ to 1 and the rest to 5 for simplicity of calculation. Figure 10.7 plots the average signaling cost against the handover probability α. Since MIPv6 only performs a binding update to HA, the $C_{\text{signaling}}$ is thus a constant and is the lowest compared to SM3 and SIPMIPv6 that have to perform binding updates to existing connections. Therefore, the $C_{\text{signaling}}$ of SM3 and SIPMIPv6 increases correspondingly with an increase in the

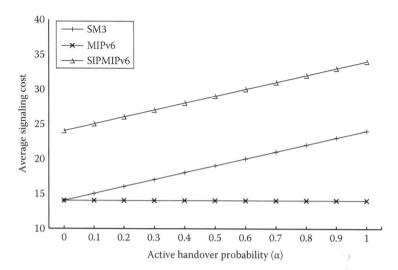

FIGURE 10.7
Average signaling cost for SM3, MIPv6, and SIPMIPv6.

α due to the additional maintenance cost, $C_{maintain}$, required to update CNs with the latest IP address.

10.6.1.2 Data Transmission Cost

In non-mobile networks, the transmission of data packets SM3 between MN and CN follows the conventional IP routing and the transmission cost is thus characterized by the number of hops between MN and CN (i.e., $H_{MN\text{-}CN}$) regardless of TCP or UDP data packets. With MIPv6, a message transmitted from CN to MN has to be routed to MN's HA first, before being tunneled to MN. Likewise, a message from MN to CN is reverse tunneled to MN's HA before being sent to CN. Being an integration of SIP-based and MIPv6 protocols, SIPMIPv6's data route is similar to SM3 for UDP packets while its TCP data routes follow that of MIPv6. C_{tx} for SM3, MIPv6, and SIPMIPv6 are shown in Equations 10.8 through 10.11.

$$\text{CSM3}_{tx} = H_{MN\text{-}CN} \tag{10.8}$$

$$\text{CMIPv6}_{tx} = H_{MN\text{-}HA} + H_{CN\text{-}HA} \tag{10.9}$$

$$\text{CSIPMIPv6}_{tx,UDP} = H_{MN\text{-}CN} \tag{10.10}$$

$$\text{CSIPMIP}_{tx,TCP} = H_{MN\text{-}HA} + H_{CN\text{-}HA} \tag{10.11}$$

For mobile nodes, besides the hop count in the physical data paths, the actual data transmission cost must also take into account the overheads incurred in acquiring a new CoA from a foreign network and in sending BU to all the relevant parties before data transmission can take place. In other words, the actual data transmission cost has to consider the signaling cost of the mobile node movement. Hence, the net data transmission cost is given as

$$C_{tx.net} = C_{move} + C_{maintain} + C_{tx} \tag{10.12}$$

The net data transmission cost is presented as a relative gain in SM3's Hop count against that of the other protocols. The Hop count is measured from the point a packet is sent from MN to CN after it has just moved into a foreign network and hence, it has to acquire its CoA and to perform binding updates before it can send out its data. The relative gain is thus the reduction in the SM3's hop count compared to another protocol. Figure 10.8 shows the Relative Gain plot in percentage, based on the condition that MN only has one CN with one handover. From Figure 10.8, SM3 begins to register a transmission gain over MIPv6 when MN sends more than two packets to CN. This is a result of the protocol characteristic of SM3 since it incurs additional maintenance cost, $C_{maintain}$, to inform CN about MN's movement into a new foreign network. Where C_{tx} is concerned, SM3 is lower given its direct data route between MN and CN as compared to MIPv6's triangular routing where packets have to go through HA as the intermediary. SM3 has a transmission gain over SIPMIPv6 for both TCP and UDP packets with an increasing gain registered for TCP packets since SIPMIPv6 incurs higher cost in C_{move} and C_{tx}. For UDP packets, SM3 yields a decreasing transmission gain over SIPMIPv6 as the latter only incurs a higher cost in C_{move} while sharing the same C_{tx}. Once MN settles itself in the new network, SM3's transmission gain over the others are then governed purely by the differences in the C_{tx} defined in Equations 10.8 through 10.11 as there is no longer any movement signaling cost.

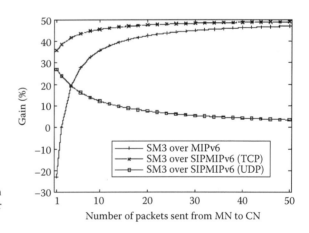

FIGURE 10.8

Relative gain in terms of data transmission cost of SM3 over MIPv6 and SIPMIPv6.

10.6.1.3 Traffic Overhead

Here we analyze the traffic overhead incurred by SM3 compared to MIPv6. First, we consider MIPv6's basic mode of operation without RO. A CBR source will generate a total of $T = R \times (8L) \times t$ bits for a duration of t s, where R is the rate (in packets/s) of the traffic source and L is the size of an IP packet. For MIPv6, the amount of traffic generated onto the network from CN to MN is given by

$$TMIPv6 = 8RtL \times HMIPv6_{CN\text{-}HA} + 8Rt\left(L + L_{tunnel}\right)HMIPv6_{HA\text{-}MN} \qquad (10.13)$$

where

L_{tunnel} is the size of the outer IP header used to encapsulate the packet
$HMIPv6_{CN\text{-}HA}$ is the number of hops between MN and HA

For SM3, the amount of traffic generated onto the network from CN to MN is given by

$$TSM3 = 8RtL \times HSM3_{CN\text{-}MN} \qquad (10.14)$$

where $HSM3_{CN\text{-}MN}$ is the number of hops between CN and MN.

The IP tunnel header in MIPv6, L_{tunnel} is 40 bytes in size. The relative gain in SM3's traffic overhead defined as (TMIPv6-TSM3)/TMIPv6 is plotted in Figure 10.9 with the following parameter setting: a 2 min VoIP session generating a flow of 89 180 byte packets per second, yielding a rate of 128 kbps, $HSM3_{CN\text{-}MN}$ is set to 15,* where $HMIPv6_{HA\text{-}MN}$ varies from 14 to 29 with $HMIPv6_{CN\text{-}HA}$ set to 1. It can be observed from Figure 10.9 that when the hop count difference is 0, i.e., $HSM3_{CN\text{-}MN} = HMIPv6_{CN\text{-}HA} + HMIPv6_{HA\text{-}MN}$,

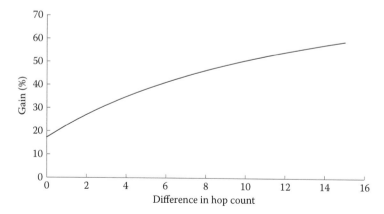

FIGURE 10.9
Relative gain in terms of traffic overhead of SM3 over MIPv6 and SIPMIPv6.

---

* Estimate based on the average hop count measurements from the Internet as reported in [32].

SM3 still registers a reduction in the traffic overhead of 17% over MIPv6. This is due to the additional IP tunnel header that encapsulates each IP data packet at the HA before it is redirected to MN. The gain in overhead traffic reduction increases as the number of hops between MN and its HA increases from 14 to 29. This plot shows that the direct route offered by SM3 eliminates the inefficiency caused by the triangular routing and the additional use of the outer IP tunnel header.

Next, consider the case of MIPv6 with RO. Before direct routing between CN and MN can be effected, MIPv6 still relies on HA's redirection of IP data packet before MN completes the binding update process to inform CN of its CoA. Thus, the initial N IP packets will still suffer from triangular routing. A Home Address option carried by the Destination Option extension header is used in an IP packet sent by an MN that is away from HA, to inform CN of the MN's permanent IP address. CN then uses a Type 2 Routing Header in its data packets to allow them to be routed directly to the CoA of MN. This header also contains MN's permanent IP address that will in turn replace the destination address of the IP packet with its permanent IP address. Both headers require an overhead of 48 bits to encapsulate MN's permanent IP address that is 128 bits. Thus, without considering the traffic incurred by the initial N IP packets and assuming that the number of hops between CN and MN are the same, each IP packet is thus 46 bytes larger in MIPv6 as compared to SM3.

10.6.2 Experiment Results

A testbed supporting multiple access technologies with a multihomed device has been set up with SM3 being implemented on the Ubuntu Linux platform kernel 2.6.17. The testbed can easily be configured to support different wireless technologies and likewise for the multihomed devices.

10.6.2.1 Mobility Support and Horizontal Handover

Figure 10.10 shows the testbed configurations to evaluate the mobility and the horizontal handover capability of SM3 as well as MIPv6. Figure 10.10a comprises two APs, a SIP-RS server, a CN, and an MN for testing SM3, while Figure 10.10b is the MIPv6 testbed. The MN is a 1.86 GHz centrino processor laptop and the CN, HA, and SIP-RS are 2.4 GHz desktop computers. The APs are Linksys WRT54G running openwrt at 54 Mbps. In this experiment, a UDP session using a traffic source from MN to CN is maintained while MN moves from the Wireless LAN A to the Wireless LAN B.

In this set of experiments, a transport session from CN to MN is maintained when MN is on the move. Figure 10.11a and b shows the number of packets received at MN at 100 ms intervals when it switches between WLAN A and WLAN B for UDP and TCP connections, respectively. In the plot for UDP, the top figure shows the UDP throughput without SM3 that breaks right after MN moves to WLAN B at $t = 120$ s. The bottom figure shows that

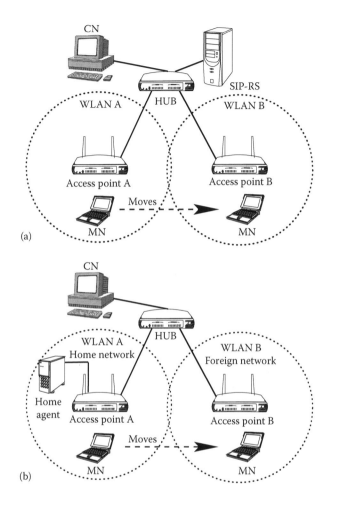

FIGURE 10.10

Testbed configuration for SM3 and MIPv6. (a) SM3. (b) MIPv6.

SM3 efficiently supports the horizontal handover with only a few hundred milliseconds data loss when MN moves to another similar network that does not have overlapped coverage. The data loss is due to the time needed for MN to acquire a new IP address and to inform CN of this new CoA. Similarly, the TCP connection suffers a small break during the handover but is able to resume its session thereafter.

In SM3, handover latency comprises three timings, namely, movement detection time, new IP address acquisition time, and SIP re-invitation processing time. The first two components are dependent on whether the handover is performed using a single wireless interface or multiple wireless interfaces. With more than one wireless interface and an overlapped wireless coverage, a "make-before-break" handover will be able to provide a

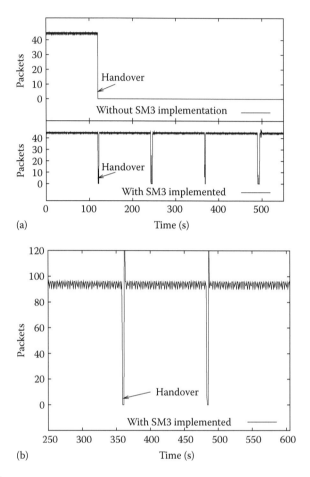

FIGURE 10.11
Received packets at MN for UDP and TCP transmissions as MN performsa horizontal handover. (a) UDP. (b) TCP.

seamless handover. This will greatly reduce if not eliminate the movement-detection latency and the new IP acquisition latency, as well as packet loss since these two processes can be carried out before breaking from the current wireless network. The time required to detect movement into another network in MN and to obtain an IP address configuration can be significant in the total handover latency when MN moves across networks. A set of steps known as DNAv4 [33] has been proposed that is aimed at decreasing the handover latency when moving between points of attachment. There is also a standard protocol proposed to alleviate the delay in the new IP addresses acquisition for DHCP using a rapid commit option [34], whereby, the usual 4-message exchange is reduced to a 2-message exchange. There are network-assisted techniques that require APs to participate in the handover process for MN to minimize packet losses in the handover. Fast handovers for MIPv6 [35] introduces a protocol to improve handover latency due to MIPv6 procedures, and HMIPv6 [12] reduces the signaling delay incurred by informing HA and/or CN of any change in the IP address of MN using the hierarchical nature of IP addressing within the visited network.

10.6.2.2 Mobility Support and Vertical Handover

Figure 10.12 is another configuration of the SM3 testbed with a CN communicating with an MN that is equipped with three interfaces: UMTS, WLAN, and WiMax. MN has access to all three networks. The SIP server resides in a public IP. CN transfers data to MN through different links as MN moves among the three networks. This means that there are periods of time when MN uses all three links, and there are periods when there is no network access at all. A streaming application using both UDP and TCP connections is deployed between MN and CN. The video streaming application is fed by a webcam at CN and streams live video to MN. The real time application requires that the video be stable and undistorted during the transmission. QoS is supported both at the application and the kernel levels.

We evaluate the performance of SM3 on three metrics, namely, end-to-end delay, packet loss and user satisfaction. Delay is a critical factor in interactive real time application. Even though streaming applications have

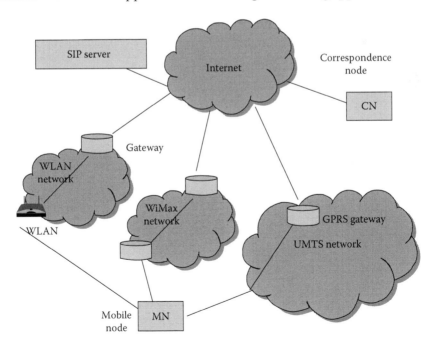

FIGURE 10.12
Testbed configuration for SM3 showing a multihomed MN with WiMax, Wi-Fi, and UMTS interfaces.

a play out buffer to store incoming packets before playing the frames, the end-to-end delay is still very substantial. Our experimental results show that the end-to-end delay cannot be lower than 2 s from CN to MN during handover. The delay is not significant when there are still active alternative links; however, it is a problem when there is no active link available or there is a great drop in bandwidth, for instance, when MN has to use 3G/UMTS that has limited bandwidth to receive data while the other two links are unavailable. Packet loss does occur especially during handover. At the network layer, all incoming packets are captured before entering the transport layer where the TCP protocol processes the packets by counting the sequence number. Out-of-order packets occur as the delay in each path for each packet varies. This delay depends on the number of hops between two ends of each path and it can severely affect large packets that will be fragmented into smaller packets. User satisfaction is collected via informal subjective rating of the audio quality, especially during the handover. A group of 20 users were asked to listen to different audio streaming. Users then rated them in terms of quality and their own judgments using a Mean Opinion Score (MOS) [36] of 1–5.

Figure 10.13 shows the end-to-end delay of the TCP transmission from MN to CN. At $t = 0$ s, MN only uses the WiMax that maintains a reliable and high-speed link. It yields an end-to-end delay of 27 ms on average. Thereafter, when MN moves into the WLAN range at $t = 15$s, it turns on WLAN to connect to CN and the end-to-end delay increases to 70 ms on average. This increase in delay is due to WLAN's shorter range resulting in more hops than WiMAX, which is only one hop away. At $t = 35$ s, UMTS is used with the

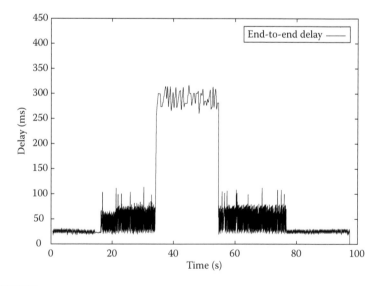

FIGURE 10.13
End-to-end delay for TCP transmission in SM3 for tri-interface multihomed MN.

other two links simultaneously, to deliver packets to the CN. Data packets are distributed in a round-robin fashion over the available links.

The average end-to-end delay is increased to 300 ms due to UMTS's lower bandwidth and more hops. At $t = 55$ s, MN disconnects from the UMTS network and instead uses only the WiMAX and WLAN links, resulting in reducing the average end-to-end delay to 70 ms. Finally, when MN reverts to using one link (WiMAX), the average delay drops to 27 ms. Given that all network links have different delays, when a packet arrives at CN from a slow link, it is queued before the transport layer, if it is not in order. The transport layer will send the re-transmission request if the timer runs out. The next time, if the packet is sent from the faster link, it will arrive in the correct order. Hence, the increase in delay. Figure 10.13 clearly vindicates the seamless connectivity and multihomed mobility support effected by SM3 as MN roams across the different heterogeneous networks. In addition, it also shows that the different interfaces can be exploited for a seamless handover using the "make-before-break" technique, whereby MN connects to a new network before relinquishing the current network.

Figure 10.14 shows the packet delivery success given by MN's vertical handovers. As TCP is used, there is no packet loss during the handovers. However, on closer examination, as amplified in the right plot of Figure 10.14, during the handover from WiMAX to WLAN around $t = 15$ s, there is a window of time when a number of packets are queued at the arrival end and cannot reach the application level. During the handover, these packets queueing before the transport layer, are grouped and delivered at the same time to the application level accounting for the observed behavior. Once again, these plots prove that SM3 can provide seamless connectivity during handover in heterogeneous networks.

For the listening tests, users listen to 10 audio streams from MN to CN and are then asked to rate the audio quality throughout the streaming session with handovers. The MN and CN are both laptops and a lower bitrate stream is used compared to that used for Figures 10.13 and 10.14. The average

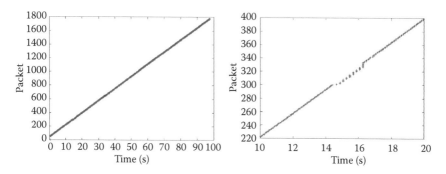

FIGURE 10.14
Received packets at MN for TCP transmission as MN performs vertical handovers.

MOS of 20 listeners stands at 3.66, which is rated as between fair (slightly annoying), and good (perceptible but not annoying). The seamless mobility support, with "make-before-break" handovers facilitated by the multihomed device, contributes to the high quality of streaming.

10.6.2.3 Traffic Distribution and Bandwidth Aggregation

Another testbed configuration is used for this set of experiments where an MN is connected to CN via heterogeneous wireless links including one WiMAX and two WIFI links. A single UDP flow is sent from CN to MN at the rate of 948 kbps. At $t = 0$ s, all packets are directed to the WiMAX link. At $t = 135$ s, WIFI interface 1 is up. Part of the traffic goes to WIFI link 1. At $t = 234$ s, the WIFI 2 interface is brought up. The traffic is further distributed among all three links. When WIFI 2 is brought down at $t = 334$ s, the traffic is redistributed to the two active links. The average measured aggregated throughput from all links is 945.8 kbps. Figure 10.15 shows the packet distribution over the three wireless links as well as the total throughput. As shown in Figure 10.15, at any point in time, the aggregated throughput is approximately equal to the data rate of the flow. This suggests that packet

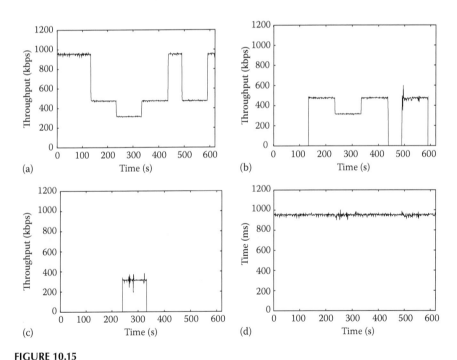

FIGURE 10.15
Throughput of an UDP flow from CN to MN. (a) Throughput of UDP flow across WiMax Link. (b) Throughput of UDP flow across Wi-Fi Link 1. (c) Throughput of UDP flow across Wi-Fi Link 2. (d) Aggregate throughput of UDP flow across all three links.

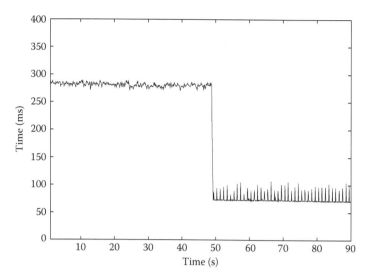

FIGURE 10.16
Delay improvement in UDP transmission after switching from WiMax Link to two Wi-Fi Links.

loss during each handover is very small, testifying to the fact that SM3 can provide seamless mobility and flow distribution for multiple heterogeneous wireless links.

In Figure 10.16, the traffic is initially carried via WiMAX with an average end-to-end delay of 270.8 ms. At $t = 49.27$ s when the two WIFI links are up, SM3 senses that the WIFI links have a lower end-to-end delay of 74.3 ms. Hence, SM3 redistributes the packets from the WIMAX link to two WIFI links, resulting in a much lower delay.

This set of experiments demonstrates that SM3 can offer a bandwidth aggregation through the simultaneous use of multiple interfaces as well as to adaptively redistribute packets across the different links. The former allows the support of the bandwidth demanding applications by low bandwidth links, while the latter facilitates efficient use of the different interfaces and improves the QoS metrics such as delay and loss. The problem of out-of-order packets arising as a result of the use of multiple links can be addressed by incorporating the In-Kernel Packet Re-ordering (IKPR) service into SM3. (Refer to discussed in Section 10.5.9.) IKPR makes use of a small buffer in the kernel space to re-order the incoming packets without violating the delay constraint and proves to be effective for UDP packets.

10.6.3 Discussion

In this section, the strengths and the limitations of SM3 are discussed with some recommendations for future work.

10.6.3.1 Tunneling and Traffic Overhead

Instead of tunnel-based approaches like Mobile IP, where packets between MN and CN are being routed via HA, SM3 uses direct routing between MN and CN as per normal conventional IP routing. The approach eliminates triangular routing in Mobile IP. The end-to-end delay is thus reduced compared to Mobile IP resulting in a lower transmission cost as well as a lower traffic overhead for SM3 data packets compared to Mobile IP.

For route optimization mode in Mobile IP, CN uses a Type 2 Routing Header in its data packets to allow them to be routed directly to the CoA of MN. This header also contains MN's permanent IP address. Both headers require an overhead of 48 bits to encapsulate MN's permanent IP address that is 128 bits. Thus, each IP packet is 46 bytes larger in MIPv6 as compared to SM3.

No new network infrastructure is needed unlike the need for a HA in MIP. The trade off for SM3 is the requirement to incorporate SM3 software to end users' devices to support the multihomed mobility and its host of extensions and merits. This may appear to be unattractive. However, this is no different from other mobility schemes. For direct communication (e.g., VoIP, file exchange) between two clients (mobile or non-mobile), SM3 provides a perfect solution for seamless communications and is the enabler for IP mobility management when the host roams between homogeneous and heterogeneous wireless networks.

10.6.3.2 Aggregate Link and Seamless Connectivity

By distributing traffic across multiple links, SM3 supports the aggregation of the link bandwidth to achieve increased total throughput without the need to modify the existing network topology or to incorporate any new protocols. When the mobile node moves to a new network, the connectivity is maintained using a soft-handoff mechanism. By interfacing to different networks with varied ranges, the multihomed support from SM3 can ensure that MN is served by overlapping wireless networks during roaming.

10.6.3.3 Network Detection and Handoff Algorithm

Handoff detection and triggering are essential in wireless mobility as they occur often when the mobile node moves among different wireless access networks. Handoff detection is decided by measurement of a relative signal strength (RSS) in the overlapped or border region of two wireless networks. Each interface uses different technologies, and manufacturers use different standards to define the signal strength and the link quality. To generate an optimized mechanism to distribute packets into different interfaces at the same time, all the measured factors or parameters have to be normalized. Each link has a different bandwidth, for instance, WLAN's bandwidth can vary from 5 Mb to 54 Mbps while WiMax can reach 70 Mbps and UMTS's bandwidth is around 50 kbps. While WLAN covers short ranges, WiMAX could reach further, up to 50 km. Hence, the signal strength/power rate for WLAN registers drastic change when MN leaves the coverage of one WLAN to another but for WiMax or UMTS, the signal strength/power remains almost constant within a small region.

10.6.3.4 Firewall and NAT Traversal

As the mobile host moves to different networks, the signaling mechanism requires that the new location be reported to the SIP server. The signaling protocol requires routing information to be provided before switching to a new network, which increases the latency when the mobile node associates with the new network. While seeking routing information from the local routers to connect to the SIP server, the mobile host might encounter firewall or Network Address Translation (NAT) [37] issues in the different subnet networks.

IP tunneling is one technique used to create a virtual private network (VPN) among the SIP server, the mobile node, and other peers. However, this approach is infeasible if the firewall is not a simple symmetric* cone type, and it reduces the scalability of the system when newly joined clients need to set up the VPN. A STUN (Session Traversal Utilities for NATs) server [38] is required to establish a tunnel to solve NAT issues, which are common in current UMTS networks where different operators possess their own private IP networks.

The STUN server enables SIP to be securely transported across private/ public network boundaries, but does not address the requirement to implement policies for extranet communication. The Session Border Controller (SBC) [39] is a proprietary network intermediary deployed at the border of a network to enforce network policies to provide a variety of functions to enable or to enhance SIP services, like VoIP. Some of its common functions include perimeter defenses like access control, topology hiding, denial-of-service prevention and detection, NAT traversal, and network management like traffic monitoring, shaping, and QoS. A SBC enables SIP signaling and media to be received from and directed to a user device behind a firewall and NAT. It achieves this by rewriting the IP addresses and ports in the call signaling headers and the session description protocol blocks attached to SIP messages. A SBC sits between the MN and the SIP registrar of the domain to perform the NAT traversal function that modifies specific SIP messages to assist an MN to maintain connection with its UA, the SIP registrar and

---

* Symmetric NATs set up separate mappings between the private IP address and the port, and the public IP address and the port, for each remote address. As a result, an endpoint cannot use STUN to determine the public address that a third-party use to route media to it through a symmetric NAT. Instead, a media relay must be used.

CN's UA. To support MN's mobility in a public network or another NAT traversal network, SM3 needs to perform port number mapping in addition to the IP address mapping for each connection. In the case when an MN moves from a NAT traversal network to a public network, its UA sends a nonstandard SIP invite request to its CNs in which it requests CNs to report the port numbers previously used by MN as recorded in CNs' state tables. With this information, MN records the port numbers accordingly and maps both the IP addresses and the ports for each incoming packet in its AMM. In the case when an MN moves from a NAT traversal network to another, the SBC in the new network passes all the SIP registers and invites requests. Each invite request message to CN includes all the ports currently used to listen to packets from CN. SBC appends a nonstandard tag onto the message that includes proposed port numbers for each MN's port to request CN's UA to map outgoing packets to MN using this port number. Upon obtaining a response from CN, SBC opens the port on the firewall. With a mapping of the outgoing packet request, CN maps the IP addresses and port numbers for each outgoing packet. Details of the operation of SM3 with SBC is given in [23].

Note that movement involving networks behind a firewall/NAT [38] requires additional IP addresses and port numbers to be included in the IP header of each IP packet.

10.6.3.5 Out-of-Sequence or Missing Packets

With SM3 allowing packet distribution across different links, for an unreliable channel using UDP, the number of out-of-sequence or missing packets are inevitable. There are several approaches to address the issues including header injection or flagged packets and buffer usage at the receiving end to re-order the packet. An In-Kernel Packet Re-Order service (IKPR) is implemented to solve the out-of-sequence packets in UDP transmission. A buffer is used to hold the packets before they enter the network layer. Each packet is marked with a flag field in the UDP header. The sequence is marked so that the receiving end could track down the order of the incoming packets. If the number of out-of-sequence packets is larger than the buffer's capacity, the oldest packet in the queue is removed and the new packet stacks up. It has been proved in simulation that the buffer is sufficient in reducing the total number of loss packets in UDP transmission [29].

For reliable TCP transmission, although there is flow control at the transport layer, the overhead from the retransmission packets can reduce the quality of the stream. As there are different paths with an unknown end-to-end, delays and handoffs may occur along the route from source to destination. The cumulative latency experienced by a packet will lead to an increase in retransmission requests since the transport layer is unaware of the presence of multiple interfaces.

Retransmissions consume valuable bandwidth and degrade link quality, resulting in the receiver having to issue a window resize advertisement to immediate routers and senders to reduce the sending packet's TCP window size. A solution to the problem could be a network detection scheme to detect any changes in the delay of each link. As the sender does not have the ability to sense the number of hops in a certain link, a trial period could be provided when MN joins a new network. Packets are sent in the link and delays measured at the receiver are compared to those of existing links. The sender is informed of the respective link delays so that the sender could change the load balancing policy in distributing packets to its physical interfaces. A slow link will get less traffic. However, it means that a signaling protocol has to be designed to signal the changes between the senders and the receivers, which complicates the deployment of the system.

10.7 Summary

In this chapter, we provided an overview of the research work on IP mobility and multihoming support. We then proceeded to discuss our proposal, i.e., SIP-based Multihomed Mobility Management (SM3). SM3 is based on the basic principles of Terminal Mobility Support Protocol (TMSP), which is an SIP-based IP mobility management scheme. SM3 is a new engineering alternative to Mobile IP (MIP, which includes both MIPv4 and MIPv6). The performance of SM3 is then compared in relation to MIP as well as another representative SIP-based protocol, SIPMIPv6, both analytically as well as via experiments conducted on the respective implemented testbeds. Results proved that SM3 is a viable method to provide seamless mobility support and QoS-aware flow distribution for multihomed mobile nodes. Compared to MIP, SM3 does not require triangular routing and hence incurs less transmission cost and traffic overhead even when compared against the MIPv6 RO mode.

SM3 utilizes the SIP protocol for session management and the localized IP-to-IP address translation helps to minimize the network traffic overhead. By using the virtual interface with the permanent address, SM3 completely hides the heterogeneity of the physical interfaces and decouples the node identifier and the node locator. This not only enables a transparent change of interface addresses but also allows the upper-layer connections to survive the breakdown of the physical interface in use. The approach adopted by SM3 does not require heavy modifications to the existing protocol architecture. SM3 only requires additional software to be bundled in the MN and its correspondent node (CN) for direct seamless communication between MN and CN across heterogeneous wireless networks. Moreover, unlike MIP, SM3 does not require new infrastructure support such as Home Agent. SM3 can therefore be easily deployed on the existing network infrastructure and it can support existing applications. SM3 does not impose any restrictions on

MN and CN. Both can reside in wired or wireless networks and can be static or mobile. During roaming, they can cross heterogeneous wireless networks seamlessly.

Besides providing full IP mobility support, SM3 supports both vertical and horizontal handovers. SM3's framework is flexible and expandable to support enhancements and refinements, such as MIH integration and support for QoS-aware data distribution. Experimental results have shown that SM3 can provide a seamless handover and QoS-aware flow distribution for WiMAX, WIFI, and UMTS links, and the traffic overhead is low.

The added ability to distribute data flows freely at either the flow level or the packet level is another enhancement incorporated into SM3. For a bandwidth intensive application that exceeds the bandwidth support of any single link, conventional flow-based distribution algorithms are unable to satisfy the requirement. However, with SM3, multiple low speed links can be aggregated to provide higher throughput to serve such bandwidth intensive applications. Distributing data flows at the packet level across multiple links leads to an out-of-sequence problem. This problem is alleviated by the proposed IKPR service, which is a receiver-based approach to complement the sender-based approach to address the out-of-sequence problem.

Abbreviations

AMM	IP-to-IP address mapping module
BU	Binding update
CoA	Care-of-address
CN	Correspondent node
FA	Foreign agent
FMC	Fixed mobile convergence
HoA	Home address
HA	Home agent
HSPA	High speed packet access
IKE	Internet key exchange
IKPR	In-kernel packet re-ordering
IKRS	In-kernel re-route service
ITU	International Telecommunication Union
IMS	IP multimedia subsystem
LPRP	Local prefix routing protocol
LTE	Long term evolution
MIH	Media independent handover
MIP	Mobile IP
MIPv4	Mobile IPv4
MIPv6	Mobile IPv6
MN	Mobile node

MNN	Mobile network node
MOS	Mean opinion score
MR	Mobile router
NAT	Network address translation
NEMO	Network mobility
NMBSP	Network mobility basic support protocol
RO	Route optimization
RRH	Reverse routing header
SBC	Session border controller
SEM	IP-swap extension module
SID	Session identity
SIP	Session initiation protocol
SM3	SIP-based multihomed mobility management
SMR	Session-to-mobility ratio
STUN	Session traversal utilities for NATs
TCP	Transmission control protocol
TDM	Traffic Distribution Module
TMSP	Terminal mobility support protocol
TNEMO	Terminal-assisted route-optimized NEMO mobility management
UA	User agent
UDP	User datagram protocol
UMTS	Universal mobile telecommunication system
URI	Universal resource identifier
VPN	Virtual private network
WMAN	Wireless metropolitan area networks

References

1. DisplaySearch, Quarterly Notebook PC Shipment and Forecast Report for Q2 2009, DisplaySearch, Austin, TX, Aug. 2009.
2. T. Teng, F. Sideco, and J. Rebello, The mobile handset industry: From unstoppable growth to foreseeable meltdown, Mobile handest market tracker, iSuppli Corporation, El Segundo, CA, 2009.
3. ITU, Worldwide mobile cellular subscribers to reach 4 billion mark late 2008, Press Release, ITU, Geneva, Switzerland, Sept. 25, 2008.
4. Infonetics Research, Mobile Broadband Cards, Routers, Services, and Subscribers Report for Q4 2008, Infonetics Research, Campbell, CA, Mar. 2009.
5. Allot Communications Ltd., Global Mobile Broadband Traffic Report Q2 2009, Allot Communications Ltd., Boston, MA, July 2009.
6. ITU, Mobility management and control framework and architecture within the NGN transport stratum, Y Series: Global information infrastructure, Internet protocol aspects and next-generation networks, ITU-T Y.2018, Geneva, Switzerland, Sept. 12, 2009.

7. ITU, Mobility management and control framework and architecture within the NGN transport stratum, Y Series: Global information infrastructure, Internet protocol aspects and next-generation networks, ITU-T Y.2808, Geneva, Switzerland, June 1, 2009.

8. J. Rosenberg. H. Schulzrinne, G. Camarillo, A. Johnson, J. Peterson, R. Sparks, M. Handley, and E. Schooler, SIP: Session initiation protocol, RFC 3261, Internet Society, Reston, VA, June 2002.

9. C. Perkins, IP mobility support, RFC 2002, Network Working Group, Oct. 1996.

10. D. Johnson, C. Perkins, and J. Arkko, Mobility support in IPv6, RFC 3775, Internet Society, Reston, VA, June 2004.

11. C. Perkins, IP mobility support for IPv4, RFC 3344, Internet Society, Reston, VA, Aug. 2002.

12. H. Soliman, C. Castelluccia, K. E. Malki, and L. Bellier, Hierarchical mobile IPv6 mobility management (HMIPv6), RFC 4140, Internet Society, Reston, VA, Aug. 2005.

13. S. Pack, X. Shen, J. W. Mark, and J. Pan, Adaptive route optimization in hierarchical mobile IPv6 networks, *IEEE Transactions on Mobile Computing*, 6(8), 903–914, Aug. 2007.

14. H. Schulzrinne and E. Wedkund, Application-layer mobility using SIP, *Mobile Computing and Communications Review*, 1(2), 47–57, July 2000.

15. E. Wedkund and H. Schulzrinne, Mobility support using SIP, *ACM WoWMoM*, Seattle, WA, Aug. 1999, pp. 76–82.

16. C. Ahlund, R. Brannstrom, K. Andersson, and O. Tjernstrom, Portbased multihomed mobile IPv6 for heterogeneous networks, *Proceedings of the 31st IEEE International Conference Local Computer Networks*, Tampa, FL, Nov. 2006, pp. 567–568.

17. T. Chen and L. Wenyu, A novel IPv6 communication framework: Mobile SHIM6 (M-SHIM6), *Proceedings of the International Conference Wireless Communications, Networking and Mobile Computing (WiCOM)*, Wuhan, China, Sept. 2006, pp. 1–3.

18. E. Nordmark and M. Bagnulo, Shim6: Level 3 multihoming shim protocol for IPv6, Internet-Draft, draft-ietf-shim6-proto-10.txt, Feb. 2008.

19. M. Bagnulo, A. Garcia-Martinez, and A. Azcorra, IPv6 multihoming support in the mobile internet, *IEEE Wireless Communications/Personal Communications*, 14(5), 92–98, Oct. 2007.

20. D. Le, J. Lei, and X. Fu, A new decentralized mobility management service architecture for IPv6-based networks, *Proceedings Third ACM Workshop on Wireless Multimedia Networking and Performance Modelling (WMuNeP)*, New York, 2007, pp. 54–61.

21. R. Atkinson, S. Bhatti, and S. Hailes, A proposal for unifying mobility with multihoming, NAT, & security, *Proceedings of the Fifth ACM international workshop on Mobility Management and Wireless Access (MobiWac)*, New York, 2007, pp. 74–83.

22. B. S. Lee, T. M. Lim, C. K. Yeo, and Q. V. Le, A mobility scheme for personal and terminal mobility, *Proceedings of the IEEE International Conference on Communications (ICC)*, Glassgow, U.K., 2007.

23. T. M. Lim, C. K. Yeo, B. S. Lee, and Q. V. Le, TMSP: Terminal mobility support protocol, *IEEE Transactions on Mobile Computing*, 8(6), 849–863, June 2009.

24. T. M. Lim, B. S. Lee, C. K. Yeo, and J. W. Tantra, A terminal-assisted route optimized NEMO management, *Proceedings of the Fifth ACM International Workshop on Mobility Management and Wireless Access*, Chania, Greece, 2007, pp. 84–90.

25. T. M. Lim, B. S. Lee, C. K. Yeo, J. W. Tantra, and Y. Xia, A terminal-assisted route optimized NEMO management, *Telecommunication Systems*, 42(3–4), 263–272, Dec. 2009.

26. V. Devarapalli, R. Wakikawa, A. Petrescu, and P. Thubert, Network mobility (NEMO) basic support protocol, RFC 3963, Internet Society, Reston, VA, Jan. 2005.

27. P. Thubert and M. Molteni, IPv6 reverse routing header and its application to mobile networks, Internet Draft, IETF, Network Working Group, Sept. 2006.

28. P. Ferguson and D. Senie, Network ingress filtering: Defeating denial of service attacks which employ IP source address spoofing, RFC 2267, Internet Society, Reston, VA, Jan. 1998.

29. D. D. Nguyen, Y. Xia, M. N. Son, C. K. Yeo, and B. S. Lee, A mobility management scheme with QoS support for heterogeneous multihomed mobile Nodes, *Proceedings of the IEEE Global Communications Conference (Globecom)*, New Orleans, LA, 2008.

30. W. C. Andi, Y. Xia, C. K. Yeo, and B. S. Lee, MIH-based SIP mobility management scheme in heterogeneous wireless networks, *Proceedings of the Second International Conference on Mobile Ubiquitous Computing, System, Services and Technologies (UBICOMM)*, Valencia, Spain, 2008.

31. W. C. Andi, C. K. Yeo, and B. S. Lee, Framework for multihomed terminal, *Computer Communications*, 33(9), 1049–1055, June 2010.

32. A. Fei, G. Pei, R. Liu, and L. Zhang, Measurements on delay and hop-count of the Internet, *Proceedings of the IEEE International Conference on Globecom Communications, (Globecom)*, Rio de Janeiro, Brazil, 1999, pp. 76–82.

33. B. Aboba, J. Carlson, and S. Cheshire, Detecting network attachment in IPv4 (DNAv4), RFC 4436, Internet Society, Reston, VA, Mar. 2006.

34. S. Park, P. Kim, and B. Volz, Rapid commit option for the dynamic host configuration protocol version 4 (DHCPv4), RFC 4039, Reston, VA, Mar. 2005.

35. E. R. Koodli, Fast Handovers for Mobile IPv6, RFC 4068, Reston, VA, July 2005.

36. ITU, Mean Opinion Score (MOS) Terminology, ITU-T P.800.1, ITU, Geneva, Switzerland, July 2006.

37. K. Egevang and P. Francis, The IP network address translator (NAT), RFC 1631 (Informational), Internet Society, Reston, VA, May 1994, obsoleted by RFC 3022 [Online]. Availablehttp://www.ietf.org/rfc/rfc1631.txt.

38. J. Rosenberg, R. Mahy, P. Matthews, and D. Wing, Session traversal utilities for NATs (STUN), RFC 5389, Internet Society, Reston, VA, Oct. 2008.

39. J. Hautakorpi, G. Camarillo, and R. Penfield, A. Hawrylyshen, and M. Bhatia, Requirements from SIP (session initiation protocol) session border control deployments, Internet Draft, Nov. 2006.

11

Vertical Handover System in Heterogeneous Wireless Networks

Yong-Sung Kim, Dong-Hee Kwon, and Young-Joo Suh

CONTENTS

11.1 Introduction

Various wireless network technologies have been developed to offer Internet access to end users. The technologies have been developed for different purposes and thus provide different services, coverage areas, network bandwidths, and so on. For example, the second-generation (2G) and third-generation (3G) cellular networks have been developed to cover large areas, whereas wireless LANs (WLANs) [1] offer smaller areas. Furthermore, 2G/3G cellular networks provide relatively low data rates up to 14.4 Mbps while WLANs offer higher data rates up to 54 Mbps. According to these heterogeneous and complementary characteristics, various wireless networks will coexist and interwork together to support the different requirements of end users such as high usability, seamless connectivity, delay-sensitive applications, etc., in next-generation (4G) wireless networks.

4G wireless networks [2] are required to support network heterogeneity and complementariness for satisfying end users' requirements. The features of 4G networks can be summarized as follows: (1) providing high-data-rate services to accommodate numerous multimedia applications; (2) supporting the mobility of outdoor pedestrians and vehicles; and (3) providing end users with high-speed, large volume, good quality, global coverage, and flexibility. According to such features, 4G networks will be adapted to various forms in several fields such as medical centers, natural environments, information industries, etc. Moreover, 4G networks will coexist with successfully deployed wireless networks because it enables a reduction in developing cost and risk. Therefore, 4G networks will be expected to have heterogeneous wireless networks and to be deployed with variable physical, networking, and architectural characteristics.

3GPP/3GPP2 has introduced two integration architectures [3,4] for 3G and WLANs as one of the first steps toward achieving seamless handovers. Since both networks have complementary characteristics and have already been well deployed in the world, the integrated architecture can provide integrated service capabilities across 3G and WLANs. One of the two integration architectures is the tightly coupled architecture, where WLANs appear as one of 3G access networks and provide 3G services to WLAN users. This architecture uses WLAN gateways in order to hide 3G protocol details from WLAN users. Figure 11.1 shows the tightly coupled architecture between

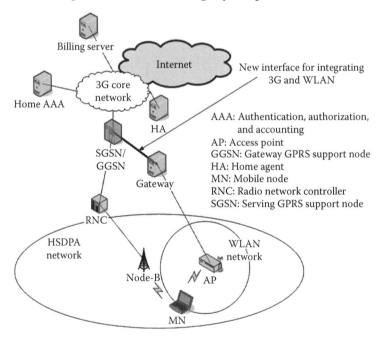

FIGURE 11.1
Tightly coupled architecture between 3G and WLAN.

3G and WLANs. A WLAN gateway is connected to SGSN/GGSN of 3G networks and provides translation services between the 3G and the WLAN protocols in order to offer 3G services to WLAN users. However, this architecture requires a new interface between the SGSN/GGSN and the WLAN gateway, which requires modifications or extensions to already deployed 3G networks. Therefore, it introduces a network deployment cost as well as a network integration cost during the initial network deployment of integrated 3G and WLANs.

The other integration architecture is the loosely coupled architecture. Contrary to the tightly coupled architecture, it integrates 3G and WLANs based on the Internet architecture. Figure 11.2 shows the loosely coupled architecture. As shown in the figure, this architecture completely separates 3G and WLANs since networks are integrated into the Internet. According to this characteristic, the loosely coupled architecture does not require any modifications or extensions to deployed 3G networks, contrary to the tightly coupled architecture. However, the loosely coupled architecture needs a mobility management protocol to support handovers between 3G and WLANs.

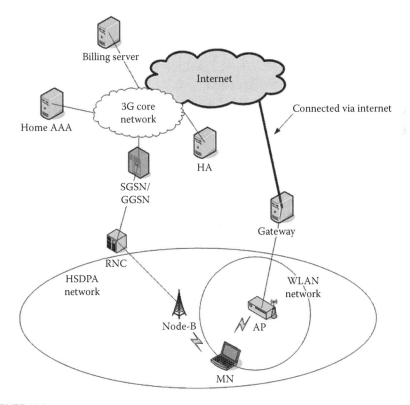

FIGURE 11.2
Loosely coupled architecture between 3G and WLAN.

4G wireless networks will consist of various wireless networks and will be integrated into IP-based networks, which further require seamless vertical handovers. Once the decision of a vertical handover has been made, the key issue for the seamless handover is a mobility management system. The handover process is divided into two processes; the Layer 2 (L2) handover process and the Layer 3 (L3) handover process. The L2 handover latency is the period between the time when the air-link with the current access router (AR) is disconnected and the time when the MN is connected to the air-link of a new AR (NAR) to which it will attach. The L3 handover latency is the sum of the agent discovery latency, the Care-of-Address (CoA) acquisition latency, and the registration latency. There are several kinds of mobility management protocols operating in different layers such as the MIPv4/v6 [5,6], the Fast Handovers for MIPv6 (FMIPv6) [7], and the Hierarchical Mobile IPv6 Mobility Management (HMIPv6) [8] in the network layer, the TCP-Migrate [9]/Stream Control Transmission Protocol (SCTP) [10] in the transport layer, and the Session Initiation Protocol (SIP) [11] in the application layer.

Mobile IP (MIP) [5] is the most widely studied solution among mobility management protocols for heterogeneous wireless networks that will be deployed on top of IP-based networks. However, MIP suffers from long handover latency due to the original handover signaling. As a result, it is not suitable for delay-sensitive applications that require lossless handovers as well as always-on Internet connectivity. Therefore, it is necessary to study a vertical handover system that provides backward compatibility to the currently deployed networks as a cost-effective way and offers seamless connectivity during handovers across heterogeneous wireless networks.

In this chapter, we design and implement a vertical handover system for heterogeneous wireless networks to provide seamless connectivity to end users without modifying or extending the network side. The system is designed as a common interface that resides between the IP/MIP layer and the actual network interfaces for an MN. The system provides a transparent service to the IP/MIP stack and seamless connectivity for inter-technology handovers by simply modifying the MN side. Moreover, the system can also be closely coupled with the MIP stack to perform a handover to its preferred network determined by handover metrics such as signal strength, QoS parameters, available services, end user preferences, etc. Thus, we address two possible designs of the vertical handover system based on inter-technology handover approaches—basic integration and tight integration. To validate the system, we have implemented the system over loosely coupled 3G/WLAN experimental test-beds. The experiment results show that the system enables an MN to handover seamlessly across different types of wireless networks without any modifications to the existing network side.

The remainder of this chapter is organized as follows. Section 11.2 describes the Mobile IP protocol and Section 11.3 introduces related work regarding the connectivity for 4G networks. In Section 11.4, we introduce a vertical handover system in detail, to support seamless handovers across heterogeneous wireless networks. Section 11.5 describes loosely-coupled architecture-based 3G/WLAN test-beds and examines the vertical handover performance on realistic environments. Finally, concluding remarks are given in Section 11.6.

11.2 Background

The IETF Mobile IP (MIP) protocol allows transparent routings of IP datagram to MNs in the Internet. An IP datagram delivery uses hierarchical addressing, where IP addresses are divided into a network ID and a host ID. Packets destined to an MN would not be delivered unless it is connected to its home network. Thus, to enable transparent routings of IP datagrams regardless of MNs' location, MIP introduces new functional entities that perform a packet forwarding, i.e., a home agent (HA) and a foreign agent (FA).

The MIP protocol performs the functions of agent discovery, tunneling, and registration. Agent discovery is intended for HAs and FAs to keep the MN informed of its current location, the availability of the agents, and the change of its point of attachment to the Internet. The Agent Advertisement (AA) of HAs and FAs is formed by including a Mobility Agent Advertisement extension in an ICMP Router Advertisement message. This ICMP message is periodically broadcast or multicast to those links to which it is connected and to which it wishes to offer routing services. AA provides the means with which an MN can locate itself. To learn if any agents are present, the MN may passively listen to AA or actively search by broadcasting an Agent Solicitation (AS) message to request AA. The other function is tunneling that is performed by HA and FA. Tunneling is generally performed by encapsulating the IP datagram with a new IP header called the tunnel header. HA encapsulates packets destined to an MN. This encapsulation point (i.e., HA) can be thought of as a tunnel entry point. On the other hand, FA decapsulates encapsulated packets and then correctly delivers the decapsulated packets to MN that it is currently serving. We can think of FA as a tunnel exit point. The tunneling decision of HA is based on the binding information that was securely formed by the registration process as dictated below. Another function is registration. When an MN is away from its home, it registers its care-of address (CoA) with its HA. Depending on its method of attachment, the MN will register either directly to its HA or through FA that forwards the registration to HA. The registration message exchanges the MN's current binding information among the MN, its HA, and (possibly) its FA. The registration creates or modifies a mobility binding at HA, associating the MN's home address with its CoA for a certain length of time.

11.3 Related Work—Vertical Handover Framework

Stemm and Katz [12] proposed a vertical handover system for infrared/ WLANs and WLAN/Ricochet networks. The system provides an end user with the best possible connectivity during a vertical handover. Siddiqui et al. [13] proposed two tightly coupled architectures by differentiating integration points, namely, SGSN and GGSN. It integrates 3G and WLANs based on the tightly coupled architecture, but it still has a problem of modifying SGSN and GGSN. Akyildiz et al. [14] studied an adaptive protocol suite for 4G wireless data networks. They noticed the importance of the adaptiveness to each layer in the protocol stack at an MN in order to address several problems including rate adaptation, congestion control, mobility support, and coding. In particular, they present the adaptiveness in layers 2 (link), 4 (transport), and 5 (application) to satisfy the demand of heterogeneity and high-speed multimedia applications in 4G wireless data networks.

Several research results address vertical handover issues by introducing an integration point among several heterogeneous networks. Buddhikot et al. [15] proposed an interworking solution for IEEE 802.11 and 3G wireless networks. They introduced a new network element called the integration of two access technologies (IOTA) gateway in IEEE 802.11 networks and a new service access software on client devices. The IOTA gateway cooperates with the client software to offer integrated 802.11/3G wireless data services that support seamless inter-technology handovers, Quality of Service (QoS) requirements, and multi-provider roaming agreements. However, they did not consider minimizing the disruption period during vertical handovers by closely integrating their approach with MIP, which is one of the issues that we want to address in this chapter. Sharma et al. [16] presented the design, the implementation, and the evaluation of a vertical handover system that enables an MN to automatically switch between the WLAN and the GPRS networks by introducing a simple extension to the existing MIP implementation. However, it needs a special agent called the GPRS foreign agent for forwarding data toward an MN while an MN is away from the home network. Chen et al. [17] showed a vertical handover solution by introducing an entity called the handover server, which provides IP tunneling techniques and serves as a home agent in the MIP protocol. It only needs a little change at the MN, but the handover server has to be deployed in the existing network, which causes a scalability overhead. This solution has been tested between 802.3 and 802.11 networks and found that it is not feasible to test vertical handovers between 802.11 and GPRS networks without deploying the handover server entity in the existing network.

Previously mentioned work has focused on a vertical handover system, but [18,19] presented the vertical handover performance with several performance metrics. Chakravorty et al. [18] conducted MIPv6 handover experiments in integrated GPRS and WLANs. They analyzed the handover impact on active TCP flows during the MN's handovers from GPRS to WLAN and vice versa. It also provided network-layer handover optimization techniques, but there was no interaction between the L2/L3 layers to detect handover timing and execution for seamless vertical handovers. Ylianttila et al. [19] presented simulation results for average throughput and handover delay obtained during vertical handovers in IEEE 802.11 and GPRS/EDGE networks. They analyzed several factors in proportion to the handover process and proposed an efficient handover decision algorithm that relieves the undesired ping-pong effect in heterogeneous wireless networks.

11.4 Vertical Handover System for Heterogeneous Wireless Networks

Based on the loosely coupled architecture, we designed and implemented a vertical handover system that allows an MN to handover across different types of wireless networks in a manner that is completely transparent to the MIP stack while providing seamless connectivity. Since our realization of the handover system is implemented as a single virtual network interface in the protocol stack of the MN, it is apparently simple and scalable compared to the modification or extension to the network side. Furthermore, the virtual network interface hides all the details about the change of network interfaces currently being used from the MIP stack. Thus, the vertical handover is handled at the virtual network interface without the MIP stack's knowledge, with the result that end users can achieve seamless connectivity no matter which technologies they currently use.

11.4.1 System Architecture

The architecture of the vertical handover system is shown in Figure 11.3. As shown in the figure, there are two main units, the handover control unit and the device control unit (DCU). The handover control unit decides which technology the MN should use and when to perform a handover with the aid of DCU. For this purpose, the handover control unit gathers the status of the network connection, reflects several kinds of handover metrics, and selects a specific radio-access technology that will be used for subsequent data communication. DCU provides interface switching among multiple network interfaces. The unit is located between the MIP stack and the network interface as a virtual network interface to provide seamless handovers to the MIP stack. This architecture offers the abstraction of a single virtual network interface to the operating system and supports multiple network interfaces according to different types of radio technologies.

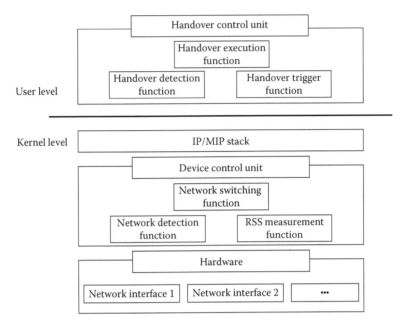

FIGURE 11.3
Vertical handover system architecture.

The functionalities of the handover control and the device control units are as follows:

- Network detection function assists the MN to learn which available technologies can be used for the MN's later data communication. It automatically detects available network interfaces when network interfaces are attached to the system and gathers network information through the attached network interfaces, such as radio access technology (e.g., cdma2000 1xEV-DO, HSDPA, WiMAX [20], WLAN, etc.).

- Received signal strength (RSS) measurement function periodically measures the current RSS of the serving BS and reports the information to the handover trigger function that will be used for a handover initiation trigger.

- Network switching function provides the end user with intra or inter-technology handovers according to the selected network interface. If this function is called, then the network interface is changed from the previous network interface to the new one by disabling the currently used network interface and enabling the requested network interface. It also configures proper parameters (e.g., MAC address, IP address, etc.) to the virtual network interface obtained from the new network interface.

- Handover trigger function gathers information from the RSS measurement function and fires a handover trigger to start a handover at the appropriate time.
- Handover detection function monitors MN's point of attachment and informs the handover execution function of the movement to a new network.
- Handover execution function actually performs the handover operation by cooperating with the handover detection function and the network switching function. When the handover detection function notifies a new link attachment, which satisfies a specific RSS threshold and a user preference, the handover execution function performs the MIP handover operation on the newly selected network interface with the aid of the network switching function.

11.4.2 Vertical Handover Signaling

Figure 11.4 shows the signaling diagram when an MN performs a handover from 3G to WLAN. When the MN detects a new available network (WLAN in this case), it measures the RSS on the WLAN. If the current RSS on the

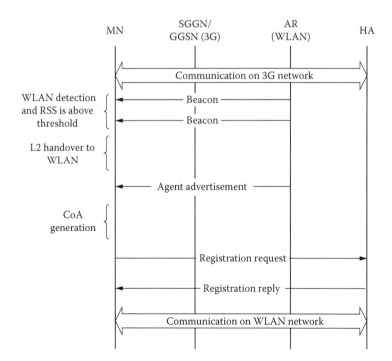

FIGURE 11.4
Handover signaling from 3G to WLAN.

3G network is below a certain threshold and the RSS on the WLAN is above a predefined threshold, the MN starts handover signaling on the WLAN. After the WLAN L2 handover operation, the MN begins its L3 handover operation when it receives an AA message from the WLAN AR. According to the newly received AA message, the MIP protocol detects its movement and configures a new CoA. Finally, it completes its MIP handover operation by exchanging Registration Request and Registration Reply messages with HA. After the registration procedure, MN resumes its data communication on the WLAN.

Figure 11.5 shows the signaling diagram when an MN performs a handover from WLAN to 3G. During communications on the WLAN, the MN detects that RSS on the WLAN is below a certain threshold and a new network is available (3G in this case). Then, the MN measures RSS on the 3G network and decides whether the measured RSS is above a predefined threshold or not. If RSS on the 3G network satisfies the threshold, then MN enables the 3G connection by exchanging PDP context request/accept messages with SGSN. When the 3G connection is enabled, the MIP detects its movement by exchanging AS and AA messages with GGSN. Next, MIP handover operations are followed. After the MIP handover signaling is finished, the MN starts its communication on the 3G network.

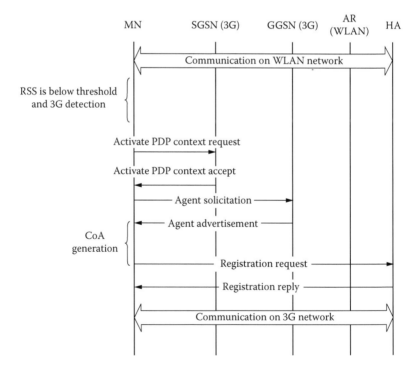

FIGURE 11.5
Handover signaling from WLAN to 3G.

11.4.3 Vertical Handover Operation

In this section, we describe the handover operation of the introduced system across heterogeneous wireless networks. The detailed operation can be described as follows:

1. *Network detection*: An MN may detect available wireless network technologies based on its network interfaces. When the MN is equipped with multiple network interfaces, the MN can obtain the information on neighboring wireless technologies as well as on one of the currently served technologies. For example, an MN equipped with 3G and WLAN interfaces moves around two wireless technologies as shown in Figure 11.2. In the figure, the MN is aware that there are two available access technologies by receiving signals from the BS and AP. From this function, the MN obtains and manages the currently available network lists, and uses the information during its handover process in the next function.

2. *Handover decision*: In heterogeneous wireless networks, there may be many handover decision metrics, such as the received signal strength (RSS), the end user preference, the QoS requirement, the service type, the communication cost, etc. Since each wireless technology has its own characteristics and services, the metric can be dependent on used wireless technologies. In this chapter, we only consider the end user preference and RSS as the most primary handover metrics, since we focus on measuring the handover performance. In order to obtain the end user preference, we simply receive the manual input from the end user at the user level. The RSS information, on the other hand, is gathered from the RSS measurement function, which constantly monitors the network status through MN's network interfaces.

3. *Handover initiation*: Once the MN learns a new available network from the handover detection function, the MN estimates the right time to start the MIP handover operation (MIP registration procedure). For the vertical handover decision, the handover trigger function uses the end user preference and RSS information. When the measured RSS of the currently serving BS (AP) drops below a predefined threshold, the detected network satisfies the end user preference, and the measured RSS of the detected network is above a certain threshold, the handover trigger function gives a handover trigger to the handover execution function.

4. *Handover execution*: When the handover execution function receives the handover trigger from the handover trigger function, it starts the MIP handover operation. When the MN starts the MIP handover operation between DCU and MIP, there are two integration design approaches, basic and tight integration designs, depending on whether DCU closely

cooperates with the MIP or not. We provide a detailed discussion about the integration designs in Section 11.4.4. In this section, we describe the handover execution process based on the basic integration design. In the basic integration design, the MN first cuts down the current network correction, and then it switches from the previous network interface to the newly selected network interface. After the new network interface is enabled, the MN starts the MIP handover signaling and its data communication on the newly selected network interface.

The vertical handover operation is summarized in Figure 11.6. First, the MN obtains the available network list from the network detection function. Then, the MN periodically monitors the RSS of currently associated BS and determines whether the RSS is below the predefined threshold or not. If the RSS is lower than the threshold, the MN selects the best BS based on the end user preference. For the best BS selection, the MN measures the RSS and gives the handover trigger to the handover execution function if the measured RSS is above the predefined threshold. As a result, the handover execution function performs MIP handover signaling based on the selected integration design between DCU and MIP. If the basic integration design is operated, the MN performs a switching from the previous network interface to the newly selected network interface and MIP handover signaling on the selected network interface. Once the MN finishes its handover signaling,

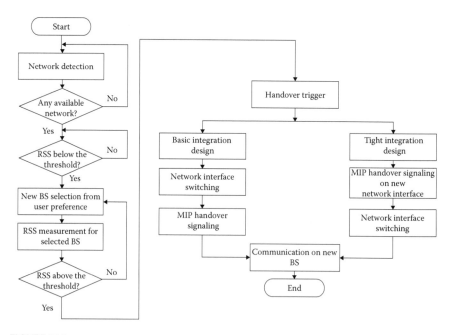

FIGURE 11.6
Vertical handover operation.

the MN resumes its communication through the newly selected BS. In the tight integration design, the MN starts its MIP handover operation before its actual network switching from the previous network interface to the new network interface that would result in less handover disruption period than the basic integration design. Since the MN uses multiple network interfaces simultaneously in the tight integration design (one for the data communication and one for the MIP handover signaling), we can expect that the MN experiences less handover latency than the basic integration design.

11.4.4 Integration Designs between DCU and MIP

In this section, we describe two integration designs between DCU and MIP in detail. The basic integration design is to make the network interface switching of DCU transparent to the MIP operation. In this case, DCU initiates a vertical handover by simply switching the network access technology (via the network switching function) among multiple network interfaces. During the interface switching, DCU enables the newly selected network interface operation while it disables the previously enabled network interface in order to prevent undesired ping-pong handovers among multiple networks. Since the MN receives several different AA messages by enabling multiple network interfaces at the same time, the MN might be under a puzzled state. Thus, the DCU enables a single network interface at a time in order to provide a simple inter-technology handover without the cooperation of MIP with the result that the MN can experience the same handover disruption period as that in horizontal MIP handovers. If some mechanism handles multiple AA messages and uses multiple network interfaces together during handovers, then the vertical handover latency can be further reduced compared to the horizontal one.

For this, we introduce another integration design, tight integration design, where DCU tightly cooperates with MIP. In this design, DCU inter works with MIP to maintain seamless connectivity during a vertical handover by triggering the MIP handover operation on the new network while maintaining the connection to the previous network. DCU reports its network switching event to the MIP stack so that the MIP stack prepares the impending handover event. With this report, the MIP stack performs its handover signaling on the newly selected network interfaces before the MN loses its current connection on the previous network. Compared to the basic integration design, this design enables simultaneous multiple network interfaces. In order to prevent unnecessary ping-pong handovers, the tight integration design introduces an AA filtering mechanism. During the AA filtering activation, DCU checks the header information of the packet received on the previous network interface. If the received packet is an AA message, then DCU filters the packet and does not forward it to the MIP stack. Since DCU only watches control messages on the previous network interface, the overhead will be small. Note that the MIP stack only sees a virtual network interface of DCU due to its transparency to the upper IP/MIP layer, which implies that the MIP stack is not aware of the fact that multiple network interfaces are currently enabled by DCU.

Due to the concurrent activation of multiple network interfaces, DCU should decide which network interface to use depending on the type of outgoing packets. If the MN is the receiver, then only the outgoing packet is the MIP registration message that should be sent on the new network interface during an impending handover. The data packets destined to the MN are received by the previous network interface until the binding cache of HA is updated. Once the binding cache is updated, the data packets to the MN are tunneled to the MN's newly visiting network. When the MN receives tunneled packets via the new network interface, DCU disables the previous network interface. If the MN is the sender, then the MN should use the previous network interface for its outgoing packets except the one for the MIP registration until the MIP registration finishes. After finishing the MIP registration, DCU disables the previous network interface. In this way, the MIP stack proceeds with its handover operation by using the new network interface in advance, while the MN maintains its connectivity to the previous network via the previous network interface. As explained above, the tight integration design does not interrupt the MN's current communication and the MN rarely experiences packet loss due to the MIP handover. It is obtained by the best utilizing multiple network interfaces and the close interaction between DCU and MIP at the MN. By closely coupling DCU with MIP, the handover performance can be improved and seamless connectivity across different types of networks can be provided.

11.5 Performance Evaluation

In this section, we present the handover performance of the introduced system for two cases: handovers between HSDPA/WLAN and cdma2000 1xEV-DO/WLAN. For each case, we measured the handover performance of the basic integration and the tight integration designs of DCU for the VoIP traffic and FTP application.

11.5.1 Handover Performance between HSDPA and WLAN

Figure 11.7 shows our experimental setup of integrated HSDPA and WLANs based on the loosely coupled architecture. The cellular HSDPA network infrastructure is the KT production network in Korea. To access the HSDPA network, the MN uses a USB type HSDPA modem, which supports a maximum data rate of 7.2 Mbps for a downlink and 384 kbps for an uplink. The WLAN

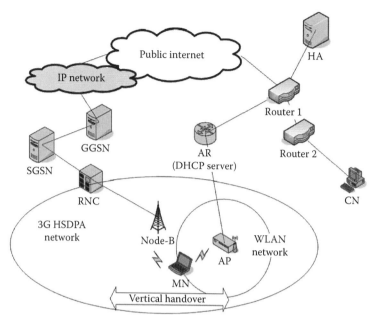

FIGURE 11.7
HSDPA-WLAN test-bed.

access is offered by IEEE 802.11b and the MN is equipped with an 802.11b PCMCIA-based card.

To support the MN's mobility between HSDPA and WLANs, HA is implemented in the experimental network. HA and the MN in the test-bed network operate on the top of the Linux operating system and use dynamics MIP that has been developed by the Helsinki University of Technology (HUT) [21]. HA is configured to send AA messages in every 1s (recommended setting in the standard). The access router (AR) also operates as a DHCP server for the MN, which implies that the MN obtains a co-located care-of address (CCoA) when the MN moves to the WLAN. Since we cannot modify some entities to support MIP in the HSDPA network, we used the CCoA mode operation of MIP during an MN's handover. Note that we only report performance results of handovers from WLAN to 3G in this experiment since handovers from 3G to WLAN are usually lossless.

Figures 11.8 and 11.9 show the performance of vertical MIP handovers for two integration designs of DCU when an MN changes its connectivity from the WLAN to HSDPA network, as shown in Figure 11.7. In both cases, we measured the end-to-end delay of each UDP packet and the number of lost packets during handovers by using Iperf [22]. The UDP packet size is 256 bytes and each packet is generated at a 20ms interval by the correspondent node (CN) to emulate the properties of VoIP traffic.

Figure 11.8 shows the snapshot of the measured end-to-end packet delivery for the basic integration design case when the CN sends UDP packets to

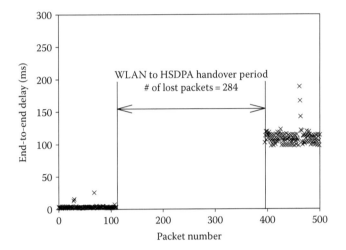

FIGURE 11.8
UDP handover performance from WLAN to HSDPA for the basic integration design.

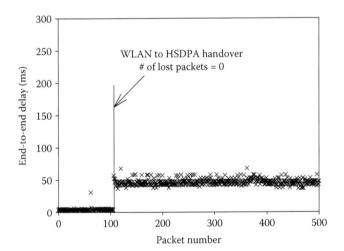

FIGURE 11.9
UDP handover performance from WLAN to HSDPA for the tight integration design.

the MN. In Figure 11.8, we observed that the MN loses 284 packets (from the 111th to the 394th packets) during the MIP vertical handover. It implies that the vertical handover latency from WLAN to HSDPA takes about 5.68 s since the packet generation interval is 20 ms in this experiment. The long handover latency of the MIP vertical handover is due to the fact that the MN must establish a HSDPA connection for L2 handover, and then perform the MIP handover operation (agent discovery, CoA acquisition, and registration procedures) for L3 handover. Since, in the basic integration design,

DCU does not simultaneously utilize multiple network interfaces, but uses only one interface at a time, it suffers from many packet losses during a vertical handover. The average end-to-end delay in the HSDPA network is around 110 ms and the packets whose delay ranges between 145 and 200 ms are mainly due to buffering [23] offered by current HSDPA networks. Most 3G networks provide a substantial amount of buffering (for every HSDPA mobile device) in their SGSN/GGSN gateways due to their low bandwidth. Thus, downlink packets are queued up at the SGSN/GGSN node and the end-to-end delay is varied during the vertical handover.

Contrary to the basic integration design, the tight integration design presents better handover performance as shown in Figure 11.9. In this case, there is no packet loss during the vertical handover from WLAN to HSDPA since DCU performs switching to the new network while maintaining the connection to the previous network (WLAN in this case). Thus, the MN can seamlessly communicate with the CN using the previous network while it completes the MIP handover operation using the new network. Until the handover finishes, DCU maintains the previous network (WLAN) as well as the new network (HSDPA network). It implies that DCU intelligently manipulates multiple network interfaces and coordinates them for heterogeneous wireless networks.

We have also studied TCP handover performance using the HSDPA-WLAN test-bed. To do so, a file whose size is about 25 MB is transferred from the CN to the MN during the MN's vertical handover. The CN uses TCP Reno and the MN uses the delayed ACK scheme in our experiments. We collected tcpdump traces of vertical handovers in the CN as well as the MN to trace individual TCP connections for each of the integration designs.

Figures 11.10 and 11.11 show the TCP performance of a vertical handover for the two integration design cases of DCU when the MN changes its

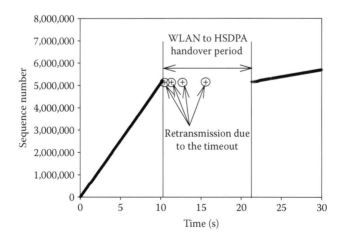

FIGURE 11.10
TCP handover performance from WLAN to HSDPA for the basic integration design.

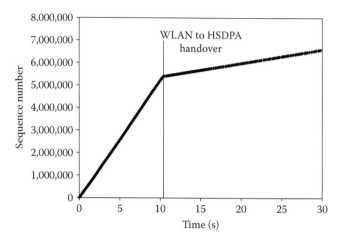

FIGURE 11.11
TCP handover performance from WLAN to HSDPA for the tight integration design.

connectivity from WLAN to HSDPA network. Each handover performance is analyzed in detail as follows. As shown in Figure 11.10, we can observe that it takes around 10 s to handover from the WLAN to HSDPA network by the basic integration design. In this case, the TCP data session backs off at the CN, retransmitting fourth before an ACK is available from the HSDPA network. The latency during the handover from WLAN to HSDPA is caused by TCP timeouts in addition to the L2 and L3 handover operations. Although the IP connection is already established in the HSDPA network (that is the MN finishes its handover operation), the MN cannot receive TCP data segments until TCP transmission timeouts expire. Thus, in the basic integration design, the MN experiences TCP segment losses during the handover from WLAN to HSDPA due to the MIP handover latency and residual TCP retransmission timeouts. Since the retransmission timeout event is critical in the TCP handover performance as described in Ref. [24], the MN suffers from low TCP handover performance in the basic integration design.

In contrast to the basic integration design, the MN does not experience TCP retransmissions during the vertical handover by the tight integration design as shown in Figure 11.11. In the tight integration design, the MN reduces MIP handover latency during the vertical handover from the WLAN to HSDPA network by simultaneously using multiple network interfaces. In the tight integration design, the MN achieves significant handover performance improvement over the basic integration design by the aid of DCU. That is, the tight integration design shows less handover latency, fewer number of TCP segments losses, and a higher TCP throughput during vertical handovers among heterogeneous wireless networks.

11.5.2 Handover Performance between cdma2000 1xEV-DO and WLAN

Figure 11.12 shows our experimental setup of integrated cdma2000 and WLANs based on a loosely coupled architecture. The cellular cdma2000 1xEV-DO network infrastructure currently in use is the SK Telecom production cdma2000 network in Korea. To access the cdma2000 network, the MN uses a USB type of the cdma2000 modem, which supports a maximum data rate of 2.4 Mbps for downlink and 153 kbps for uplink. Except the cdma2000 1xEV-DO, other entities are the same as those used in the integrated HSDPA and WLAN test-bed.

In this experiment, the UDP packet size is 256 bytes and each packet is generated at 50 m intervals by the correspondent node (CN). Figure 11.13 shows the end-to-end delay performance for UDP traffics of the handover case from WLAN to cdma2000 network. We can see that the MIP vertical handover performance is almost the same as that from WLAN to HSDPA. The basic integration design still shows lots of packet losses during the handover as shown in Figure 11.13. There are 21 packet losses (from the 73rd to the 93rd) and the handover latency is around 1.05 s since the packet generation interval is 50 ms in this experiment. Since the MN does not simultaneously use multiple network interfaces for the vertical handover as discussed earlier, the MN experiences packet losses during the handover in the basic integration design.

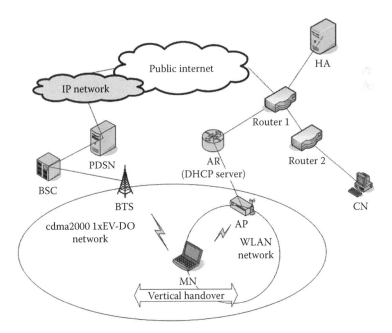

FIGURE 11.12
cdma2000-WLAN test-bed.

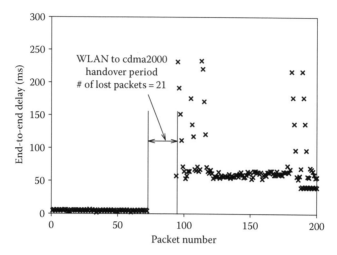

FIGURE 11.13
UDP handover performance from WLAN to cdma2000 for the basic integration design.

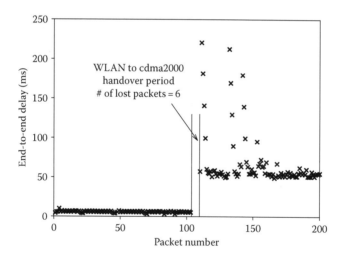

FIGURE 11.14
UDP handover performance from WLAN to cdma2000 for the tight integration design.

On the other hand, the tight integration design shows better handover performance compared to the basic integration design as shown in Figure 11.14. Since the tight integration design supports multiple network interfaces simultaneously with the aid of DCU, the MN uses its previous network interface (WLAN interface in this case) for data communication, while the MN performs L2 and L3 handover operations using the cdma2000 network interface. In the tight integration design, we expect no packet loss during

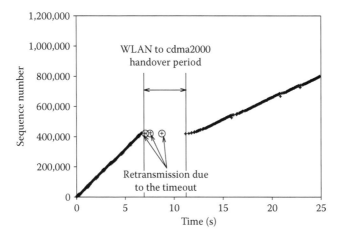

FIGURE 11.15
TCP handover performance from WLAN to cdma2000 for the basic integration design.

the vertical handover. However, there are some packet losses, around six packets (from the 104th to 109th packets) during the handover, and the handover latency is around 300 ms. When the MN switches to the cdma2000 interface after it finishes the MIP handover operation, the MN loses the buffered packets in the previous WLAN interface. Since DCU manages multiple network interfaces transparently, the buffered packets in each network interface are not managed and handled in DCU for low layer transparency. However, it still achieves better handover performance compared to the basic integration design.

Figure 11.15 shows the TCP handover performance of the basic integration design between WLAN and cdma2000 network. As shown in Figure 11.15, we can find that it takes about 4.5 s to finish the MIP handover operation and receive TCP segments. Due to the short RTO value and the long handover latency of the basic integration design, the MN experiences long handover latency and many TCP segment losses. In this experiment, the retransmission timeout occurs three times during an MN's vertical handover. Compared to the basic integration design, the MN experiences less TCP segment losses during the vertical handover by the tight integration design. Figure 11.16 shows the TCP handover performance of the tight integration design. In. Figure 11.16, it achieves significant handover performance improvement over the basic integration design with the aid of DCU, even though the MN experiences TCP timeouts during the vertical handover due to different network characteristics between WLAN and cdma2000. That is, the tight integration design shows less handover latency, fewer number of TCP segments losses, and a higher TCP throughput during vertical handovers among heterogeneous wireless networks.

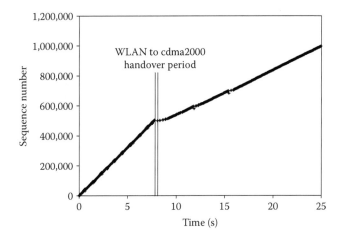

FIGURE 11.16
TCP handover performance from WLAN to cdma2000 for the tight integration design.

11.6 Conclusion

Various wireless network technologies have been developed and they have distinct characteristics. Next-generation wireless networks will consist of a diverse set of wireless networks. One of the key issues in next-generation wireless networks is the mobility management scheme, which maintains seamless connectivity during handovers over heterogeneous wireless networks.

In this chapter, we introduced a client software module to support the integration of multiple wireless technologies, such as HSDPA, WiMAX, WLAN, etc. We also introduced a vertical handover system that offers seamless vertical handovers to end users across heterogeneous wireless networks without modifying or extending the network side. The benefit of modifying only the client side, rather than modifying or extending the network side, is a system that is simpler, scalable, and cost-effective. We described the design and the implementation of the system and measured its handover performance. The system is designed to provide transparent services to its upper layers, IP/MIP layers, and manipulates different types of wireless network interfaces in a single virtual network interface. Besides the basic integration design, we introduced the tight integration design where the operations of DCU and MIP are closely coupled.

To analyze the handover performance of the system, we presented an experimental internetwork mobility between HSDPA and WLAN and between cdma2000 1xEV-DO and WLAN. The experimental results show that the tight integration design provides seamless mobility during vertical handovers. In our experience, the tight integration design can mitigate MN's traffic flow with least disruptions during vertical handovers across heterogeneous wireless networks.

References

1. IEEE Std. 802.11-1999, IEEE standard for local and metropolitan area networks, Part 11, IEEE, New York, 1999.
2. S. Y. Hui and K. H. Yeung, Challenges in the migration to 4G mobile systems, *IEEE Communications Magazine*, 41(12), pp. 54–59, December 2003.
3. A. Salkintzis, C. Fors, and R. Pazhyannur, WLAN-GPRS integration for next-generation mobile data networks, *IEEE Wireless Communications*, 9(5), 112–124, October 2002.
4. M. M. Buddhikot, G. Chandranmenon, S. Han, Y.-W. Lee, S. Miller, and L. Salgarelli, Design and implementation of a WLAN/CDMA2000 interworking architecture, *IEEE Wireless Communications*, 41(11), 90–100, November 2003.
5. C. Perkins, IP mobility support, RFC 2002, IETF Standard Track, October 1996.
6. D. Johnson, C. Perkins, and J. Arkko, Mobility support in IPv6, RFC 3775, IETF Standard Track, June 2004.
7. R. Koodli, Fast handovers for mobile IPv6, RFC 4608, IETF Standard Track, July 2005.
8. H. Soliman, C. Castelluccia, K. El Malki, and L. Bellier, Hierarchical mobile IPv6 mobility management (HMIPv6), RFC 4140, IETF Standard Track, August 2005.
9. A. C. Snoren and H. Balakrishnan, An end-to-end approach to host mobility, in *Proceedings of the ACM MobiCom*, Boston, MA, August 2000.
10. L. Ma, F. Yu, and V. C. M. Leung, A new method to support UMTS/WLAN vertical handover using SCTP, *IEEE Wireless Communications*, 11(4), 44–51, August 2004.
11. J. Rosenberg, H. Schulzrinne, G. Camarillo, A. Johnston, J. Peterson, R. Sparks, M. Handley, and E. Schooler, SIP: Session initiation protocol, RFC 3261, IETF Standard Track, June 2002.
12. M. Stemm and R. H. Katz, Vertical handoff in wireless overlay networks, *ACM Mobile Networking and Application (MONET)*, 3(4), 335–350, 1998.
13. F. Siddiqui, S. Zeadally, and E. Yaprak, Design architectures for 3G and IEEE 802.11 WLAN integration, *Lecture Notes in Computer Science (LNCS)*, 3421, 1047–1054, April 2005.
14. I. Akyildiz, Y. Altunbasak, F. Fekri, and R. Sivakumar, AdaptNet: An adaptive protocol suite for the next-generation wireless Internet, *IEEE Communication Magazine*, 42(3), 128–136, March 2004.
15. M. Buddhikot, G. Chandranmenon, S. Han, Y. W. Lee, S. Miller, and L. Salgarelli, Integration of 802.11 and third-generation wireless data networks, in *Proceedings of the IEEE Conference on Computer Communications (INFOCOM)*, San Francisco, CA, April 2003.
16. S. Sharma, I. Baek, Y. Dodia, and T. Chiueh, OmniCon: A mobile IP-based vertical handoff system for Wireless LAN and GPRS Links, in *Proceedings of the IEEE International Conference on Parallel Processing Workshops (ICPPW)*, Montreal, Canada, August 2004.
17. L.-J. Chen, T. Sun, G. Yang, and M. Gerla, USHA: A simple and practical seamless vertical handoff solution, in *Proceedings of the IEEE Consumer Communications and Networking Conference (CCNC)*, Las Vegas, NV, January 2006.

18. R. Chakravorty, P. Vidales, K. Subramanian, I. Pratt, and J. Crowcroft, Performance issues with vertical handovers—Experiences from GPRS cellular and WLAN hot-spots integration, in *Proceedings of the IEEE Conference on Pervasive Computing and Communications (PERCOM)*, Orlando, FL, March 2004.

19. M.Ylianttila, M. Pande, J. Mäkelä, and P. Mähönen, Optimization scheme for mobile users performing vertical handoffs between IEEE 802.11 and GPRS/EDGE networks, in *Proceedings of the IEEE Global Telecommunications Conference (GLOBECOM)*, San Antonio, TX, November 2001.

20. Std. 802.16–2001, IEEE standard for local and metropolitan area networks, Part 16, IEEE, New York, 2001.

21. B. Andersoon, J. Malinen, D. Forsberg, H. Kari, and J. Hautio, Dynamics mobile IP, available from http://dynamics.sourceforge.net/?paget=main.

22. NLANR/DAST, Iperf, available from http://sourceforge.net/projects/iperf.

23. TIA, TIA–835-E: cdma2000 Wireless IP network standard, 2010.

24. R. Caceres and L. Iftode, Improving the performance of reliable transport protocols in mobile computing environments, *IEEE Journal on Selected Areas in Communications*, 13(5), 850–857, June 1995.

12

A Framework for Implementing IEEE 802.21 Media-Independent Handover Services

Wan-Seon Lim and Young-Joo Suh

CONTENTS

12.1 Introduction

The integration of different communication technologies is one of the key features in next generation networks. In the integrated network environments, it is expected that users can access the Internet on an "anytime, anywhere" basis, and with better quality of service (QoS) by selecting the most appropriate interface according to their needs. Although network integration enhances user experiences, it raises several challenging issues such as candidate network discovery, call admission control, secure context transfer, and power management for multimode terminals.

There have been several standard group activities to handle those issues in integrated heterogeneous networks. For example, the integration of 3GPP and non-3GPP accesses (e.g., CDMA 2000, WLAN, WiMAX) has actively been studied by the 3GPP consortium [1]. In addition, the integration of 3GPP2-WLAN and HyperLAN-3G systems has been addressed in [2] and [3], respectively. However, these solutions designed based on specific networks cannot be directly adapted to different types of networks.

The IEEE 802.21 standard provides *media-independent* mechanisms that facilitate handovers in heterogeneous networks including both IEEE 802 and non-IEEE 802 networks [4]. The core entity of IEEE 802.21 is the media-independent handover function (MIHF), which acts as an intermediate layer between media-specific link layers and upper layers. MIHF is located in both mobile nodes (MNs) and network entities, and provides media-independent handover (MIH) services to upper layers. By using the MIH services, various user applications and mobility management protocols such as mobile IP (MIP) and session initiation protocol (SIP) can improve their performance in heterogeneous networks.

In this chapter, we first provide an overview of the IEEE 802.11 standard, and introduce an MIH framework for implementing MIH services. Although the IEEE 802.21 standard presents a general MIH architecture, it neither addresses the implementation details nor provides an indication of preferred implementations. In the introduced MIH framework, a centralized module called connection manager (CM) utilizes MIH services for seamless vertical handovers and efficient network discoveries. To evaluate the introduced MIH framework, we implemented it in integrated IEEE 802.11/802.16e networks.

12.2 Overview of IEEE 802.21

12.2.1 MIH Architecture

The IEEE 802.21 MIH services have been designed for enabling seamless service continuity among different types of networks including 3GPP, 3GPP2, and the IEEE 802 standard family. Figure 12.1 shows an example of network model with MIH services. In the model, an MIH-capable MN supports multiple access technologies and Access Network-3 couples the MN to the home core network. The MIH point of service (MIH PoS) is a network entity that exchanges MIH messages with the MN. (A network entity that does not exchange MIH messages with MN is referred to as MIH non-PoS.) MIH PoS may include an information server in order to provide useful network information such as costs, provider identification, provider services, and priorities. Note that the PoS location varies according to the deployment scenario and technology-specific reasons. MIH point of attachment (PoA) is the network side endpoint of a layer 2 link that includes an MN as the other

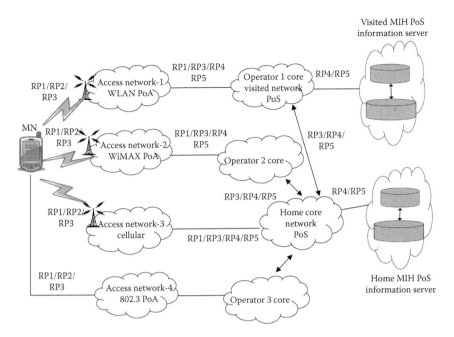

FIGURE 12.1
Example of network model with MIH services.

endpoint. The model also shows the five types of reference points (RPs) as follows:

- RP1: between the MN and an MIH PoS on the network entity of the serving PoA.
- RP2: between the MN and an MIH PoS on the network entity of the candidate PoA.
- RP3: between the MN and an MIH PoS on a non-PoA network entity.
- RP4: between an MIH PoS and an MIH non-PoS instance in different network entities.
- RP5: between two MIH PoS instances in different network entities.

Figure 12.2 illustrates the internal MIH architecture of MN, IEEE 802 network, 3GPP network, and other core networks. As shown in the figure, there is a central entity called MIHF in each node. MIHF is logically located above various media-dependent interfaces and provides abstracted services to upper layers through a media-independent interface. MIHF users, typically mobility management protocols, are abstractions of the functional entities that use the MIHF functionality. MIHF communicates with other entities such as MIHF users and lower layers through service access points (SAPs).

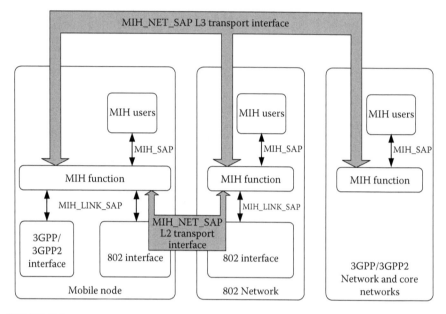

FIGURE 12.2
Internal MIH architecture.

In particular, Figure 12.2 shows three types of SAPs defined in the IEEE 802.21 standard as follows:

- MIH_SAP: it provides a media-independent interface for higher layers to control and monitor heterogeneous access links.
- MIH_LINK_SAP: it provides a media-specific interface for MIHF to control and monitor media-specific links.
- MIH_NET_SAP: it supports the exchange of MIH information and messages with a remote MIHF.

By using SAPs, MIHF provides a common view to MIH users across different media-specific links. MIH_LINK_SAP enables MIHF to obtain media-specific information, and the information can be propagated to MIH users through MIH_SAP.

12.2.2 MIH Services

The IEEE 802.21 standard defines three main services that facilitate handovers between heterogeneous links, which are as follows:

- Media-independent event service (MIES): it provides event classification, event filtering, and event reporting corresponding to dynamic changes in link characteristics, link status, and link quality.

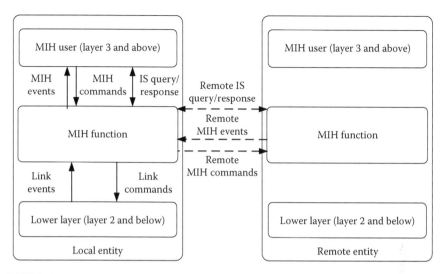

FIGURE 12.3
Communication flow of MIES, MICS, and MIIS between local and remote entities.

- Media-independent command service (MICS): it enables MIH users to manage and control link behaviors relevant to handovers and mobility.
- Media-independent information service (MIIS): it provides details on the characteristics and services provided by the serving and neighboring networks in order to enable effective system accesses and effective handover decisions.

Figure 12.3 shows a flow of MIES, MICS, and MIIS between local and remote entities. MIES is generated asynchronously, while MICS and MIIS are generated synchronously by a query/response mechanism. Now we present a detailed description of each service.

12.2.2.1 Media-Independent Event Service

MIES defines events that indicate or predict changes in state and transmission behavior of physical and link layers. Events may be originated from the local node or remote entity in the network. Remote events traverse across the network medium from one MIHF to a peer MIHF through the reference points RP1, RP2, and RP3. The event model is based on a subscription/delivery mechanism. Therefore, MIH users interested in an event type should register to the event. Local events are subscribed by the local MIHF within a single node, while remote events are subscribed by a remote entity and are delivered over a network.

There are two categories of events: link events and MIH events. Both link and MIH events traverse from a lower to a higher layer, and the relationship

TABLE 12.1

MIH Events

MIH Event Name	(L)ocal (R)emote	Description
MIH_Link_Up	(L, R)	L2 connection is established and link is available for use.
MIH_Link_Down	(L, R)	L2 connection is broken and link is not available for use.
MIH_Link_Parameters_Report	(L, R)	Link parameters have crossed a specified threshold and need to be reported.
MIH_Link_Going_Down	(L, R)	Link conditions are degrading and connection loss is imminent.
MIH_Link_Handover_Imminent	(L, R)	L2 handover is imminent based on either the changes in the link conditions or additional information available in the network.
MIH_Link_Handover_Complete	(L, R)	L2 link handover to a new PoA has been completed.
MIH_Link_PDU_Transmit_Status	(L)	Indicate transmission status of a PDU.

between them is shown in Figure 12.3. Link events are defined as events that originate from lower layers below MIHF, and they are delivered to MIHF. Within MIHF, link events propagate further to MIH users that have subscribed for specific events. MIH events are defined as events that originate from MIHF or link events that are propagated to MIH users via MIHF. Table 12.1 shows main MIH events where (L), (R), or (L, R), indicate whether the event can be subscribed by a local MIH user, a remote MIH user, or both, respectively.

12.2.2.2 Media-Independent Command Service

MICS provides commands to determine the status of links and control the physical and data link layers for optimal performance. Commands are sent from higher layers to lower layers, and can be invoked either locally or remotely by MIH users or by MIHF itself. The recipient of a command can be located within the protocol stack that originates the command or within a remote entity. Local commands are sent to MIHF by MIH users, and then are propagated from MIHF to lower layers. Remote commands may propagate from the local MIHF to MIHF in a remote entity via the R1, R2, R3, and R5 reference points. The information provided by MICS is dynamic information composed of link parameters such as channel quality and achievable data rate, whereas the information provided by MIIS is composed of less dynamic or static parameters such as network type, MAC or IP address, and higher layer service information. Information obtained via MICS and MIIS could be used in combination by MN and network to facilitate the handover.

TABLE 12.2

MIH Commands

MIH Command Name	(L)ocal (R)emote	Description
MIH_Link_Get_Parameters	(L, R)	Get the status of a link.
MIH_Link_Configure_Thresholds	(L, R)	Configure link parameter thresholds.
MIH_Link_Actions	(L, R)	Control the behavior of a set of links.
MIH_Net_HO_Candidate_Query	(R)	Network initiates handover and sends a list of suggested networks and associated PoA.
MIH_MN_HO_Candidate_Query	(R)	MN obtains handover-related information about possible candidate networks.
MIH_N2N_HO_Query_Resources	(R)	Sent by the serving MIHF entity to the target MIHF entity to allow for resource query.
MIH_MN_HO_Complete	(R)	MIHF of the MN indicates the handover completion to the target or source MIHF.
MIH_N2N_HO_Complete	(R)	Either source or target MIHF indicate the handover completion to the other.

There are two categories of commands, MIN commands and link commands, where the communication flow is shown in Figure 12.3. MIH commands are generated by MIH users and sent to MIHF in the local protocol stack or to MIHF in a peer protocol stack through MIHF transport protocol. For the best available network selection of MIH users under varying network conditions, a set of commands defined in the IEEE 802.21 standard. Table 12.2 shows main MIH commands. Link commands originate from MIHF, on behalf of the MIH user, and are directed to lower layers in order to control the behavior of the lower layers. Link commands are local only and should be implemented by technology-dependent link primitives.

12.2.2.3 Media-Independent Information Service

MIIS provides a unified framework to acquire neighboring network information within a geographical area to facilitate network selections and handovers. By using MIIS, an MN is able to acquire a global view of all heterogeneous networks in the area of interest of the MN. In contrast to the asynchronous push model of MIES, MIIS is based on a query/response mechanism. The information may be present locally in the MN, but it is usually in an external information server. MIIS provides a generic mechanism to allow a service provider and a user to exchange information on different handover candidate access networks, such as IEEE 802 networks, 3GPP networks, and 3GPP2 networks. Through MIIS, the MN can obtain information of various access networks from any single network. For example, by using a 3GPP interface, it may be possible to obtain information not only on all other 3GPP networks but also on all IEEE 802 and 3GPP2 networks in a given region without the need to power up each individual interface. This capability of MIIS allows optimal power saving in heterogeneous networks.

For MIIS, the standard defines information structures called information elements (IEs) and a query/response mechanism for information transfer. In particular, both resource description framework (RDF) [5] and type-length-value (TLV) format can be used to specify a media-independent way of representing information across different access technologies. IEs can be divided into the following groups:

- General information: These IEs provide a general overview of different networks providing coverage within an area.
- Access network-specific information: These IEs provide specific information for each access technology and operator.
- PoA-specific information: These IEs provide information about different PoAs for each of the available access networks.
- PoA-specific higher layer service: These IEs provide higher layer services and individual capabilities of different PoAs.
- Other information such as service specific or vendor/network specific can be added.

Table 12.3 shows main IEs defined in the standard.

TABLE 12.3

Information Elements

IE Name	Description
General information elements	
IE_NETWORK_TYPE	Link types of access networks that are available in a given geographical area.
Access network-specific information elements	
IE_NETWORK_ID	Identifier for the access network.
IE_COST	Indication of cost for service or network usage.
IE_NETWORK_QOS	QoS characteristics of the link layer.
PoA-specific information elements	
IE_POA_LINK_ADDR	Link-layer address of PoA.
IE_POA_LOCATION	Geographical location of PoA such as coordinate-based location information, civic address, and cell ID.
IE_POA_CHANNEL_RANGE	Spectrum range supported by the channel for that PoA.
PoA-specific higher layer service information elements	
IE_POA_IP_ADDR	IP Address of PoA.
Other information elements	
Vendor-specific IEs	Vendor-specific services.

The communication between different entities in order to gather information related to MIIS may be performed through all reference points defined earlier. Information can be stored in a network element referred to as MIIS server, which is located at the network side. The server maintains information on given access networks in its local database and responds to an information request.

12.3 Design of IEEE 802.21 MIH Framework

The standardization of IEEE 802.21 has been finalized recently. However, only little work has been performed for the performance evaluation of IEEE 802.21 services. There have been some studies on utilizing MIH services to enhance the performance of mobility management protocols such as mobile IPv6 (MIPv6) [6], fast mobile IPv6 (FMIPv6) [7], and proxy mobile IPv6 (PMIPv6) [8]. However, in those studies, the performance has been evaluated by a simple analysis, and thus there are no experiment results in real test beds. In this section, we introduce an MIH framework for utilizing MIES, MICS, and MIIS in real test beds. Then, we present use cases of the MIH framework, seamless vertical handovers, and efficient 802.11 AP discoveries in integrated 802.11/802.16e networks.

12.3.1 Overview of MIH Framework

Figure 12.4 shows an overview of the introduced MIH framework, which includes an MIH-capable MN, 802.11PoS, 802.16e PoS, and MIIS server. Note that although the framework is based on 802.11 and 802.16e networks, it can be easily extended to other networks such as 3GPP or 3GPP2. The MIHF module of the MN supports MIH services by interworking with device drivers

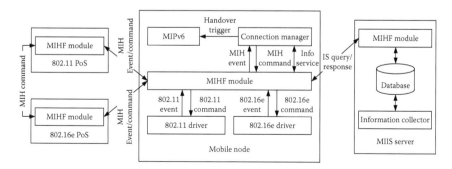

FIGURE 12.4
Introduced MIH framework.

in the local system, 802.11/802.16e PoS, and an external MIIS server. In the framework, both local and remote MIES/MICS are available. For supporting the local MIES and MICS, the MIHF module cooperates with device drivers for each network interface. When an MIH user sends a request of MICS to the MIHF module, it converts the MICS into a link-specific command message for 802.11 or 802.16e and sends it to an appropriate driver. When an event is generated from the device driver, the MIHF module converts the received link-specific event into an MIH event and sends it to the MIH user. For supporting the remote MIES and MICS, the MIHF module of MN communicates with the MIHF module of PoS. In this chapter, we focus on the local MIES and MICS and do not describe an internal structure of MIH PoS for supporting remote MIES and MICS.

In our approach, the device driver of each interface should be modified to support link events and link command services. One simple way for supporting event and link command services is using standard system calls provided by OS (e.g., Linux or Windows). For example, in the architecture proposed in [9], monitoring modules are installed in each network interface, which continuously monitors a specific interface through the system calls of Linux. If a link event is observed, the monitoring module sends it to the upper layers. This simple approach, however, causes the polling overhead and supports only limited events such as link up, link down, and change of signal strength. Moreover, via standard system calls provided by OS, most of the command services cannot be supported. Therefore, we have decided to modify the device driver of 802.11 and 802.16e network cards for supporting various link event and link command services. The modified driver can announce events actively without an explicit polling mechanism and support most of link commands defined in the standard.

MIIS is supported by interworking with an MIIS server that maintains the information on managed networks and provides neighboring network information to an MN. For MIIS, the IEEE 802.21 standard defines MIH_ Get_Information request and MIH_Get_Information response messages. When an MIH user generates a MIH_GET_Information request message, the MIHF module of the MN transfers the message to that of the MIIS server and waits for a MIH_GET_Information response message. Upon receiving a MIH_GET_Information request, the MIIS server retrieves the requested information from its local database. Then, the MIIS server sends a MIH_GET_Information response message to the MN. Since the method to gather information at the MIIS server is not defined in the standard, it is implementation dependent. We introduce an information collector (IC) module that obtains the information of the managed networks by periodic polling. The obtained information is saved to the local database of the MIIS server.

As a next step, we design the MIH user that exploits MIH services. As shown in Figure 12.4, we introduce a centralized module, called CM. In our framework, only CM becomes an MIH user. Even though each MIH user can directly use the MIH service as shown in Figure 12.3, this centralized approach can increase the flexibility of the system and minimize modifications on existing applications that want to utilize the MIH service.

In the proposed framework, CM has two main roles: seamless vertical handover and efficient 802.11 AP discovery. For seamless vertical handovers, CM discovers available networks and chooses the best candidate by utilizing MIH services. After deciding to perform a handover to a new access network, CM generates a handover trigger and sends it to the MIPv6 module. Note that although CM cooperates with MIPv6 in the example of Figure 12.4, other applications and mobility management protocols such as FMIPv6, PMIPv6, and SIP can also exploit the handover trigger. For an efficient 802.11 AP discovery, CM obtains the channel information of neighboring 802.11 APs via MIIS, and manages the scanning and switching operations of the 802.11 interface.

12.3.2 Seamless Vertical Handover

In heterogeneous networking environments, an MN that has multiple network interfaces may perform not only a horizontal handover (handover between cells of the same technology) but also vertical handover (handover between cells that are using different network technologies). For an efficient vertical handover, various factors, such as the received signal strength (RSS), user's preference among different networks, supported date rates, QoS mapping, and cost should be considered. Generally, users prefer a faster and cheaper network when the service coverage of multiple networks overlaps. For example, if both 802.11 and 802.16e networks are accessible, many users may want to use the 802.11 network.

Vertical handovers can be divided into two cases. Handovers from a network which supports smaller coverage with high data rates to a network which supports larger coverage with low data rates, and vice versa. Figure 12.5 shows a typical deployment of integrated 802.11/802.16e networks. The first case occurs when an MN moves out of the coverage of an 802.11 AP and the latter case occurs when it moves into the coverage of an 802.11 AP. Since these two types of handovers have distinct characteristics, we handle them differently in vertical handover operations.

Although MIPv6 is a mobility management protocol that has originally been designed for supporting horizontal handovers, it also has been widely accepted as a solution for vertical handovers. However, long latencies caused by MIPv6-based vertical handovers result in packet losses. To overcome this problem, cross-layer-based handover schemes that enable the MIPv6 stack to utilize the lower layer information have been proposed [10,11]. Although such cross-layer-based schemes show good performance in terms of handover latency, the remaining problem is that there is no predefined interface between the MIPv6 protocol stack and lower layers.

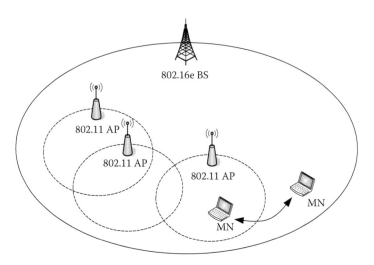

FIGURE 12.5
Vertical handover scenarios in the integrated IEEE 802.11/802.16e networks.

MIH services, which provide a media-independent interface to upper layers from various link specific lower layers, can be a good solution for solving the problem.

Now, we describe the vertical handover operation in conjunction with MIPv6 and explain why it causes long handover latency. Then, we introduce a new MIPv6 handover operation that utilizes MIH services. Our description of MIPv6 is based on MIPL, which is a widely used software program supporting MIPv6 in Linux systems [12]. Note that MIPL uses triggers from the MAC layer such as "link up" and "link down." These triggers are provided by the Linux kernel, and the usage of them is not specified in the MIPv6 standard.

Figure 12.6 shows the MIPv6-based handover operation from 802.11 network to 802.16e network. As shown in the figure, a handover operation is started when the 802.11 link is broken due to the MN's movement. In such case, a link down trigger is generated from the 802.11 MAC layer. If the MN detects a break of the 802.11 link through the link down trigger, then it tries to be connected to 802.16e link. After activating the 802.16e link, the MN generates a new care-of-address (CoA) based on a network prefix of the 802.16e network and performs the duplicate address detection (DAD) procedure. Note that if the MN maintains connectivity to the 802.16e link even when it uses the 802.11 interface, the link connection and the CoA generation procedure can be omitted. If a CoA is generated successfully, then the MN sends a binding update (BU) message to the home agent and waits to receive a biding acknowledgment (BA) message. After exchanging BU and BA messages, the MN can use the 802.16e network for data communications with the corresponding node (CN).

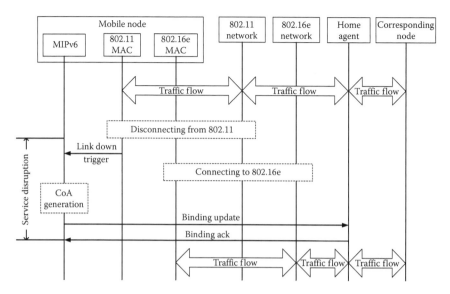

FIGURE 12.6
MIPv6-based handover from 802.11 to 802.16e.

The link down trigger, which is not considered in the MIPv6 standard, helps the MIPv6 module detect the disconnection of the 802.11 link faster. In the MIPv6 standard, the movement detection is achieved by receiving router advertisement (RA) messages that are periodically sent from the access router. Hence, the MN notifies that the 802.11 link is not available when it fails to receive a new RA within the lifetime of the past RA. Since the link down trigger is usually generated earlier than the expiration of RA, the handover operation can be performed faster. However, even if the MIPv6 module uses such MAC layer triggers, the packet loss during a handover cannot be eliminated completely. From the time of disconnecting the 802.11 network to the time of receiving the BA via the 802.16e network, the MN neither sends nor receives any data packets.

Figure 12.7 shows the MIPv6-based handover operation from 802.16e network to 802.11 network. When an MN moves into the service coverage of an AP, it tries to be connected to the AP by the association process defined in the 802.11 standard. If it finishes the association process to a new AP, then a link up trigger is generated from the 802.11 MAC layer. The following operations are similar to the operations in Figure 12.6. The MN configures a new CoA and exchanges BU and BA messages with its home agent. Note that if the 802.16e network completely covers the coverage of the AP just as the example of Figure 12.5, the MN can continuously use the 802.16e interface during the construction of new path. In such case, the MN can perform seamless handovers.

Now we introduce a new handover operation that utilizes MIH services. The aim of the MIH service–assisted handover is to guarantee service continuity without packet losses regardless of the handover's direction.

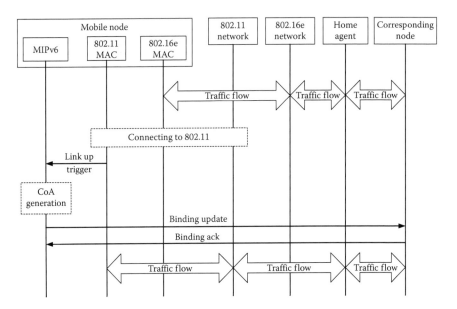

FIGURE 12.7
MIPv6-based handover from 802.16e to 802.11.

Figure 12.8 shows an MIH service–assisted handover operation from 802.11 network to 802.16e network. Note that all messages in the figure, except for BU and BA, are defined in the 802.21 standard. An MIH_Link_Parameters_Report is generated when one of link parameters has crossed the threshold set by an MIH_Link_Configure_Thresholds command. When CM receives an MIH_Link_Parameters_Report, it sends a MIH_Get_Information request frame to the MIIS server to obtain information on neighboring networks. The MIH_Get_Information request is sent to the MIIS server via the MIHF module. Upon receiving the MIH_Get_Information request, the MIIS server sends a MIH_Get_Information response frame that contains the neighboring network information such as type of network, PoA address, subnet prefix, IP configuration method, and data channel range. The MN can perform this information gathering operation anytime before an actual handover happens.

To eliminate packet losses during a handover, an MN should start a handover operation prior to the disconnection of the 802.11 link. As stated earlier, the 802.21 standard defines a MIH_Link_Going_Down event that is generated when the link condition is degraded and a connection loss is imminent. Upon receiving the MIH_Link_Going_Down at CM, the MN establishes a L2 connection to the 802.16e network and configures new CoA. Here, the network information obtained from the MIIS server can be used for a fast CoA configuration. After that, the MN exchanges BU and BA with its home agent just as in the legacy MIPv6 operation. If the MN receives BA before disconnecting from the 802.11 network as shown in Figure 12.8, then packet losses can be eliminated.

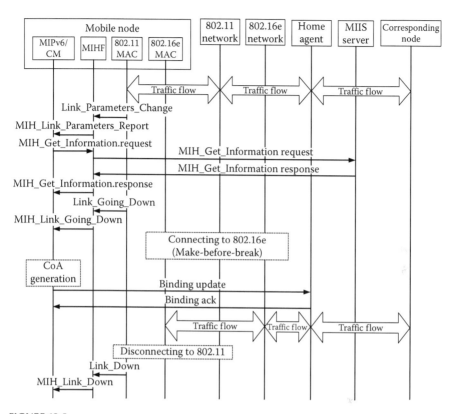

FIGURE 12.8
MIH service–assisted handover from 802.11 to 802.16e.

Figure 12.9 shows the operation of the MIH service–assisted handover from 802.16e network to 802.11 network. Since an MN can always access the 802.16e network due to its large service coverage, the packet loss during a handover is not an issue in handovers from 802.16e network to 802.11 network. The main differences between Figures 12.7 and 12.9 are that the MN obtains the information on the 802.11 AP from the MIIS server and tries to be connected to the 802.11 network by a new AP discovery scheme. Detailed operations of the new discovery scheme are introduced in Section 12.3.3.

12.3.3 Efficient AP Discovery

AP discoveries in 802.11 networks can be realized by means of the scanning operation. In the 802.11 standard, two types of scanning modes are defined: passive scanning mode and active scanning mode. In the passive scanning mode, an MN listens to each channel of the physical medium one by one for receiving beacon frames. After listening to all channels, the MN chooses the most appropriate AP according to the RSS of beacon frames. For the passive scanning mode, the MN waits for at least a period of the beacon interval,

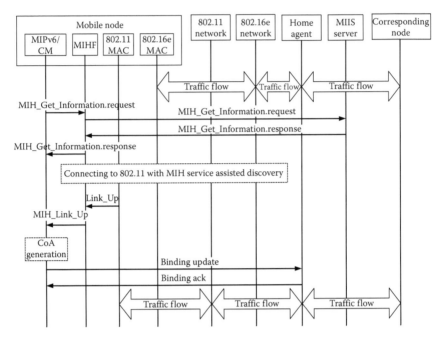

FIGURE 12.9
MIH service–assisted handover from 802.16e to 802.11.

before it switches to the next channel. Therefore, the passive scanning mode incurs significant delay.

On the other hand, in the active scanning mode, an MN transmits a probe request frame and waits for probe responses from APs. If no response has been received by *MinChannelTime*, then the next channel is scanned. Otherwise, if one or more responses are received by *MinChannelTime*, then the MN waits for more responses at most *MaxChannelTime*. Once all channels have been scanned, the MN chooses the most appropriate AP according to the RSS of probe responses.

It is typically assumed that an MN always turns on its WLAN interface and continuously performs scanning to find a new AP. Since APs are not always accessible due to their small service coverage as shown in Figure 12.5, such continuous scanning may cause an unnecessary use of energy. For an efficient power management of MN that is usually battery-powered, keeping the 802.11 interface turned on should be avoided. The most intuitive way to reduce such unnecessary power consumption is to switch on the 802.11 interface at a certain predefined time interval. When the 802.11 interface is turned on, all the available channels are scanned sequentially by passive or active scanning mode, and then it is turned off until the next wake-up time. This periodic switching scheme is quite simple but a more efficient discovery scheme can be developed with additional information by MIIS.

In the literature, several AP discovery schemes have been proposed which obtain information on 802.11 APs via protocol-specific broadcast messages or external servers. Choi and Choi proposed that an 802.16e base station (BS) periodically broadcasts the information on the density of 802.11 APs within its cell coverage [13]. By using the density information, the MN decides the scan interval for the 802.11 interface. Cao et al. proposed an AP discovery scheme that utilizes the 3GPP network to broadcast the channel information on WLANs [14]. Protocol-specific approaches in [13,14] need to modify the message format, and thus may cause compatibility problems with existing devices.

In this section, we introduce an efficient AP discovery scheme based on MIIS of the IEEE 802.21 standard. Compared to existing schemes in [13,14], one of the advantages of using MIIS is that an MN can obtain information on APs via a common interface regardless of the currently connected network type. Moreover, modifications on existing protocols such as the extension of broadcast messages can be avoided.

As discussed earlier, the MIIS server can provide information on neighboring APs to an MN through PoA-specific IEs in the MIH_Get_Information response. One useful IE for an AP discovery is TYPE_IE_POA_LOCATION, which contains the geographical location of PoA. If the MN obtains the location information, then it can turn on its 802.11 interface only when it moves into the coverage of APs. Even though this location-based discovery approach can significantly reduce unnecessary scanning operations, it needs an extra device such as global positioning system (GPS) to learn the current location of the MN. In the introduced scheme, an MN exploits the channel information via TYPE_IE_POA_CHANNEL_RANGE that represents the channel range of PoA.

Now we describe how an MN can discover APs efficiently by exploiting MIIS. As shown in Figure 12.9, while the MN uses the 802.16e network, CM of the MN sends the MIH_Get_Information request to the MIIS server. Upon receiving the MIH_Get_Information response from the MIIS server, which contains the information on neighboring APs, CM makes a list of valid channels. Then, the MN starts to discover a new AP based on the valid channel list.

By using a valid channel list, an MN can perform scanning for selected channels (selective scanning) rather than scanning for all channels (full scanning). Such selective scanning can be adapted to all existing discovery schemes for enhancing their performance. Here, we select a periodic switching scheme, which is simple and does not need extra cost. Figure 12.10 illustrates the timing diagram of the conventional periodic switching scheme. When the MN turns on its 802.11 interface, it scans all channels by active scanning or passive scanning mode. If there is no detected AP after a scanning phase, then the WLAN interface is turned off during the remaining time of a switching interval T. If one or more APs are detected, then the MN associates to a new AP and uses the 802.11 interface for data communications. As shown in Figure 12.10, we can observe that if we reduce the scanning time in a given switching interval T, then the 802.11 interface of MN will

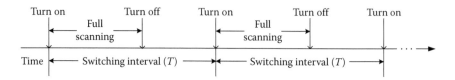

FIGURE 12.10
Timing model of conventional periodic switching scheme.

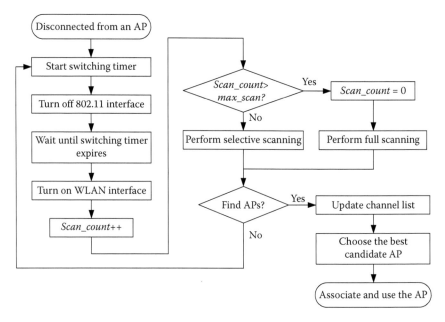

FIGURE 12.11
MIIS-assisted AP discovery scheme.

remain in the turn off state for a longer time and thus the MN can reduce the energy consumption. In addition, a shorter scanning time leads to faster AP detection when the MN moves into the service coverage of APs.

Figure 12.11 shows the overall operation of the introduced AP discovery scheme. After disconnecting from an AP, MN starts the "switching timer" with switching interval T and turns off its 802.11 interface. When the switching timer expires, the MN restarts the switching timer and increases the value of *Scan_count*, which is set to zero initially. After that, the MN turns on the WLAN interface and decides whether to perform selective scanning or full scanning. In case that *Scan_count* does not exceed the predefined threshold *Max_scan*, the MN performs selective scanning. After the scanning operation, if one or more APs are found, then the MN uses the WLAN interface for data transmissions and stops the switching timer. Otherwise, it turns off its WLAN interface again and waits until the next switching timer expires. If the MN fails to discover APs until *Scan_count* exceeds *Max_scan*, then it resets *Scan_count* to zero and performs a full scanning. This allows the MN to find APs on channels that are not announced by the MIIS server. If the MN finds APs on such channels, then it updates the channel list. It is worth noticing that the MIIS server cannot manage all running APs within its service area since numerous 802.11 APs are individually deployed and managed. The updated channel list is used in a later selective scanning.

12.4 Performance Evaluation

12.4.1 Implementation

In order to evaluate the performance of MIH services, we have implemented the proposed framework on a real test bed. An MN runs on the Linux 2.6.16 by applying a kernel patch for MIPL, and it has dual PCMCIA slots. It equips with both 802.11 interface and 802.16e interface cards. The 802.11 card used in the experiment is Proxim 8481-WD (802.11a/b/g/combo type) and the 802.16e card is Samsung SPH-H1100. The implementation of MN is composed of four main parts, the implementations of the MIHF module and CM, and the extensions of device drivers and the MIPv6 module.

The Proxim 8481-WD model supports the MadWifi driver, an open Linux kernel device driver for 802.11 chipsets from Atheros [15]. We have extended the MadWifi driver v0.9.2 to support various MIH events and commands services defined in the IEEE 802.21 standard. In addition, we also have extended MadWifi driver to support the selective scanning operation. Unfortunately, in contrast to the case of 802.11, there is no open source driver for the 802.16e. Therefore, we used the MIH-enabled 802.16e driver developed by Samsung Electronics, one of the major vendors developing 802.16e products.

Although MIHF is logically located between L2 layer and L3 layer, we have implemented the MIHF module at the user level for ease of implementations. It has multiple interfaces to communicate with other modules: CM, device drivers, and the MIIS server. Since CM runs on the user level, the MIHF module opens the UNIX domain socket for CM in order to provide the interface for MIH services. For interworking with 802.11 and 802.16e device drivers located at the kernel level, the MIHF module has the netlink socket interface and the ioctl (input/output control) interface that are provided by the Linux system. For supporting MIIS, the MIHF module has an IPv6 socket interface opened to the MIIS server. MIH_Get_Information request and response messages are exchanged through this IPv6 socket interface.

CM is a user-level application and it interfaces with the MIHF and MIPv6 modules. If CM decides to perform a handover according to the introduced handover operation, then it should send a handover trigger to the MIPv6 module. For this, the UNIX domain socket between CM and the MIPv6 module is opened. In addition, CM sends a valid channel list to the MadWifi driver through the ioctl interface for supporting the new AP discovery scheme. CM also handles the periodic switching of the 802.11 interface.

We used MIPL v2.0.2, the user-level implementation of MIPv6 as the MIPv6 module [12]. According to the introduced MIH framework, the MIPv6 module does not need to recognize or use MIH services, but it just receives a handover trigger from CM. Note that the legacy MIPL starts the handover procedure when the lifetime of a RA expires or the Linux kernel generates link up/down triggers. We have extended MIPL to recognize a handover trigger generated by CM. Upon receiving the handover trigger from CM, MIPL starts to exchange binding messages according to the normal MIPv6 operation.

Now we describe our implementation of MIIS server. The MIIS server has been implemented based on the Linux 2.6 kernel. In the MIIS server, MySQL-5.0.18 is installed for the database. The MIIS server opens two IPv6 sockets for the MIHF module and IC module. The MIHF module of the MIIS server uses the IPv6 socket for exchanging MIH_Get_Information messages with the MIHF module of MN. For an information gathering, IC module periodically sends an information request message to the managed entities in the 802.11 and 802.16e networks through the IPv6 socket. With such a polling mechanism, the MIIS server periodically updates the network information maintained in the database. To send information request messages to network entities, the MIIS server should have the basic information on the managed network entities, such as IPv6 address and polling interval. For this, we have implemented a management application that helps the server dynamically learn the basic information during runtime.

12.4.2 Experiments

First, we measured the performance of the introduced vertical handover operation. The network topology is shown in Figure 12.12. The 802.11 network consists of an AP and access router (AR), and the 802.16e network consists of a BS and access control router (ACR). The coverage of the AP is relatively small and it is completely included in the coverage of the 802.16e BS. During the experiments, the MN does not move at all. Instead, in order to simulate the movement of the MN, we gradually changed the transmission power of the 802.11 AP. When the transmission power becomes high, the MN detects that the signal from the AP becomes strong as it moves towards the AP. In contrast, when the transmission power becomes low, the MN detects that the signal from the AP becomes weak as it goes away from the AP. Therefore, the MN performs vertical handovers between 802.11 and 802.16e networks according the AP's transmission power level. For adjusting the transmission power level of AP, we used the host AP driver, a Linux driver for Prism2/2.5/3 chipsets [16]. Unlike the AP, the transmission power of the BS is fixed. Therefore, the MN always maintains the connectivity to the 802.16e network during the experiments.

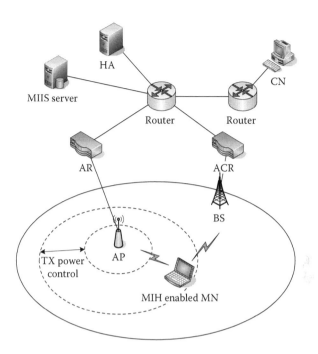

FIGURE 12.12
Network topology for vertical handovers.

We performed experiments to measure the number of packet losses in the legacy MIPv6 handover and the introduced MIH service–assisted handover. During the experiments, the MN sends a ping request message every 200 ms to the CN in the fixed network. If the MN maintains the L3 connection successfully, then it can receive a ping response message sent by the CN. We counted the number of lost ping response messages. In the handover from 802.16e network to 802.11 network, there is no loss regardless of using MIH services. However, when the MN performs a handover in the reverse direction, we can observe differences between the legacy MIPv6 handover and the MIH service–assisted handover. In the legacy MIPv6 handover, the MN loses about five messages, while there is no loss with the MIH service–assisted handover.

Now we analyze the handover results in detail. In the handover from 802.16e network to 802.11 network, when the MN is connected to the 802.11 AP, it waits for receiving a RA message from the AR in order to generate CoA. Upon receiving the RA, the MN generates a global scope address based on the network prefix contained in the RA and its MAC address. Then, to validate the uniqueness of the generated address, the MN broadcasts a neighbor solicitation (NS) message. If the MN fails to receive a neighbor acknowledgment (NA) message within a predefined time (the default value is 1 s), it uses the address as CoA, and sends a BU to the HA. When the HA replies to the MN by sending BA, the MN starts to use 802.11 network for

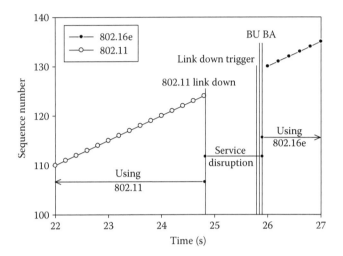

FIGURE 12.13
MIPv6 handover from 802.11 to 802.16e.

data communications. Since the MN can use the 802.16e interface until it completes the path set up to the 802.11 network by receiving the BA, the handover latency is almost zero regardless of whether the MN uses MIH service or not.

Figure 12.13 shows the measured result of the MIPv6 handover from 802.11 network to 802.16e network. We can observe relatively long handover latency, which is composed of the time to detect the link disconnection and the time to receiving BA from the HA. When the MAC layer of the MN detects the disconnection of the 802.11 link, it sends a link down trigger to MIPL. Then the MN exchanges BU/BA messages with the HA and starts to use the 802.16e network. Here, a CoA generation is not needed because the MN has been connected to the 802.16e network. It is important to note that there is a time gap between the link disconnection time and the detection time of the link disconnection at the MAC layer. Since the MAC layer of the MN determines that the 802.11 link is not available when it misses several beacon frames consecutively (e.g., 10 beacon frames in the Madwifi driver), it takes about 1 s to detect the link disconnection. In addition, the delay of exchanging BU and BA is about 100 ms since we performed experiments in a small-size test bed. As a result, the MIPv6 handover latency of the experiment is about 1.1 s, as shown in Figure 12.13. Note that in a large-scale network where the distance between the MN and the HA is far away, the service disruption time of the legacy MIPv6 handover may increase due to the long delay required for exchanging BU and BA.

Figure 12.14 shows the result of the MIH-assisted handover from 802.11 network to 802.16e network. Unlike Figure 12.13, the MIH_Link_Going_ Down event is generated when the link condition of 802.11 is degrading and packet losses are imminent. With the help of the MIH_Link_Going_Down

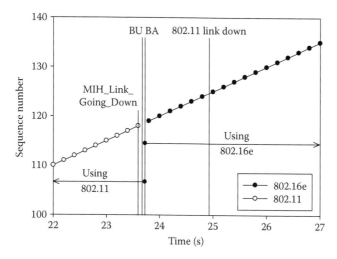

FIGURE 12.14
MIH service–assisted handover from 802.11 to 802.16e.

event, the MN can exchange BU and BA before the 802.11 link down. As a result, the handover latency becomes almost zero and packet losses are eliminated. In the MIH-assisted handover, it is important that the threshold for the MIH_Link_Going_Down event is adjusted properly according to the characteristics of the wireless channel and the movement of the MN. If the MIH_Link_Going_Down event is generated too late, then the MN cannot exchange BU and BA before the link disconnection, and thus it will experience the service disruption. Otherwise, if the MIH_Link_Going_Down event is generated too early, then the MN will stop using the current AP too quickly.

We have also evaluated the performance of the conventional periodic switching scheme described in Figure 12.10 and the introduced MIIS-assisted AP discovery scheme. The network topology for experiments of AP discovery schemes is shown in Figure 12.15. For ease of discussion, in this section, we call the two schemes as PSS and MAS, respectively. There are three 802.11b APs that operate in different channels. Although the MN's 802.11 interface is Proxim 8481-WD, which is 802.11a/b/g/ combo type, we set the channel mode of the MN to 802.11b mode since all APs are working with the mode. Therefore, the MN scans the channels for 802.11b (2.4 GHz) only.

We first measured the scanning time for the full scanning, which is used in PSS, and selective scanning which is used in MAS. Table 12.4 shows the experimental results. The selective scanning reduces scanning time significantly in both passive scanning and active scanning modes. In the case of full scanning, the MN scans all channels of 802.11b, and thus the scanning time increases significantly. In contrast, the selective scanning takes much shorter time since the MN scans only three channels used by the APs with the help of MIIS. We can also see that the active scanning takes shorter

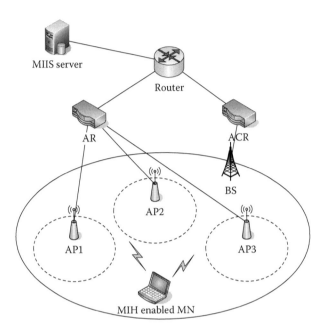

FIGURE 12.15
Network topology for AP discovery.

TABLE 12.4

Comparison of Scanning Time

	Full Scanning (ms)	Selective Scanning (ms)
Passive scanning mode	1771	331
Active scanning mode	762	133

time than the passive scanning. This is because the active scanning mode allows the MN to transmit a probe request frame immediately rather than passively waiting for the beacon frame from APs.

To compare the energy-consumption level of PSS and MAS, we placed the MN out of the service coverage of APs and measured the consumed energy according to the AP discovery. The wake-up interval for PSS and MAS is set to 10 s and the experiment is carried out during 1 h. Figure 12.16 shows the energy consumption of the MN. The results are obtained from the proc file of the Linux system located at "/proc/acpi/battery/BAT0/state." In both the scan modes, MAS consumes smaller energy than PSS due to the reduction in the number of scanned channels. Note that the experimental results include wasted time by turning on/off the interface. Therefore, the difference of energy consumption between PSS and MAS is less significant compared to the difference of the scanning time between the full scanning and selective

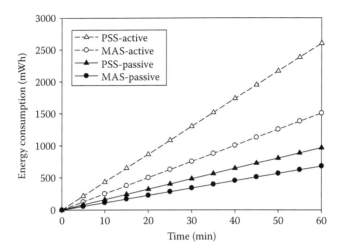

FIGURE 12.16
Comparison of energy consumption between PSS and MAS.

scanning in Table 12.4. However, this result is a remarkable enhancement for portable devices such as mobile phones and personal digital assistants (PDAs), which have limited battery capacity. An interesting result is that the energy consumption is advantageous to the passive scanning. The reason is that, although the passive scanning has longer scanning time than the active scanning, it just listens to the channel without sending a request frame.

We also measured the AP discovery time in the case where an MN enters into the service coverage of AP. To measure the AP discovery time, we fixed the MN within the coverage of one of three APs (target AP). Initially, all APs turn off their radio. After a random amount of time, the target AP turns on its radio to make the MN detect the signal. We measured the time from when the target AP turns on its radio to when the MN completes the association with the target AP. Figure 12.17 shows the average AP discovery time of 10 measurements as a function of the switching interval. We can see that MAS makes the MN to associate with a new AP more quickly than PSS.

12.4.3 Simulations

In order to evaluate the performance of the AP discovery scheme in various environments, we performed simulations by using the ns-2 simulator. The simulation topology is $3 \times 3\,km^2$ and one 802.16e BS covers the entire area. The number of APs varies from 5 to 25 and the MIIS server manages 80% of the APs. In other words, information on 20% of APs is not saved in the database of the MIIS server. The transmission range of APs is 250 m and channel numbers used by the APs are 1, 6, and 11. An MN has two types of network interfaces, 802.11 and 802.16e. During the simulations, the MN always uses the passive scanning mode. The total simulation time is 4 h and

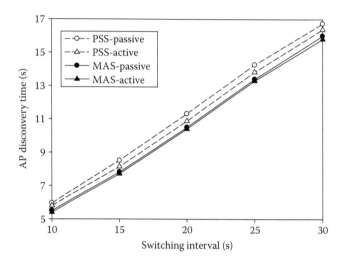

FIGURE 12.17
Comparison of AP discovery time between PSS and MAS.

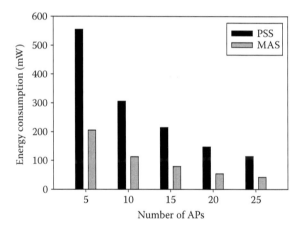

FIGURE 12.18
Average energy consumption by varying the number of APs.

the MN moves as the random way point model with pause time of 10 s and the average speed of 15 m/s. The switching interval is set to 10 s, and the value of *Max_scan* is set to 10. For simulations, we use two performance metrics: average energy consumption before connecting to the AP and the average AP discovery time which is defined as the time from entering the service coverage of an AP to discovering it.

Figure 12.18 shows the average energy consumption of PSS and MAS by varying the number of APs. In both PSS and MAS, the average energy consumption decreases with the increase of the number of APs. With the help of the selective scanning, MAS can save much more energy than PSS, especially

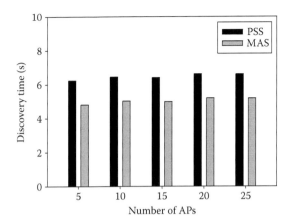

FIGURE 12.19
Average AP discovery time by varying the number of APs.

in case of 5 APs. Figure 12.19 shows the simulation results of the average AP discovery time. Similar to the results of Figure 12.17, MAS can reduce the average discovery time compared to PSS.

12.5 Conclusion

The IEEE 802.21 standard aims at enabling the optimization of handovers between heterogeneous IEEE 802 networks and facilitates handovers between IEEE 802 networks and non-IEEE 802 networks. The standard defines MIHF, which is located above various media-dependent interfaces and provides a single media-independent interface to upper layer entities, referred to as MIH users. MIHF provides three types of MIH services, MIES, MICS, and MIIS to MIH users. MIH users can exploit the services for the mobility management and handover processes. Even though the standard defines the MIH services and presents reference models, it does not provide implementation details. Moreover, in the literature, only a limited work has been carried out for the evaluation of the 802.21 framework in real test beds.

In this chapter, we provided an overview of the IEEE 802.11 standard. We also introduced an MIH framework for an implementation of the 802.21 standard, and then evaluated its performance in integrated 802.11/802.16e networks. The introduced MIH framework is composed of MIH-enabled MN, MIH PoS, and the MIIS server. In order to increase the flexibility of the system and to minimize modifications on existing modules and applications, we introduced a new entity named CM. CM is an MIH user of MN, which utilizes various MIH services. In this chapter, we introduced two functions of CM, seamless vertical handover and efficient 802.11 AP discovery.

We evaluated the introduced MIH framework in integrated 802.11/802.16e networks. The results showed that, with the help of MIH services, an MN can start its handover operation before the disconnection of the old link and thus packet losses and the handover latency can be reduced. The new AP discovery scheme of CM enhances the performance of MNs in terms of the AP discovery time and energy consumption.

References

1. 3GPP TS 23.402 ver.8.5.0, Architecture enhancements for non-3GPP accesses, Mar. 2009.
2. 3GPP2 S.R0087-0 v1.0, 3GPP2-WLAN interworking, state 1 requirements, July 2004.
3. ETSI, Requirements and architecture for interworking between HYPERLAN/3 and 3rd generation cellular systems, Technical Report, European Telecommunications Standards Institute, Sophia Antipolis, France, August 2001.
4. IEEE 802.21-2008, IEEE 802.21 standard and metropolitan area networks: Media independent handover services, IEEE, New York, Jan. 2009.
5. W3C Recommendation, RDF/XML syntax specification, http://www.w3.org/TR/rdf-sparql-XMLres
6. Q. B. Mussabbir and W. Yao, Optimized FMIPv6 handover using IEEE802.21 MIH services, *ACM MobiArch*, San Francisco, CA, Dec. 2006.
7. Y. Y. An, B. H. Yae, K. W. Lee, Y. Z. Cho, and W. Y. Jung, Reduction of handover latency using MIH services in MIPv6, *AINA*, Vienna, Austria, Mar. 2006.
8. L. A. Magagula and H. A. Chan, EEE 802.21 optimized handover delay for proxy mobile IPV6, *IEEE MILCOM*, San Diego, CA, Nov. 2008.
9. F. Cacace and L. Vollero, Managing mobility and adaptation in upcoming 802.21 enabled devices, *WMASH*, Los Angeles, CA, Sep. 2006.
10. M. Bernaschi, F. Cacace, and G. Iannello, Vertical handoff performance in heterogeneous networks, *International Workshop on Mobile and Wireless Networking*, Aug. 2004.
11. B.-J. Chang and S.-Y. Lin, Mobile IPv6-based efficient vertical handoff approach for heterogeneous wireless networks, *Wireless Communications and Mobile Computing*, 6(5), 691–709, Aug. 2006.
12. MIPL—Mobile IPv6 for Linux, http://www.mobile-ipv6.org/
13. Y. Choi and S. Choi, Service charge and energy-aware vertical handoff in integrated IEEE 802.16e/802.11 networks, *IEEE INFOCOM*, Anchorage, AK, May 2007.
14. Z. Cao, J. Jiang, and P. Fan, WLAN discovery scheme delay analysis and its enhancement for 3GPP WLAN interworking networks, *IEICE Transactions on Communications*, 90-B(6), 1523–1527, June 2007.
15. MadWifi—A Linux kernel device driver for wireless LAN chipsets from Atheros, http://madwifi.org/
16. Host AP driver for Intersil Prism2/2.5/3, http://hostap.epitest.fi/

13

Converged NGN-Based IPTV Architecture and Services

Eugen Mikoczy and Pavol Podhradsky

CONTENTS

The fixed mobile convergence (FMC) not only transforms technologies for delivery of the digital television but also helps to change the user from being a passive consumer of unidirectional broadcasted media toward using TV for active, interactive, mobile, and personalized bidirectional multimedia communication. The users expect that they are enabled to access any content, anytime, anyhow, anywhere, and on any device that they wish to be entertained with. The Next Generation of Networks (NGN) has been considered as a fully converged architecture that can provide a wide spectrum of multimedia services and applications to end users. The Internet Protocol Television (IPTV) represents a specific group of multimedia services including the television services, which are not only in the sphere of interest of the telecommunication technical community but also in that of the subscribers. This chapter is based on the FMC aspects of NGN-based IPTV architecture as the new directions for providing converged IPTV service. First and foremost, we provide an overview about evolution trends, related standardization, and architectures for NGN-based IPTV (called NGN integrated IPTV and IMS-based IPTV). The concluding part of this chapter focuses on the convergence of Mobile TV with IPTV or hybrid IPTV architecture (hybrid IPTV with terrestrial/satellite/cable broadcasting), which will lead us to the potential concept of Converged NGN–based IPTV [38].

13.1 Evolution Trends of Digital Television Toward IP Networks and IPTV

Digital television has evolved as recently as from the early 1990s. The history of television start of course earlier, probably from the time when humans had invented telegraphy over wire or radio transmition and from the time when they started to think of means of transmitting static images between locations. When inventors first achieved in transmitting the first images in the nineteenth century by trying to transmit them over wireline or subsequently wireless, they focused on delivering of moving pictures and, in fact, in bringing cinema technology to homes and living rooms. The TV service started in the last century as an analog terrestrial broadcasting service. Besides the first real television broadcasting, which started in the early twentieth century, there exist only a few inventions and technologies, which have changed our world and formed modern civilization. The digitalization of the communication, not only changed the voice services but also television. TV service, which started as terrestrial or satellite digital television, can now be delivered over new fixed and mobile broadband networks. The Internet as the other important invention of the last century affects the evolution of communication between people and in the delivery of multimedia content of Internet Protocol networks. The deployment of the IPTV over different broadband access networks was made possible because of the new types of broadband access networks and improved media coding algorithms [1]. The IPTV could be considered as the second generation of the digital television. ITU-T defines the IPTV through the following definition [2]: "IPTV are multimedia services such as television/video/ audio/text/graphics/data delivered over IP-based networks managed to support the required level of QoS/ QoE, security, interactivity and reliability."

The NGN-based IPTV recently developed by industry, academic community, and standardization institutions may be considered as the third generation of the digital television [1,3].

13.2 IPTV Standardization Overview

The first real IPTV solution was developed in the late 1990s mainly as streaming solutions over IP networks (using basic streaming control and transport protocols over IP networks) and later with deployment of IPTV-like services also by several operators. It was based on proprietary components and IPTV middleware. Although some important protocols such as RTSP, RTP, or IGMP existed from the beginning (which provide the basis for IPTV solutions). There has been no standardization of IPTV architecture and end devices or services that has been accepted worldwide. The main drawback for such solutions was that there had been limited possibilities for operators to combine different vendors and equipment to the complex IPTV architecture. The operator had to evaluate from the list of headend, middleware, content protection systems, networks equipment, and end devices, which could interwork and be integrated to the existing infrastructure.

The NGN have been originally defined as multiservice architecture, which could deliver television services, but in reality the first releases of NGN specification have been without the IPTV services. Section 13.2.1 describes most relevant standardization activities leading to NGN-based IPTV standardized architecture.

13.2.1 DVB

The Digital Video Broadcasting [DVB] Project is an industry-led consortium of broadcasters, manufacturers, network operators, software developers, regulatory to design open interoperable standards for the global delivery of digital media services (most of the specifications are published later as ETSI standards).

The DVB work, related to the IPTV is done on DVB-IPTV standards [DVB] (formerly DVB-IPI as DVB—Internet Protocol Infrastructure), could be found in the DVB-IPTV BlueBook [4], which provides a set of technical specifications to cover the delivery of DVB MPEG 2–based services over bidirectional IP networks, including specifications of the transport encapsulation of MPEG 2 services over IP, and also provide the protocols to access such services. Another important issue is solved in the DVB-IPTV specification [4] for the Service Discovery and Selection (SD&S) mechanism for DVB MPEG 2 based Audio/Video (A/V) services over bidirectional IP networks to define the service discovery information, its data format, and the protocols.

13.2.2 ITU-T

International Telecommunication Union [ITU-T] is in force on topics from service definition to network architecture and security, from broadband DSL to Gb/s optical transmission systems to NGN and IP-related issues, all together the fundamental components of today's information and communication technologies (ICTs).

ITU-T Focus Group IPTV evaluates the IPTV on various issues such as services, architecture, IPTV middleware, and security. ITU-T IPTV FG has been transformed to ITU-T IPTV SGI (Global Standards Initiative) [ITU-T SGI] and also another relevant study group as, for example, SG13 (responsible primarily for NGN). The IPTV documents are evaluated and finalized for inclusion to the next ITU-T NGN recommendations. The first standard published from the IPTV SGI is related to the NGN-based IPTV architecture: Recommendation ITU-T Y.1910 IPTV functional architecture [2].

13.2.3 ETSI

The European Telecommunications Standards Institute [ETSI] produces globally-applicable standards for Information and Communications Technologies (ICT), including fixed, mobile, radio, converged, broadcast, and Internet technologies. ETSI is officially recognized by the European Commission as a European Standards Organization.

Since its creation in 2003, ETSI TISPAN (Telecommunications and Internet converged Services and Protocols for Advanced Networking) [TISPAN] has been the key standardization body in creating the NGN specifications. NGN Release 1 was finalized in December 2005, which provided the robust and open standards for the first generation of NGN systems. The NGN Release 1 specifications adopt the Third Generation Partnership Project IP Multimedia Subsystem (3GPP IMS) standard for SIP-based applications, but also add further functional blocks and subsystems to handle the non-SIP applications. Initially, TISPAN worked on harmonizing the IMS core for both wireless and wireline networks (TISPAN NGN R1 with 3GPP R7 and TIPSAN NGN R2 with 3GPP R8). However, in early 2008, the common IMS specifications were transferred back to 3GPP so that only one single standard organization could take the responsibility for providing a Common IMS fitting of any network (fixed, 3GPP, CDMA2000, etc.). The NGN Release 2 was finalized in early 2008, and added a key element to the NGN such as the IMS and non-IMS-based IPTV, Home Networks and devices, as well as the NGN interconnected with Corporate Networks.

In TISPAN NGN there are several specifications that address the IPTV regarding service requirements [5] in stage 1 and architectures in stage 2 with non-IMS IPTV subsystem [6] as well as the IMS-based IPTV [7]. In stage 3, TISPAN IPTV specification deals with protocol and implementation details. We will focus on ETSI TISPAN NGN–based IPTV architecture and present more details in Section 13.4.

13.2.4 3GPP

The 3GPP unites telecommunications standard bodies worldwide (including ETSI, ATIS, ARIB, CCSA, TTA, and TTC). The original scope of 3GPP was to produce Technical Specifications and Technical Reports for a 3G Mobile System–based on evolved GSM core networks and the radio access. The 3GPP has published several releases (R99, R4, R5, R6, R7, R8, R9, and most recently R10), which describe some aspects of UMTS services, architectures, and protocols. The an IP Multimedia Subsystem (IMS) [8] (subsystem specified as part of 3GPP UMTS core network domain for providing multimedia service) that has already been accepted by the industry as the unified service control architecture for NGN (some other bodies like ITU-T or ETSI TISPAN also bring the concept of IMS-based IPTV, where IMS is also used for providing IPTV services). The 3GPP is now working on Long Term Evolution (LTE), which will be built up on UMTS, as the Industry looks beyond 3G as well as related concepts of Service Architecture Evolution (SAE) and Evolved Packed Core (EPC). 3GPP will contribute to the ITU-R toward the development of the IMT-Advanced via its proposal for the LTE-Advanced.

The 3GPP also specifies MBMS (Multimedia Multicast/Broadcast Services) specifications [9] mainly to define an efficient way to deliver and control multicast and broadcast services over 3G networks. MBMS are limited at this moment to Mobile TV channel bandwidth, which is up to 256 kb/s per MBMS Bearer Service. The Advantage of MBMS in comparison with the DVB-H is that the same IP-based common infrastructure is used for Mobile TV as for 3G data services.

13.2.5 IETF

The Internet Engineering Task Force [IETF] is a protocol engineering and development arm of the Internet. Most of the IPTV standards are in using the existing protocols defined by the IETF from the Internet Protocol (IP) version 4 or 6 [10,11] to the application protocols like the Session Initiation

Protocol (SIP) for session control [12], the Real Time Stream Protocol (RTSP) [13] for media control of CoD (content on demand) services as well as the Internet Group Management Protocol (IGMP) [14] for the IPv4 or Multicast Listener Discovery (MLD) [15] for the IPv6 multicast-based services. The Real Time Transport Protocol (RTP) is used for media delivery [16]. The Hypertext Transfer Protocol (HTTP) [17] is used for transfer of metadata or presentation of client graphical interface using the web browser technologies.

There are also several Internet drafts regarding the IPTV channel description to enable a unified IPTV service identification within the IETF.

13.2.6 Alliance for Telecommunications Industry Solutions

The Alliance for Telecommunications Industry Solutions [ATIS] is a standardization organization that develops technical and operational standards for the communications industry and is accredited by the American National Standards Institute (ANSI).

The ATIS initiates the IPTV Interoperability Forum (IIF) that develops the standards to enable the interoperability, interconnection, and implementation of the IPTV systems and services, including video on demand and interactive TV services [18]. The ATIS analyzes several important aspects of the IPTV within the IIF initiative and defines IPTV logical domains, IPTV reference architectures (IMS and non-IMS-based IPTV, and their coexistence) [19], content delivery concepts with quality of experience, digital rights management (DRM) requirements, interoperability standards as well as testing requirements for components, reliability, and robustness of service components.

13.2.7 CableLabs

Cable Television Laboratories, Inc. [CableLabs] is a non-profit research and development consortium founded by the cable operating companies dedicated to pursue new cable telecommunications technologies. The PacketCable is one of CableLabs-led initiative to develop interoperable interface specifications for delivering advanced, real time multimedia services over a two-way cable plant. Built on top of the industry's highly successful DOCSIS cable modem infrastructure (actually 3.0, with European adaptation EuroDOCSIS), PacketCable networks use Internet protocol (IP) technology to enable a wide range of multimedia services, such as IP telephony, multimedia conferencing, interactive gaming, and general multimedia applications. In the architecture of the PacketCable 2.0 [20] the IMS concept is utilized, too.

13.2.8 Open Mobile Alliance

The Open Mobile Alliance [OMA] introduces the concept service enablers to provide standardized components in order to create an environment in which services may be developed and deployed. The OMA enables the decomposition into these components and the interactions among them comprise the OSE (OMA Service Environment) framework architecture [21].

From the Mobile TV perspective, most relevant work in the OMA deals with the Mobile Broadcast Services Enabler [22] to address functional issues, which are generic enough to be common to many Broadcast Services that can be defined and implemented in a bearer-independent way. These functional issues are Service Guide, File Distribution, Media Stream Distribution, Service Protection, Content Protection, Service Interaction, Service Provisioning, Terminal Provisioning, Notification, and so on. Generally, it is expected that Mobile Broadcast Services should enable the distribution of rich, interactive, and bandwidth consuming media content to a large number of mobile audiences.

13.2.9 Open IPTV Forum

The Open IPTV Forum [OIPF] is a pan-industry initiative with the purpose of producing end-to-end specifications for the development of end-to-end solution to allow any consumer's end-device, compliant to the Open IPTV Forum specifications, to access enriched and personalized IPTV services [23] either in managed or a non-managed networks (non-managed is having no guarantee of QoS and Over Open Top—over Internet). The Open IPTV Forum focuses on standardizing the user-to-network interface (UNI) both for managed and non-managed network with their NGN-based architecture [24].

13.2.10 Home Gateway Initiative

The Home Gateway Initiative [HGI] is an open forum launched by Telecommunication providers with the aim to release specifications of the home gateway. In addition to telecommunication providers, several manufacturers have joined the alliance and actually provide specification for NGN capable of home gateways [25].

13.3 NGN-Based IPTV Architecture and Services

The major players in any IPTV delivery chain comprise content providers, service providers, network providers, and end-users. A content provider is a source of content, for example, TV stations, studios, content aggregators, etc. The IPTV platform, usually owned by service provider, has to provide all functions necessary for control and delivery of IPTV services over network infrastructure (network provider) to the end user.

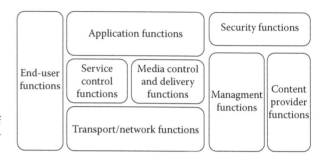

FIGURE 13.1
High-level architecture of NGN-based IPTV functional.

Main blocks NGN-based IPTV platforms are as the following points (Figure 13.1)

- Application functions
- Service control functions and User profiles
- Media control and delivery functions
- Supporting, management and security functions
- End user function

Application functions can include several service logics of the IPTV services, mechanisms for service discovery, and selection to find the right services and content as well as help to interact with other application and external systems.

Service control functions provide functionality for authentication and authorization of service requests. This function is also responsible for the setup and control of all the IPTV services. It can also reserve resources toward transport control functions.

User profiles contain user data and user profiles related to user's services.

Media control and delivery functions has received content and media streams from content provider and then control and provide media processing, media delivery, content storing, transcoding, and relaying of content.

End-user functions represent home network and user equipment, for example, end devices (e.g., TV with set-top-box, mobile, etc.) but also home networking part, including Home Access Gateways.

The greatest advantage of NGN-based IPTV architecture is the possibility to integrate IPTV services with other NGN services, reused existing NGN capabilities, better utilized resources, personalization of services, and mobility.

The NGN/IMS existing functions can be reused for providing following capabilities to the IPTV systems:

- User registration and authentication
- User subscription management
- Session management, routing, service triggering, numbering

- Interaction with the existing NGN service enablers (presence, messaging, group management, etc.)
- QoS and bearer control
- Mobility, FMC capability
- Charging and billing
- Security and management mechanisms

There are IPTV specific functions, which have to be additionally described as the following:

- Service discovery and selection, presentation, e.g., EPG
- Service and Content protection, e.g., DRM and CAS
- Service and Content management, managing the services and contents in the Content Provider domains, and/or the Service Provider domains
- Content distribution, delivery, and locating control
- Multicast support and control
- VCR control, e.g., play/pause/fast-forward/rewind

The producing of specification is usually defined by standardization bodies in three stages [2], [ETSI]:

1. Collect Service and system requirements, service use cases.
2. Define functional entities and architecture, reference points, service procedures.
3. Specify the implementation, signaling flows, protocols details.

In this section, the stress is given to ETSI TISPAN NGN Release 2 and 3 specification analyses. The ETSI TISPAN working group produces several stage 1 and 2 specifications for NGN-based IPTV:

- IPTV service requirements in ETSI TS 181 016 [5]
- NGN integrated IPTV subsystem architecture in ETSI TS 182 028 [6] (formally called NGN dedicated IPTV in R2)
- IMS support for IPTV architecture in ETSI RTS 182 027 [7]

The specifications focused on the implementation of the IPTV functions, interfaces, procedures, protocol recommendations (stage 3) have also been prepared for both the IPTV architectures:

- NGN dedicated/integrated IPTV in ETSI TS 183 064 [26]
- IMS-based IPTV in ETSI TS 183 063 [27]

TABLE 13.1

NGN-Based IPTV Architecture and Mapping of Functional Entities

Functional Block/Entities	TISPAN NGN Integrated IPTV [6]	TISPAN IMS–Based IPTV [7]	Converged NGN-Based IPTV
Applications functions	Client facing IPTV application (CFIA), service discovery and selection (SD&S)	Service control (SCF), service discovery (SDF), service selection (SSF)	IPTV Converged application function (ICAF), service discovery and selection (SDSF)
Service control functions	IPTV-control (IPTV-C)	IP multimedia subsystem (IMS)	IP multimedia subsystem (IMS), IPTV service control (ISCF)
Media control and delivery functions	Media control function (MCF), media delivery function (MDF)	Media control function (MCF), media delivery function (MDF)	IPTV MCF (IMCF), interconnect, proxy, serving MDF (I-/P-/S-IMDF)
User profiles	User profile server function (UPSF), IPTV user data function (IUDF)	User profile server function (UPSF)	User profile server function (UPSF)

The ITU-T NGN-based IPTV architecture is specified in recommendation of ITU-T Y.1910, namely, IPTV functional architecture [1]. Also other institutions like ATIS or Open IPTV forum are working on NGN-based IPTV. TISPAN describes in technical report [28] the mapping and interconnection with other NGN-based IPTV systems (Table 13.1).

13.3.1 NGN-Based IPTV Services

The main goal of NGN-based IPTV is to provide to the end user a comprehensive list of converged IPTV services. New generation IPTV services have to provide service personalization, interactivity, blending of services, user targeting, enhanced accessibility, and mobility (Table 13.2).

IPTV services can be split into three groups:

1. Basic IPTV services
2. Advance IPTV services
3. Converged IPTV services

1. Basic IPTV services comprise minimal set of IPTV services, which are expected from NGN-based IPTV service provider:
   a. Broadcast TV (with or without trick modes)—delivery of linearly broadcasted TV channels.
   b. Trick Modes—enable control playback and pause, forward, rewind content.

TABLE 13.2

Selected NGN-Based IPTV Services in ETSI TISPAN

Service and Feature	IPTV Requirements		NGN Dedicated/ Integrated IPTV Subsystem		IMS-Based IPTV	
ETSI TISPAN Specification	TS 181 016 [5]		TS 182 028 [6]		RTS 182 027 [7]	
TISPAN Release	R2	R3	R2	R3	R2	R3
Linear/broadcast TV (BC)	M	M	M	M	M	M
BC with trick play	M	M	M	M	O	O
Content on demand (CoD)	M	M	M	M	M	M
Push CoD	NA	O	M	M	NA	O
Near COD	M	M	M	M	NA	NA
Network PVR	M	M	M	M	M/O	M/O
Client PVR	NA	O	NA	O	NA	O
Service information (EPG)	M	M	M	M	M	M
Parental control	M	M	M	M	M	M
Pay-per-view	NA	M	M	M	NA	O
Interactive TV	NA	M	M	M	NA	O
Time shifted TV	O	O	O	O	O	O
User generated content	NA	M	NA	O	NA	O
Communications and messaging	NA	O	O	O	NA	O
Notifications	NA	O	O	O	NA	O
IPTV presence	O	O	O	O	O	O
Profiling and personalization	O	O	O	O	O	O
Content recommendation	NA	O	NA	O	NA	O
Targeted advertising	NA	O	NA	O	NA	O
Bookmarks (Content marking)	NA	O	NA	O	NA	O
Personalized channel	NA	O	NA	O	NA	O
Personalized service composition	NA	O	NA	O	NA	O
Remote control of IPTV service	NA	O	NA	O	NA	O
Games	NA	O	NA	O	NA	O
Emergency information	NA	O	NA	O	NA	O
Incoming call management	NA	O	O	O	NA	O
Interaction with third party application (e.g., Parlay)	NA	O	NA	O	NA	O
Interaction with Internet services	NA	O	NA	O	NA	O
Interaction with NGN services	O	O	O	O	O	O
Service portability	NA	O	NA	O	NA	O
Service continuation between IPTV UEs	NA	O	NA	O	NA	O
Service continuation fixed mobile	NA	O	NA	O	NA	O

Source: Draft ETSI TR 182 030 V<0.0.3> (2009-09), TISPAN: NGN based IPTV mapping or interconnect between IPTV systems, ETSI, Sophia Antipolis, France, 2009.

Note: M – Mandatory, O – Optional, NA – not available or not specified.

    c. Pay Per View (PPV)—user pays, for example, only for particular show or time period and not for all the TV channels or TV package.

    d. Content on Demand (CoD)—user requests content consumption on demand (e.g., Video on Demand or Music on Demand).

    e. Personal Video Recording (PVR)—user can record content in network (network or n-PVR) or locally in STB (client or c-PVR).

    f. Electronic Program Guide (EPG)—provides service selection information required by the viewer to find and select the programs to be watched.

    g. Parental Control—protection mechanism to limit access to television content for children's in age-below rating of program.

2. Advance IPTV services

    a. Profiling and personalization—features that enable personalized IPTV services based on user preferences and user profile. Provider can also use the information about user's behavior and content consumptions.

    b. User Generated Content (UGC)—content produced by the end user with the intention to share it with other users.

    c. Content Recommendation (CR)—service advisory for favorite shows based on user's preferences and behaviors.

    d. Time Shift TV (TsTV)—user can browse and play from past broadcasted content prerecorded by provider.

    e. Personalized channel (PCh)—user specific list of programs that are scheduled as playlist for personalized preview.

    f. Targeted advertising (TAI)—advertising mechanism, which is targeted to specified group of user based on his user profiles.

    g. Interactive TV (iTV)—service providing interactivity between provider/broadcaster and end user or between several users.

3. Converged IPTV services

    a. Convergence of the IPTV and other NGN Service and their interaction (e.g., presence-based game, incoming call notification, sharing the remote control).

    b. Interaction with third Party application (e.g., Parlay) enable interworking with third party developed applications.

    c. Interaction with Internet Services (e.g., convergence of IPTV with web 2.0 services and social media).

d. Service Continuation between IPTV UEs allows park service on one device and pick up and continue with service consumption on another one.

e. Service Continuation between Fixed Mobile and support for mobility/roaming, service accessibility across different access networks and terminals.

f. Remote control of IPTV services, for example, control of recordings or content recommendations.

g. Hybrid IPTV services (combination of Satellite/Terrestrial delivery with IPTV).

13.3.2 NGN-Integrated IPTV

The NGN integrated (previously in TISPAN Release 2 called dedicated) IPTV subsystem (as shown in Figure 13.2.) has to provide the basic integration of IPTV functions to NGN architecture and other NGN subsystems especially with central NGN database with user data and user profiles (UPSF—User

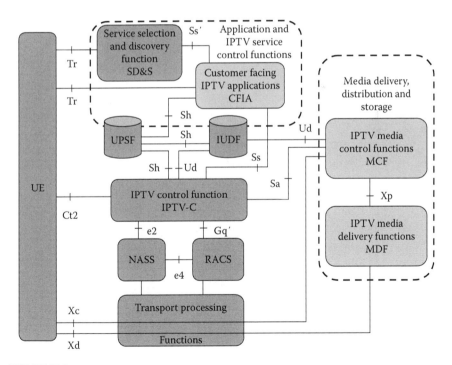

FIGURE 13.2
Simplified TISPAN NGN dedicated IPTV subsystem functional architecture. (From ETSI TS 182 028 V2.0.0 (2008-01) in R2, ETSI TS 182 028 V3.3.1 (2009-10) in R3 TISPAN, IPTV Architecture: NGN integrated IPTV subsystem Architecture, ETSI, Sophia Antipolis, France, 2009.)

Profile Server Function) and transport control subsystems for network attachment control (NASS—Network Attachment Subsystem) as well as resource control (RACS—Resources Admission Control Subsystem) [6].

The users can have (using with his user equipment—UE, like set-top-box STB) access to the service description (e.g., Electronic Program Guide) via SD&S service selection and discovery procedures that follow DVB IPTV specification [4] and use the http protocol over Tr reference point. The same Tr interface can be used by UE for accessing the user interface and service selection over Customer Facing IPTV application (CFIA). CFIA provides over the Tr http based interface IPTV service provisioning, selection, and authorization. The IPTV control (IPTV-C) is enabled over Ct2 interface (http or RTSP control). Media (e.g., content on demand—CoD) can be streamed by unicast or multicast over Xd from Media Delivery Function (MDF). For media control, for example, the trick play command uses the RTSP protocol via Xc by Media Control Function (MCF).

13.3.3 NGN IMS–Based IPTV Architecture

The second alternative and/or next evolution step is in replacing the IPTV control by IMS-based subsystems (based on [5,7]) and in introducing session-based IPTV service control based on SIP protocol (Figure 13.3). The service discovery is performed by SIP application servers called Service Discovery Function (SDF). The SDF provides information about Service Selection Function (SSF), which contains necessary service information (e.g., TV program guides) for initiate IPTV services. Service Control Function (SCF) is responsible to serve SIP-based service initiation requests delivered over Gm interface via IMS core for any IPTV services required from UE. MCF and MDF functionalities are similar to those in NGN dedicated IPTV with a difference that SIP request can be forwarded from SCF to MCF via IMS core and y2 interface. The resource allocation and reservation is performed over existing IMS interface to RACS (Gq'). Specific combination of SIP and RTSP protocols are designed for CoD service initiation where SIP INVITE message is used for session setup, but control of media stream is handled via Xc using RTSP.

The IMS-based IPTV has a number of advantages because IMS can act as a unified service control subsystem for all NGN services instead of establishing an additional specialized subsystem (case of NGN dedicated IPTV subsystem). Additionally, IMS can more naturally support mobility, interaction with NGN service enablers (like messaging or presence), service personalization, or quadruple play services (voice, data, video, and mobile).

IMS-based IPTV is a new concept where the first implementations based on ETSI TISPAN R2 have been already tested by MSF in 2008. ETSI plugtest specification [29] allows vendors and operators to perform a real interoperability test of IMS-based IPTV servers and platforms (based on

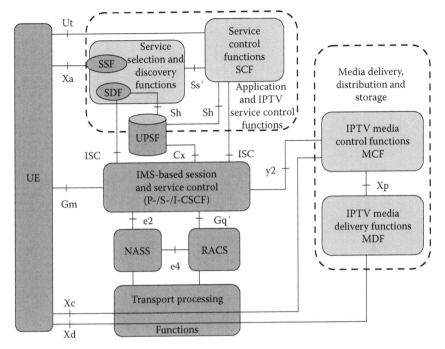

FIGURE 13.3
Simplified TISPAN IMS–based IPTV functional architecture. (From ETSI TS 182 027 V2.0.0 (2008-02) R2, Draft ETSI RTS 182 027 V3.3.1 (2009-08) in R3, TISPAN: IPTV Architecture, IPTV functions supported by the IMS subsystem, ETSI, Sophia Antipolis, France, 2009.)

TISPAN R2 [7]). The most comprehensive set of new IPTV services will be implemented based on TISPAN Release 3 NGN-based IPTV [5].

13.3.4 Technical Comparison between TISPAN IMS and Non-IMS-Based IPTV Systems

When IMS-based IPTV [6] is compared with NGN integrated IPTV [7], the main differences between both solutions are as follows (Table 13.3):

- Separation of service selection and discovery in IMS-based IPTV (SIP-based discovery).
- Similar service selection specs, IMS-based IPTV additionally support OMA BCAST ESG.
- SIP-based service initiation and service control.
- Support of the direct RTP encapsulation.
- Related differences in interfaces and protocols (from Tr to Ut, Xa and change from http-based Ct, Ss, Sa to SIP-based Gm, ISC, y2 interfaces).

TABLE 13.3

Comparison of Characteristics for Non-IMS and IMS-Based NGN IPTV Concepts

General Characteristics	IMS-Based IPTV Architecture (NGN IMS Based)	NGN Integrated IPTV Architecture (NGN Non-IMS)
ETSI TISPAN specification	ETSI TS 183 063 [26]	ETSI TS 183 064 [27]
1. SD&S	ETSI TS 102034 based SD&S model—separate SDF, SSF SIP based (Mandatory), HTTP (Optional), DVBSTP (Optional) via Xa to SSF—HTTP based	ETSI TS 102034 based SD&S model—single SD&S HTTP based (Mandatory) DVBSTP (Optional) via Tr to SD&S—HTTP based
2. Service selection information (e.g., program guides)	DVB SD&S (ETSI TS 102034)	DVB SD&S (ETSI TS 102034)
	DVB BCG (ETSI TS 102 539)	DVB BCG (ETSI TS 102 539)
	OMA BCAST ESG	
	TISPAN XML	
3. Multicast control—IGMP	SIP-based initiation	Pure IGMP based
	IGMP join to ECF/EFF	IGMP join to ECF/EFF
	IGMPv3, MDLv2	IGMPv3, MDLv2
4. Unicast control—RTSP methods	SIP-based initiation	RTSP based on ETSI TS 102034
	Mixture RTSP control (RFC 2326), partially ETSI TS 102034 based	Coupled, decoupled mode
	Method 1—new coupled SIP/RTSP	
	Method 2—SIP and RTSP separated	
5. Media Delivery	MPEG2TS over RTP	MPEG2TS over RTP
	MPEG2TS over UDP Direct RTP encapsulation	MPEG2TS over UDP
6. Service control (initialization, modification, teardown)	SIP-based service control using IMS [9]	HTTP resp. RTSP based
	Session-based control	
7. Service configuration	Ut–XCAP	Tr–XCAP
8. Resource allocation and reservation	Via core IMS	IPTV-C
	Gq' to RACS	Gq' to RACS
9. User profile, user data	Distributed	Distributed
	UPSF (SSP located)	UPSF (SSP located), IUDF
	SCF (SIP AS)	CFIA (http AS), IPTVC
	SSP used (SSF, SDF via Sh)	SSP used (SD&S, IPTVC)

Source: Mikóczy, E., Next generation of multimedia services—NGN-based IPTV architecture, in *15th International Conference on Systems, Signals and Image Processing (IWSSIP 2008)*, Bratislava, Slovak Republic, June 25–28, 2008.

13.4 Convergence of IPTV with Mobile TV Technology and Hybrid IPTV Concepts

Beside TISPAN, NGN-based IPTV have been designed to access network independently and are enabled to provide IPTV services over fixed or mobile networks. There are several technologies that can really support multicast/ broadcast over mobile technology in an efficient way for providing TV on mobile. Other type of convergence is the hybrid IPTV where unidirectional terrestrial/satellite delivery is used for media delivery of Live TV, but for interaction and delivery of other IPTV services bidirectional IP networks are used.

13.4.1 IPTV Services over 3GPP IMS–Based MBMS/PSS

The 3GPP Packet Switch Streaming (PSS) [31] provides a framework for Internet Protocol (IP)-based unicast streaming applications, while the 3GPP Multimedia Broadcast and Multicast Service (MBMS) provide a framework for broadcast and multicast streaming and download applications in 3GPP networks, supporting the MBMS bearer service. The MBMS Bearer Service [32] includes a multicast and broadcast mode that uses IP Multicast addresses for efficient delivery of the IP flows. The advantage of the MBMS Bearer Service compared to legacy UMTS bearer services is the sharing of resources and multicasting, which enables the efficient usage of radio-network and core-network resources, with an emphasis on radio interface efficiency.

The 3GPP IP Multimedia Subsystem (IMS) enables the deployment of IP multimedia applications. PSS and MBMS User Services are IP multimedia services but they were specified before IMS. But IMS brings enablers and features to operators and subscribers that can enhance the experience of PSS and MBMS User Services.

The purpose of IMS-based PSS/MBMS specification [33] is to initiate and control PSS and MBMS User Service. This should enable deployment of PSS and MBMS user services as IMS services. IMS-based PSS/MBMS specification uses components of the 3GPP PSS, 3GPP MBMS, ETSI TISPAN IPTV, and Open IPTV Forum standards specifications.

ETSI TISPAN analyzes the concept for providing selected TISPAN IPTV services to mobile UE over 3GPP mobile access networks supporting unicast-based streaming services (using PSS) or multicast-based streaming services (using MBMS). In Figure 13.4 the principle of mapping of how IMS-based IPTV from TISPAN can be combined with 3GPP IMS-based PSS and MBMS is shown [33].

13.4.2 OMA BCAST

The OMA Mobile Broadcast [OMA BCAST] Services Enabler refers to the term "Mobile Broadcast" that can be understand as a broad range of Broadcast Services, which jointly leverages the unidirectional one-to-many broadcast

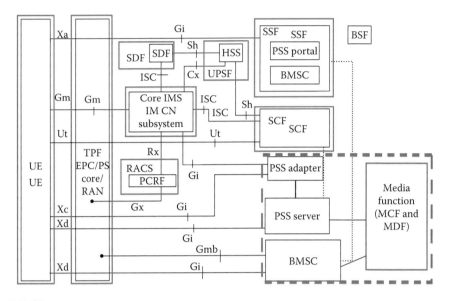

FIGURE 13.4
Conceptual mappings of 3GPP IMS–based PSS and MBMS with TISPAN IMS-based IPTV.
(From Draft ETSI TR 182 030 V<0.0.3> (2009-09), TISPAN: NGN based IPTV mapping or inter-
connect between IPTV systems, ETSI, Sophia Antipolis, France, 2009.)

paradigm and the bidirectional unicast paradigm in a mobile environment,
and covers one-to-many services ranging from classical broadcast to mobile
multicast (generally called Broadcast Distribution Systems—BDS).

The OMA BCAST Enabler 1.0 addresses some usual IPTV functionalities,
which are generic enough to be common to many Broadcast Services, and
which can be defined and implemented in a bearer-independent way for Mobile
TV. These functionalities are as the follows: Service Guide, File Distribution,
Stream Distribution, Service Protection, Content Protection, Service Interaction,
Service Provisioning, Terminal Provisioning, and Notification.

BCAST Enabler uses an underlying Broadcast Distribution Systems (BDS)
for distribution of streams over existing broadcast/multicast networks. BDSs is
mainly considered in the current version of BCAST architecture, which include
3GPP MBMS, 3GPP2 BCMCS, and DVB IPDC, not excluding any other future
BDSs (e.g., for BCAST 1.1 DVB-SH, WiMAX, and FLO-IP are also considered).

13.4.3 IMS-Based IPTV with DVB-H Delivery

DVB-H is a transmission system [34] using DVB standards, for providing an
efficient way of carrying multimedia services over digital terrestrial broadcasting networks to handheld terminals. IP Datacast over DVB-H is the additional technology that allows the delivery of IP datagram over an end-to-end
broadcast system. An inherent part of the IPDC over DVB-H system is that
it comprises an unidirectional broadcast path that could be used to deliver

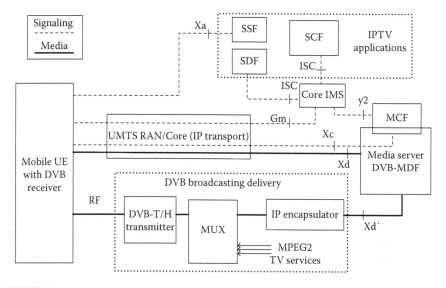

FIGURE 13.5
Conceptual delivery of selected IMS-based IPTV services over DVB-H. (From Draft ETSI TR 182 030 V<0.0.3> (2009-09), TISPAN: NGN based IPTV mapping or interconnect between IPTV systems, ETSI, Sophia Antipolis, France, 2009.)

some of the IP-based noninteractive streaming services (like Live TV broadcast, Near VOD, or Broadcast-based Push CoD), which may be combined (Figure 13.5) with a bidirectional mobile/cellular or fixed interactivity path (also other IPTV services have to be delivered over IP networks like unicast or interactive services). IP Datacast over DVB-H is thus a platform that can be used for enabling the convergence of services from broadcast/media and telecommunications domains (e.g., mobile/cellular).

13.4.4 IPTV over Cable

Cable operators have longer history in providing TV services through cable. If they would like to offer more complex triple play services and interactive IPTV services they have two main options. First, to deliver TV streams over RF as DVB-C and the other one is in providing services over data services through cable. Second option is to provide full IP-based IPTV services over advanced high speed cable technology, for example DOCSIS 3.0 [CableLabs].

13.5 Concept of Converged NGN-Based IPTV

The TISPAN in Release 3 (in informative annex of [7]) proposed possible migration scenarios among non-NGN, NGN non-IMS, and NGN IMS–based architecture. The last evolution step called converged NGN-based

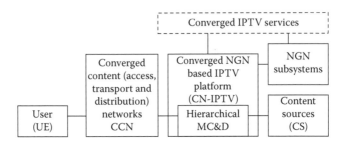

FIGURE 13.6
High level view to IPTV domains.

IPTV is also mentioned (this concept is not described in Release 3 because it is beyond scope of this chapter). The ITU-T has been proposing the converged application framework where the non-IMS IPTV merges with the IMS-based IPTV architecture. But in this case, just as purely, unity of all elements supports all interfaces from both architectures that make no sense from the complexity perspective. Both architectures are able to provide similar services. Sections 13.5.1 through 13.5.3 describe the author's view about the potential concept of such architecture with the additional goal to also describe the method to adapt the multiple types of content sources, with multiple access and distribution networks converged to a single functional architecture.

The concept of converged NGN-based IPTV can be split into several subsystems and domains, where each one plays an important role to provide the converged approach of multiservice NGN architecture (Figure 13.6)

- Content sources
- Converged NGN-based IPTV architecture—overall concept, description of functionalities
- Concept of hierarchical content control and delivery subsystem
- Content transport and distribution networks
- Converged IPTV services

13.5.1 Functional Architecture for Converged NGN–Based IPTV

The proposed Converged NGN–based IPTV architecture (CN-IPTV) is an evolution of the IMS-based IPTV from ETSI TISPAN Release 3. We combined IPTV service control function (ISCF) for service control, which can use the IMS for specific cases; but not all signaling traffics need to pass the core IMS (which improve the performance and shorter delays). The concept includes media control and delivery organized in hierarchical MC&D architecture that is extended with specialized types of MDFs [35]. Additionally, an elaborate [3] of the possible integration with some DVB,

3GPP UMTS, and OMA BCAST component that can extend the mobility capabilities for IPTV services are also provided. DVB specified how DVB services could be provided over IP networks in specification called DVB-IPTV (formally DVB-IPI) and also discussed the possible compatibility with TISPAN NGN–based IPTV [36]. OMA defined service enabler is primarily for mobile related broadcasted/multicasted services (OMA BCAST), but it could also be used as a base for adaptation of several types of access technologies to our proposed concept for converged NGN–based IPTV. OMA BCAST 1.1 explains the adaptation of the DVB-H, 3GPP MBMS, WiMAX access technologies.

The proposed converged NGN–based IPTV architecture consists of the following elements (Figure 13.7):

ICAF—IPTV Converged Application Functions
SDSF—Service Discovery and Selection Functions
ISCF—IPTV Service Control Function
UPSF—User Profile Server Function (existing one)
IMCF—IPTV Media Control Function
IMDF—IPTV Media Delivery Function (P—Proxy, S—Serving, I—Interconnection)
IMS—IP Multimedia subsystem (P-/S-/ICSCF)
NASS—Network Attachment Subsystem
RACS—Resource and Admission Control Subsystem

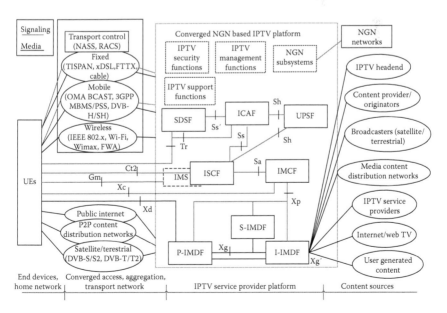

FIGURE 13.7
Proposed conceptual high level architecture for Converged NGN–based IPTV.

Any IPTV architecture is not fully completed without other functionalities, which we have also included in the proposed concept:

- IPTV supporting function (e.g., content preparation and manipulation)
- IPTV management functions (e.g., content management)
- IPTV security functions (e.g., content protection, IPTV service protection)
- IPTV charging (based on NGN charging for online/offline charging but enhanced for IPTV)
- Interworking or interfacing with other NGN subsystems

Three types of MDFs are introduced by hierarchical MC&D. MDFs are split into multiple specific functionalities with three architecture components described [35] and extended [3] as follows:

Interconnection—IPTV Media Delivery Function (I-IMDF): this element handles the media import and ingress of content from multiple content sources (ingress of media, metadata, content provider information, and interconnection to external domains):

- IPTV Headend or from content providers/originators or broadcasters.
- From other IPTV service providers in case of interconnection or roaming or as offer of the content from service provider playing a role of content aggregator or Media Content Distribution Network (MCDN).
- From the Internet sources like the Web-based TV or from the end users like user generated content.
- The IMDF need to hide the IPTV service provider infrastructure for external domains but also provide necessary functionality to interconnect to heterogeneous content sources (which can hold a variety of coding, transport, signaling schemas) and convert to content/metadata/signaling to formats supported by Converged NGN–based IPTV.

Serving-IPTV Media Delivery Function (S-IMDF): this element handles the processing of contents (e.g., encoding, content protection, and transcoding) and it is also responsible for the storage of contents and metadata as well as the propagation of content information. S-IMDF provides centralized oriented services such as Content on demand for long tail content (less popular content), or recording/storing of user independent content (n-PVR or Time shifted TV or Near CoD).

Primary-IPTV Media Delivery Function (P-IMDF): this element is the primary contact point of the users, which also provides the streaming and downloading functionalities for all IPTV services according to the required quality, format, and the type of casting (multi-/uni-/broad-casting) for particular user's end device and network accessing to P-IMDF (personalization of user specific content delivery). This element could also store the most frequently accessed CoD assets or user's specific contents (specific user n-PVRs, user generated content). The P-IMDF could be responsible for the adaptation of IPTV service delivery to other access technologies or distribution networks:

- Preferred way is that the P-IMDFs are located as near as possible to UE, for example, near to edge of the network. These elements can be combined with specific elements of access network (e.g., in case of also using OMA BCAST and 3GPP MBMS with integrated control elements, in case of MBMS, for example). P-IMDFs can also integrate Broadcast-Multicast Service Centre (BM-SC). The BM-SC provides functions needed for user service provisioning and content delivery in MBMS capable UMTS. P-IMDF integration with 3GPP MBMS and PSS helps to provide IPTV services delivery over exiting UMTS mobile network.
- P-IMDF can also support mobility and seamless handover between different technologies.
- P-IMDF may required to transcode or adapt the content to the required bitrate, codec, or content encapsulation to specific transporting technologies.
- Can be used as security elements for the content protection (e.g., digital rights management and content encoding).

The element responsible for the service discovery and selection functions (SDSF) could support multiple formats and mechanisms. But the main enhancement in proposed architecture is the SDSF's potential to aggregate metadata information from multiple sources (e.g., content provider, electronic program guide provider, Internet, broadcasted service information), and provide them everywhere to the UE in personalized manner and allows them to integrate with other relevant information (presence, statistics, recommendations, etc.), too.

The last but not the least element is (for sure) the IPTV converged application function (ICAF), which could provide combinational services and converged services with enhanced service logic and service orchestration. The ICAF can interact with other NGN application servers and subsystems and can be used for personalized service behavior based on user's preferences and settings.

The enhanced service control entity called IPTV Service Control Function (ISCF) is responsible (in conjunction with core IMS) for service control and for providing basic IPTV services, service interaction, and also supports mobility on service control and application layers. Similar to other NGN-based IPTV, the converged one need to ensure relevant resource allocation and QoS handling (using existing interfaces to transport control). Main goal of Converged NGN–based IPTV is to provide the flexible service provider platform for IPTV services (and any NGN services as well) that may deliver personalized services over multiple access network with nomadic or seamless mobility. But we have to differentiate the IPTV service provider access infrastructure with guaranteed quality (fixed technology as xDSL or FTTx, mobile technology like UMTS with IMS-based MBMS/PSS or wireless access technologies like WiFi or Wimax) from other additional distribution possibilities, like public Internet (without QoS and unpredictable conditions), terrestrial/satellite distribution (mainly unidirectional for broadcasting with other technologies used for interaction, back channel signaling or unicast services), or the P2P content distribution network.

13.5.2 Protocol Stack for Converged NGN-Based IPTV

The protocol stack for such a complex system as Converged NGN–based IPTV platform includes almost all existing NGN protocols (Figure 13.8).

Common layer for all of them is IP, which should be carried over different physical and data link technologies. Additionally, to IP-based NGN protocol family, we can transmit the media over other type of broadcasting networks like DVB-T or DVB-S (where adaptation and other ways for interactive services and back channels for signaling may be needed) in case of hybrid IPTV services.

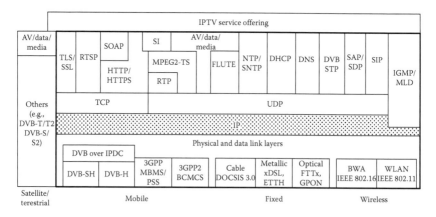

FIGURE 13.8
Protocol stack for Converged NGN–based IPTV platform. (From Mikóczy, E. and Podhradský, P., Evolution of IPTV architecture and services towards NGN, in *Recent Advances in Multimedia Signal Processing and Communications*, Series: Studies in Computational Intelligence, Grgic, M. et al. (eds.), vol. 231, Springer, Berlin/Heidelberg, Germany, 2009, ISBN: 978-3-642-02899-1.)

13.5.3 High-Level Procedures for Converged NGN-Based IPTV

The concluding section (Section 13.6) provides high level procedures for Converged NGN–based IPTV architecture and also shows protocols used (Figure 13.9).

Initially, the UE is required to start or boot (i.e., a set-top-box, PC, mobile or any other device with an IPTV client) and perform the network attachment to obtain network parameters (i.e., an IP address, etc.).

1. After the network attachment, the UE is required to start the service initiation steps.

2. The UE shall perform service provider discovery in order to enable SDSF procedure followed by the IPTV service selection and attachment as defined in [5]. SDSF can use HTTP over Tr or DVBSTP over Tr.

3. Then the UE shall perform the service selection procedures with SDSF via Tr (using HTTP over Tr) to receive service selection information.

4. At this stage, the IPTV UE needs to acquire and use collected service selection information to establish an appropriate selected service via ICAF (in step 5).

5. The ICAF can control service behavior via ISCF. The ISCF is also capable of initiating a resource reservation and allocation process

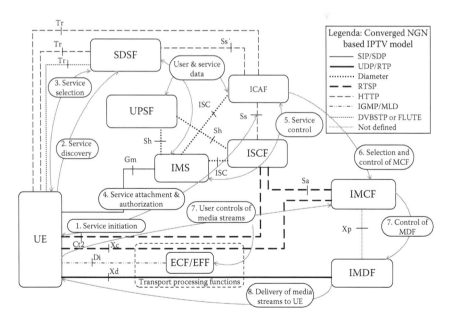

FIGURE 13.9
Converged NGN-based IPTV—protocol model with FE relation.

for network resources needed by the IPTV service, according to the capabilities of the UE and access network (using standardized transport control functions of NGN RACS available).

6. Following a successful session initiation, the ISCF informs the MCF via Sa reference point (or UE in some cases, i.e., BC) about identification of selected content from the Media Delivery Function (or ECF/EFF for BC services) to initiate delivery of the selected multimedia content (CoD, nPVR).

7. The UE may interactively control CoD media stream over the Xc reference point (between the UE and the MCF) via RTSP protocol. The UE may control BC media stream over the Di reference point (between the UE and the ECF/EFF) with IGMP/MLD protocol.

8. The IMDF performs media delivery over the Xd interface using UDP/RTP stream delivery and several transport variants.

13.6 Conclusions and Future Evolution of NGN-Based IPTV

The future evolution of IPTV architecture and services will depend on the acceptance of the NGN-based IPTV concept by operators and vendors. But finally, only the satisfied end users as paying subscriber are crucial for any commercially successful service and therefore not only used technology but definitely simple usability and rich set of IPTV services/content will determine which solutions will dominate on the market. The changes in the ICT (Information and Communication Technology) world are also changing user behavior. Some trends we can already see, during these days, where the Internet is converging the world (provide application abstraction layer for fixed mobile convergence where services are accessible for any device with Internet connection) and also provide unlimited source of content (but still higher quality content providers prefer TV delivery instead of Internet). The role of classical network operators, which invest a lot on infrastructure is questionable because they are pushed out from providing services in to the role of pure bit pipes. The technologies of next generation access network (like FTTH, LTE, DOCSIS 3.0) will dramatize this situation and expand the opportunity for Internet application provider deliver their services "over the top" to end users without owning any piece of network infrastructure (just few server farms/clouds). We can recognize how social media and communities are changing communications habits, sharing personal content, and watching videos on web. But operators could use their strengths and position their IPTV products as last window opportunity to provide added value service with required quality, reliability, and personalization. The converged IPTV can also provide the best from Internet world directly to TV or PC/mobile.

We need to understand that majority of people still will prefer comfort of their living rooms and simplicity by using just remote control from their TV (set-top-box). Beside the group of Internet user growing most rapidly, still TV is primarily used as a source of "trusted" information for most of world population (and we could not forget billions without TV or Internet). Full digitalization of TV delivery and wide accessibility of IPTV can mean that users will be sooner or later adapted to using more advance IPTV features (EPG, Pause TV, interactive TV, etc.) as provided with previous generation of analog TV. We can expect that the amount of IPTV user will grow significantly in the next years. Then IPTV will change from only premium product to mandatory service of any integrated/converged operator. We can expect changes also in our homes, where all equipment will be "on net" and you can consume content or control your home remotely with interconnection of your TV with smart home technologies. We can easily identify several topics and future trends, which will most probably be concrete for NGN-based IPTV and will be the hottest topics in NGN standardization [37]:

1. Evolution of NGN-based IPTV in context of Future Networks requirements (re-design of Internet Architecture discussed in scientific community as Future Internet)

2. New IPTV services (e.g., enhanced Release 3 services, TV communities, TV commerce, multiscreen approach, public interest services, fully personalized IPTV,etc.)

3. Support for new media (Ultra-HD, 3D Content, Virtual Realities, Networking/Social Media)

4. Hybrid IPTV models (partial delivery of TISPAN IPTV services over "non-TISPAN" networks e.g., DVB-H/T/S/C/SH, OMA BCAST, 3GPP MBMS/PSS, DOCSIS3.0, unmanaged networks/over-the-top)

5. IPTV interconnections (roaming support, interconnection with Media Content Delivery/Content Providers/Media Sources) and integration with Content Delivery Networks, Peer to Peer

6. Home network support for managed/unmanaged models, integration of IPTV services with future home networks (smart homes services, metering, near field communication, etc.)

7. Convergence of end devices (converged end devices for IMS/non-IMS IPTV or hybrid models)

8. Enhanced IPTV security (Service and Content Protection in converged and open environment, content mobility)

9. IPTV management (content distribution management, interconnection aspects)

10. Interoperability issues, consolidation in standardization of NGN-based IPTV architecture

Acknowledgments

The first author is also a senior designer in Slovak Telekom where he has been actively participating on ETSI TISPAN NGN Release 2/3 standardization with more than 100 agreed contributions included in several TISPAN IPTV–related specifications (e.g., IPTV-related work items WI0005, WI2048, WI2049, WI3127, WI3137, WI7029, WI1059, WI2070, WI2074, WI 2079, WI3208). Some parts of this chapter are based on contributions of author within ETSI TISPAN.

This chapter also presents some of the results from participation in various research project at Slovak University of Technology in Bratislava such as NGNlab project [NGNlab], European CELTIC EUREKA project Netlab [Netlab], [38], Slovak National research projects: AV project 4/0019/07: Converged technologies for next generation networks (NGN), Slovak National basic research projects VEGA 1/0720/09 and VEGA 1/4084/07.

References

1. López, D., Mikoczy, E., Moreno, J., Cuevas, A., and Vázquez, E., IPTV modeling and architecture over IMS, in *IP Multimedia Subsystem (IMS) Handbook*, M. Ilyas and S. A. Ahson (Eds.), Chapter 22, CRC Press, Boca Raton, FL, 2008, pp. 443–471, ISBN: 978-1420064599.

2. Recommendation ITU-T Y.1910 (09/2008), IPTV functional architecture, ITU-T, Geneva, Switzerland, 2008.

3. Mikóczy, E. and Podhradský, P., Evolution of IPTV architecture and services towards NGN, in *Recent Advances in Multimedia Signal Processing and Communications*, Series: Studies in Computational Intelligence, M. Grgic, K. Delac, and M. Ghanbari (Eds.), vol. 231, Springer, Berlin/Heidelberg, Germany, 2009, ISBN: 978-3-642-02899-1.

4. ETSI TS 102 034 V1.4.1 (2009-08) Technical Specification, DVB: Transport of MPEG 2 based DVB services over IP based networks, ETSI, Sophia Antipolis, France, 2009.

5. ETSI TS 181 016 V2.0.0 (2007-11) for Release 2, ETSI TS 181 016 V3.3.1 (2009-07) in Release 3, TISPAN; Service layer requirements to integrate NGN services and IPTV, ETSI, Sophia Antipolis, France, 2007.

6. ETSI TS 182 028 V2.0.0 (2008-01) in R2, ETSI TS 182 028 V3.3.1 (2009-10) in R3 TISPAN; IPTV Architecture: NGN integrated IPTV subsystem Architecture, ETSI, Sophia Antipolis, France, 2009.

7. ETSI TS 182 027 V2.0.0 (2008-02) R2, Draft ETSI RTS 182 027 V3.3.1 (2009-08) in R3, TISPAN; IPTV Architecture, IPTV functions supported by the IMS subsystem, ETSI, Sophia Antipolis, France, 2009.

8. 3GPP TS 23.228 V7.7.0 (2007-03), IP Multimedia Subsystem (IMS), Stage 2, 3GPP, Sophia Antipolis, France, 2007.

9. 3GPP Technical Specification, MBMS 3GPP TS 23.246 V8.2.0 (2008-06), Multimedia broadcast/multicast service (MBMS), architecture and functional description (Release 8), Third Generation Partnership Project, Sophia Antipolis, France, 2008.
10. RFC 791—Internet protocol, IETF, Fremont, CA, 1981.
11. RFC 2460—Internet protocol, Version 6 (IPv6) Specification, IETF, Fremont, CA, 1998.
12. RFC 3261—SIP: Session initiation protocol, IETF, Fremont, CA, 2002.
13. RFC 2326—Real time streaming protocol (RTSP), IETF, Fremont, CA, 1998.
14. RFC 3376—Internet group management protocol, Version 3, IETF, Fremont, CA, 2002.
15. RFC 3810—Multicast listener discovery version 2 (MLDv2) for IPv6, IETF, Fremont, CA, 2004.
16. RFC 3550—RTP: A transport protocol for real-time applications, IETF, Fremont, CA, 2003.
17. RFC 2616—Hypertext transfer protocol—HTTP/1.1, IETF, Fremont, CA, 1999.
18. Alliance for Telecommunications Industry Solutions (ATIS), IPTV Architecture Requirements (ATIS-0800002), ATIS IPTV Interoperability Forum (IIF), ATIS, Washington, DC, 2006.
19. Alliance for Telecommunications Industry Solutions (ATIS), IPTV High Level Architecture Standard (ATIS-0800007), ATIS IPTV Interoperability Forum (IIF), ATIS, Washington, DC, 2007.
20. CableLabs document PKT-TR-ARCH-FRM-V05-080425, PacketCable™ 2.0, Architecture Framework Technical Report, Cable Television Laboratories Inc., Louisville, CO, 2008.
21. Open Mobile Alliance Document OMA-AD-Service-Environment-V1_0_4-20070201-A, OMA Service Environment 1.0.4, OMA, San Diego, CA, 2007.
22. OMA draft OMA-AD-BCAST-V1_0-20081209-C, BCAST mobile broadcast services architecture, candidate Version 1.0—12/2008, Open Mobile Alliance, San Diego, CA, 2008.
23. Open IPTV Forum—Service and Platform Requirements—V 2.0, Open IPTV Forum, Sophia Antipolis, France, 2008.
24. Open IPTV Forum—Functional Architecture—V 1.2, Open IPTV Forum, Sophia Antipolis, France, 2008.
25. Home Gateway Technical Requirements: Release 1, Version 1.0, Home Gateway Initiative, Sophia Antipolis, France, 2006.
26. ETSI TS 183 064 V2.1.1 (2008-10) in R2, Draft ETSI TS 183 064 V3.1.0 (2009-08) in R3, TISPAN; NGN Integrated IPTV subsystem Stage 3, ETSI, Sophia Antipolis, France, 2009.
27. TS 183 063 V2.1.0 (2008-06) in R2, Draft ETSI TS 183 063 V3.2.0 (2009-10) in R3, TISPAN; IMS based IPTV Stage 3, ETSI, Sophia Antipolis, France, 2009.
28. Draft ETSI TR 182 030 V<0.0.3> (2009-09), TISPAN; NGN based IPTV mapping or interconnect between IPTV systems, ETSI, Sophia Antipolis, France, 2009.
29. Draft ETSI TS 186 020 V2.0.5 (2009-08), TISPAN; IMS-based IPTV interoperability test specification, ETSI, Sophia Antipolis, France, 2009.
30. Mikóczy, E., Next generation of multimedia services—NGN based IPTV architecture, in *15th International Conference on Systems, Signals and Image Processing (IWSSIP 2008)*, June 25–28, 2008, Bratislava, Slovak Republic.

31. 3GPP TS 26.234, Transparent end-to-end packet-switched streaming service (PSS), protocols and codecs, v.8.4, 3GPP, Sophia Antipolis, France, 2009.
32. 3GPP TS 23.246, Multimedia broadcast/multicast service (MBMS): Architecture and functional description, v.8.4, 3GPP, Sophia Antipolis, France, 2009.
33. 3GPP TS 26.237 V8.3.0 (2009-09), IP multimedia subsystem (IMS) based packet switch streaming (PSS) and multimedia broadcast/multicast service (MBMS): User service, protocols (Release 8), Sophia Antipolis, France.
34. ETSI EN 302 304 V1.1.1 (2004-11), DVB: Transmission system for handheld terminals (DVB-H), ETSI, Sophia Antipolis, France.
35. Mikoczy, E., Sivchenko, D., Xu, B., and Moreno, J., IPTV services over IMS—Architecture and standardization, *IEEE Communication Magazine*, 47(5), 128, May 2008, ISSN: 0163-6804, 2008.
36. DVB Document A128, DVB-IP Phase 1.3 in the context of ETSI TISPAN NGN, DVB, Geneva, Switzerland, 2008.
37. Mikoczy, E., Discussion on future topics as ETSI TISPAN contribution 22bTD113, *ETSI TISPAN 22bis meeting*, Sophia Antipolis, France, November 2–6, 2009.
38. Mikóczy, E., Kadlic, R., and Podhradský, P., Concept for mobility and interconnection aspects in Converged NGN based IPTV architecture, *International Journal of Digital Multimedia Broadcasting*, Special issue: Convergence of Digital TV Systems and Services, Hindawi Publishing, New York, 2010, ISSN: 1687-7578, e-ISSN: 1687-7586.

Web Pages

[DVB]   Digital Video Broadcasting web site, http://www.dvb.org/technology/standards/
[ITU-T]   International Telecommunication Union web page, http://www.itu.int/ITU-T/
[ITU-T SGI] ITU-T web page of Internet Protocol Television Global Standards Initiative, http://www.itu.int/ITU-T/gsi/iptv/
[ETSI]   ETSI web page, http://www.etsi.org
[TISPAN]   ETSI TISPAN web page, http://www.etsi.org/tispan/ http://portal.etsi.org/docbox/tispan/Open/NGN_LATEST_DRAFTS/
[3GPP]   Third Generation Partnership Project web page, http://www.3gpp.org/
[OMA]   Open Mobile Alliance web page, http://www.openmobilealliance.org/
[IETF]   Internet Engineering Task Force web page, http://www.ietf.org/
[ATIS]   Alliance for Telecommunications Industry web page, http://www.atis.org/
[CableLabs] CableLabs web page, http://www.cablelabs.com
[OIPF]   Open IPTV Forum web page, http://www.openiptvforum.org/
[HGI]   Home Gateway Initiative web page, http://www.homegatewayinitiative.org/
[NGNlab]   NGNlab—NGN laboratory at Slovak University of Technology in Bratislava, project web page http://www.ngnlab.eu
[Netlab]   NetLab: Use Cases for Interconnected Testbeds and Living Labs, project web page, http://www.celtic-nitiative.org/Projects/NETLAB/default.asp

14

Interconnection of NGN-Based IPTV Systems

M. Oskar van Deventer, Pieter Nooren, Radovan Kadlic,
and Eugen Mikoczy

CONTENTS

As Metcalfe's Law [1] estimates "The value of a telecommunications network is proportional to the square of the number of connected users of the system"; although this may be debatable from the mathematical point of view [2], it takes place. The scalability and service accessibility have been the main drivers for the interconnection of telecommunication networks. The telephony network and the Internet are two highly interconnected services that achieve their value by connecting any user to any other user, and by providing access to services and content worldwide.

This chapter looks into the interconnection of content distribution networks and, more specifically, interconnection of NGN-based IPTV systems. Such types of interconnections would be beneficial to the consumers. The interconnection would enable consumers access to a wider range of content, namely, content available in other fixed and/or mobile networks. Roaming and mobility capabilities supported by the interconnection would also enable consumers access to contents from a wider range of access points, namely, from access points belonging to other fixed and/or mobile networks. In addition, it would provide the consumer with a consistent, personalized, content-rich, and service-rich user experience from any place and at any time.

Section 14.1 considers some business role models and requirements for the interconnection of content distribution networks from the perspective of different players in the content-delivery value chain. It concludes that any successful business model for content delivery network (CDN) interconnection and IPTV roaming interconnection should be beneficial to the content providers, as well as to network operators.

Section 14.2 presents the service scenarios for IPTV roaming interconnection. It analyzes the impacts when specific IPTV services would also be available in a roaming context [3].

Section 14.3 presents the technical scenarios for IPTV roaming interconnection. It shows that there are various levels of interconnection, varying from the basic IP transport interconnection, via IMS session interconnection to two variants of IPTV service interconnections. Finally, Section 14.4 provides the conclusion.

14.1 Business Roles and IPTV Requirements for Interconnection of Content Distribution Networks

Video content services are key in the competition between network operators in their role as triple-play service providers. As telephony and Internet are becoming commodity services, video content services are the only way to differentiate a triple-play service offer. This explains why network operators are currently fiercely investing in video content services like high-definition TV, digital TV, mobile TV, and IPTV, and in advanced value-added services like electronic service guides, content-on-demand, and many more.

However, the competitive landscape around content delivery services is much more complex than just competition between network operators. Using business role models, this section explains some of the business drivers. These business drivers are applied to interconnection of network-operator-based CDNs to derive commercial and technical requirements.

14.1.1 Classic Content Delivery Value Chain under Pressure

Figure 14.1 shows a business role model for a classic content delivery value chain. The model distinguishes the following roles:

- The Content Provider creates, or at least owns the content. "Hollywood," national studios and news agencies are classic examples of content creators.

- The Advertiser provides advertisements that are to be distributed with the content. TV commercials are the classic examples.

- The Content Aggregator aggregates content and advertisements into a sellable form. Television channels are the classic example of content aggregators.

- The Network Operator delivers content services to consumers over its access and core networks. Classic examples are telephony and cable operators.

- The Consumer consumes the content. In the classic model, consumers are the subscribers of television services.

This classic content delivery chain is financed in part by fees to the consumer and in part by the advertiser.

It can be observed that this model is an abstract representation of the real world. These abstract business roles do not necessarily map one-to-one with the actual companies or parties. Some parties may assume multiple business roles,

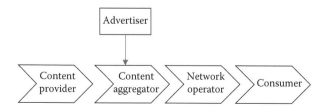

FIGURE 14.1
Business role model for a classic content delivery chain.

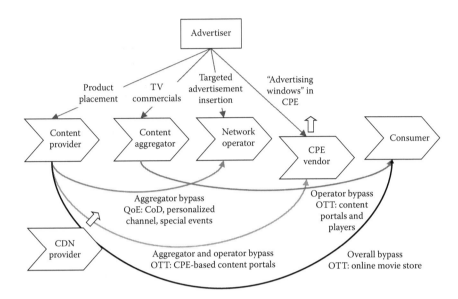

FIGURE 14.2
Classic value chain under pressure.

e.g., content creator and content aggregator (e.g., Disney Channel, BBC). In addition, some business role models make the distinction between the role of providing a network access service ("Bit pipe Provider") and a content delivery service ("TV Service Provider"), but in practice, these roles are usually combined.

The classic content delivery chain is under stress. Among others, the following movements can be recognized, see Figure 14.2:

1. Advertisers are exploring alternative routes to place advertisements. Product placement with the content creator, targeted advertisement insertion with the Network Operator, and advertisement opportunities with CPE vendors (see next bullet) are examples of such alternatives to classic TV commercials.

2. Customer Premises Equipment (CPE) Vendors are finding a role in the content delivery value chain by providing over-the-top (OTT) portals to Content Providers. Philips NetTV, Microsoft Xbox, and Apple iMovie are all examples of such CPE-based content portals.

3. Content Aggregators are bypassing Network Operators with OTT content portals. BBC iPlayer [4] and Uitzending Gemist [5] are both portals to watch TV programs at a later time and also some other content on demand.

4. Network Operators are bypassing Content Aggregators with enhanced quality-of-experience (QoE) services like Content-on-Demand (CoD), personalized television channels, network-based personal video recording, and special-events channels.

5. Content Providers are exploring many routes to bypass other players in the content delivery value chain.

6. CDN Providers are building overlay networks to support parties to deliver content OTT, independent of Network Operators' content service offerings. Examples are Akamai, Limelight, and Level-3, but also peer-to-peer (P2P) networks like Bittorrent.

Clearly, two business roles are the most under pressure: Content Aggregator and Network Operator.

The value of the classic Content Aggregator role diminishes due to the explosion of the number of television channels, possibilities for content on demand, network-based personal video recording, and alternative delivery routes for content and advertisements. Whereas, earlier most commercial TV channels paid Network Operators to deliver their advertisement-loaded channel, some of them are now forced to charge the Network Operators. Even though the role of the Content Aggregator may never disappear as consumers will always need guidance to find contents, it will no longer be pivotal in the value chain.

The largest threats to Network Operators are the various forms of OTT content delivery, reducing the role of Network Operator to a bit pipe. Section 14.1.2 explains how this threat can be seen as a business driver for Network Operators to interconnect NGN-based IPTV systems.

14.1.2 Business Drivers for Network Operators to Interconnect NGN-Based IPTV Systems

In order to respond to the OTT threat, Network Operators are enhancing their video content services, as already highlighted in the introduction of this section. Interconnection and roaming would provide the following business opportunities to Network Operators:

- Content Providers would benefit from the interconnected operator CDNs by having a one-stop shopping for a wide geographical coverage combined with a managed QoE content delivery, resulting from local caching and other Quality-of-Service (QoS) control methods.
- Network Operators would also benefit from the additional wholesale business from mutual interconnection and roaming agreements.
- Consumers would benefit from access to a wider range of content, namely, content available in other fixed and/or mobile networks, and access to content from a wider range of access points, with a high degree of personalization and a managed QoE.

Interconnection and roaming have little impact on the business of Advertisers, as most aspects of personalization and targeted advertisement insertion are

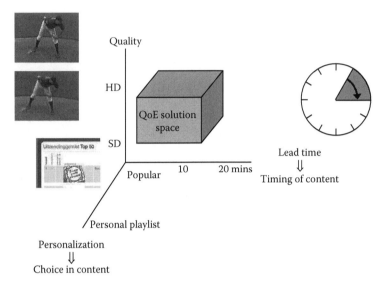

FIGURE 14.3
Combining CDN capability with IPTV services provides higher QoE flexibility.

out of the scope of interconnection and roaming. To the Content Aggregators, interconnection and roaming could be part of the threat of being bypassed, as described in Section 14.1.1.

The Eureka Celtic Rubens project [6] has demonstrated the value of combining CDN technology with IPTV services. This combination offers a higher flexibility in the control of the QoE perceived by the end users. Figure 14.3 illustrates the three axes of flexibility:

- The first axis is the technical quality of the content itself. If there is sufficient bandwidth available, then the user can select the high definition (HD) version, whereas at times of bandwidth scarcity, the user can best switch back to standard definition (SD).

- The second axis is the time that the content is played. If at the time of the user request there is insufficient bandwidth to deliver a piece of content in HD format, then the user may be offered the choice to start watching at a later time. The content is then preloaded while the user is doing something else. User tests in the Rubens project showed that many users would use and appreciate this possibility.

- The third axis is the degree of personalization. The CDN could be configured such that at peak times only the most popular content is delivered, while "long tail" content is only delivered at times that there is sufficient bandwidth in the network. In addition, this predictability of "which content is available when" will enhance the QoE perception of users.

Adding interconnection to this picture, the result is a one-stop shopping with a wide geographical coverage for the Content Provider, combined with a managed and highly flexible QoE for the Consumers.

14.1.3 Requirements on Interconnections NGN-Based IPTV Systems

The analysis in Section 14.1.2 shows that Content Providers play a key role in the business success or failure of interconnection of NGN-based IPTV systems. Content Providers require strict control over distribution and delivery of their content, which they can enforce with business agreements and copyright law. "Content is king" in the business relationship between Content Provider and Network Operator. The biggest fear of content providers is uncontrolled "leakage" of content, with the feat of the music industry, because of the illegal downloads sharply in mind. Whereas monolithic CDNs like Akamai can enforce strict control over content delivery, this is not automatically guaranteed by an interconnected CDN. In addition, roaming involves an intrinsic risk of content leakage to other domains. Interconnection of NGN-based IPTV systems has been heavily debated within ETSI TISPAN in 2008 [7]. The main argument against defining a standardized interconnection point at that time was the lack of confidence that Content Providers would allow such interconnection [8]. Therefore, any content service interconnection solution must satisfy the following requirement:

> The [NGN-based ITV] interconnection solution shall provide full control on to which users a particular piece of content is delivered.

Support of a cross-domain Digital Rights Management solution may be an ingredient to fulfill this requirement.

Figure 14.4 shows a basic business role model for IPTV/CDN interconnection. For simplicity reasons, the other business roles are not shown. This business role model supports a variety of business models.

1. CDN interconnection, for example
   a. The Upstream and Downstream Operators act jointly as a CDN to the Content Provider.
   b. The Upstream Operator acts as a CDN for the Downstream Operator.
2. Roaming interconnection, for example
   a. A roaming Consumer from the Upstream Operator consumes content provided by the Downstream Operator.
   b. A roaming Consumer from the Upstream Operator consumes content from its own services, distributed by the Downstream Operator.

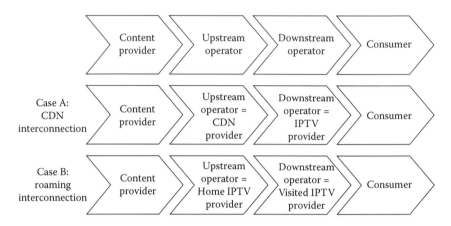

FIGURE 14.4
Business role model for IPTV/CDN interconnection.

This illustrates that the "other operator" is another key role in the business success or failure of interconnection of NGN-based IPTV systems. Interconnection and/or content brokering only happens when it is beneficial for all parties involved: the Upstream Operator, the Downstream Operator, and always the Content Provider. This means that there should be sufficient cross-domain traffic to justify an interconnection. This is equally true for telephony (enough demand for cross-domain calls) as for content (enough demand for cross-domain content). In telephony, existing interconnect solutions (SS7, SIP-based, etc.) all support a variety of business models, like terminating-fee models, originating-fee models, free-phone models, premium-rate models, and transit-fee models. For content interconnection (CDN, IPTV, etc.), a similar variety of business models can be envisioned depending on the type of content, like advertisement-supported content, user-generated content, user-owned content, premium content, local content, remote content, and roaming content.

> The [NGN-based IPTV] interconnection solution shall support a wide variety of Operator-to-Operator business models, respecting the rights of the Content Provider.

A critical issue with interconnection is the control of resources in the own and other operator domain, which include both transport and caching resources. Technically, there are two control models.

- Pull model: the Downstream Operator pulls content (streaming, cached) from the Upstream Operator domain, and controls its delivery to the Consumer.

- Push model: the Upstream Operator pushes content into Downstream Operator domain, and controls its delivery to the Consumer.

> The [NGN-based IPTV] interconnection solution shall support a variety of technical models to control the content delivery between Upstream and Downstream Operators, including a push and a pull model.

Finally, it should be noted that some contents have a real time streaming character and is typically transported using multicast and unicast, whereas other contents have an off-line character and is typically cached and delivered using unicast.

> The [NGN-based IPTV] interconnection solution shall support cross-domain control of unicast, multicast, and caching resources, respecting the integrity (capacity, resource control) of the other operator's domain.

Sections 14.2 and 14.3 explain the IPTV roaming interconnection scenarios in more detail. CDN interconnection is for further study [9].

14.2 IPTV Roaming Interconnect from Service Perspective

As highlighted in the Section 14.1, one of the applications of IPTV interconnection is roaming interconnection, enabling a roaming consumer access to IPTV services. This section and Section 14.3 focus on roaming interconnection for IPTV.

14.2.1 IPTV Service from Roaming Perspective

Roaming is a service originally used for mobile/wireless devices [10]. The concept of roaming is that the user can use all or subset of services provided by his home provider in the network of other providers. A home provider or network is that provider or network, where the subscriber would have signed a contract. The term "home network" is used here in the context of roaming, i.e., the network that is associated with the subscription of the user. This provider also sends invoices to subscribers. A visited provider or network is one where a subscriber is only temporarily located, when visiting other locations or countries. The first implementation of Roaming is originated in GSM system. Nowadays, we can imagine roaming to be useful for any device that can be easily moved (STB, Desktop PC, etc.). In this section we will describe how roaming can be implemented for IPTV platforms. Several issues are related to roaming, for example:

- Legal (law, provider relations, political, etc.)
- Economical (market specifics, rates, billing, etc.)
- Technical (technology, connectivity, etc.)

In this chapter, we focus mostly on technical issues and solution.

For IPTV roaming, we can take into the account not only the mobile devices, but also the stable devices used in living rooms.

Sections 14.2.2 through 14.2.7 discuss technical scenarios for selected IPTV services like Linear TV, Live TV with Trick play, Time-shift TV, and Content on Demand (COD). We analyze the scenarios and advantages from both the operator and user perspectives, especially when Visited network resources could be reused during roaming. Additionally, two specific use cases for Advance Personal Video Recorder (PVR) and User Generated Content (UGC) are explained as potentially attractive user services in roaming case.

14.2.2 Linear TV, Broadcast TV

Every channel of both operators must have a common channel identifier or at least synchronized metadata in both networks to provide service discovery and sufficient service selection information. If the users can access some live channels that are present in the visited network in the same quality then the Visited Network must be able to provide the required service selection. The provider of the Visited Network can agree with the Home Network provider the channels that can be offered to the subscriber from the Visited Network (providing more channels); however, they are not a part of subscriber offer in the Home Network (e.g., some local channels provided for free or paid additionally to the monthly fee). There should be other policies and addressing for multicast-based services in Home and Visited network; otherwise, interconnect should not support multicast. In these cases, specialized media servers from home network can adapt service to unicast and media servers in Visited transform address back to multicast or deliver media to end devices by unicast.

14.2.3 Trick Play of Live TV Stored in Network

In case of Live TV channel with trick play it is reasonable to make use of the local resource from the Visited Network. Otherwise, the operators must agree on the network that should be used to store the data needed for trick play (Visited/Home). If the Visited Network does not have enough space for trick playing another channel or the operator of the Visited Network do not want to spend their own resources, it is necessary to offer the service from the Home Network. This solution spends the bandwidth on the interconnect interface. The media servers responsible for interconnection in Visited Network asks the Home Network to deliver the stream for the trick play recording and then the trick play session will start to serve the unicast stream from the Visited media servers to the end devices.

14.2.4 Time-Shift TV Service Served by Network

In case that for the channel is provided time-shift then the channel is stored in network for user consumption but is usually planned by service provider (compared to user initiated recording in case of nPVR). If a channel is available and time-shifted in Visited Networks then it is possible to use this resource from the Visited Network (caching/time-shifted in Visited proxing media servers). Because the operator of the Visited Network cannot store a huge amount of content from all the time-shifted channels from all roaming partners and their Home Network offer's, what can be unfeasible and wasting of resources and a subject for legislative problems. Therefore, it is more probable that this service will be offered mainly from the Home Network. The Home Network will need to spend the bandwidth on the interconnect interface, and this could mean that the user request for time-shifting in roaming scenario will be refused.

14.2.5 CoD Service

Content on Demand (CoD) is very simple from the technical point of view, because it is a de-facto catalogue and database of content (e.g., movie files/assets) and pure unicast delivery. If the user search in CoD catalogue (e.g., in electronic content guides) and chooses the content that is accessible from the Visited Network, the user equipment (UE) will get it from this resource and do not need to consume the interconnect bandwidth. Otherwise, the content will be provided from the Home Network interconnection media servers. If there are many visiting users from the same network and there is some popular content in it, it is useful to cache it anyway in the Visited Network. If there is any need for some adaptation of the content, this must be done by transcoding on media servers in prior to sending media to the user equipment over Visited Network.

14.2.6 Advanced PVR

If the user requests recording (PVR—personal video recording) of the content, which is already stored in the Visited Network storage, it should be served from this resource. If such type of content is not available, the UE can get the content from the Home Network.

If the UE requests to record a channel, it should be done in the Home Network, because there is a high probability of subscriber's presence in the Home Network and therefore its playback in this network.

In some special cases, there are channels offered only in Visited Network, here the n-PVR should be prepared in the Visited Network and the playback should be possible only in this network. In the case of client-based PVR (c-PVR), it is independent from roaming because every time it is stored only in UE storage, internally.

We also specified some specific use cases, where the combination of both PVRs (c-PVR with combination of n-PVR) could be provided, which is called Advance or Hybrid PVR.

There exist multiple scenarios or situations when the client PVR is disabled to record show (e.g., end device is disconnected in time of scheduled recording or the network parameters are not sufficient to stream or there could be a missing capacity of storage in local device), IPTV solution should record in the network and later deliver to end device. In the case of different end device capabilities (resolution, encoding, etc.), the record by n-PVR can be prepared in several formats but an appropriate record will be distributed to the end device storage for later preview.

14.2.7 User Generated Content—UGC

Nowadays, it is possible to buy a camera in any shop for a good price. Many people or families have their own (personal) cameras. People like to make short movies of "everything" like holidays, natural beauties, night sky, etc. Some of these movies are really interesting for entertainment, educational, and many other reasons. This content is becoming very popular in on-demand segment of IPTV. It is good for operators from a business point of view to provide a space for this content and make it available to other users who might be interested in watching it.

All user contents must be stored in the Home Network, because home operator is offering the service to the subscriber and hold the space for him. If the user wants to upload or upstream some content, the UE will open the upload/upstream channel to his home storage on media servers in the Home Network and have it stored in that location.

Some arguments why it is better from technical point to store content in home network are

- Users wants to show their content to their friends, who are mostly from the same country (maybe same operator, so same home network).
- User wants to show this content, when he is at home.
- Content is often only in one mother language—area of interest may be from same country. The playback of this content in Visited network should be done from the Home Network, too.

14.3 Roaming Scenarios for NGN-Based IPTV Services

Based on the type of visited network, we can see different possible scenarios. Device can roam only from network, which offers these requirements minimally

- IP-based network with required network elements/technology
- Offering minimal bandwidth and connectivity, which is needed for provided services
- Support required protocols/interfaces and media transport/encapsulation

Basic IPTV roaming scenarios that can be identified [13,16] for IMS-based IPTV interconnections are following:

- Scenario A—Visited network is not IMS based.
- Scenario B—Visited network is based on IMS, but provider has no IMS/Converged IPTV platform.
- Scenario C—Visited network is based on IMS, but all IPTV services are offered purely from IPTV platform in home network.
- Scenario D—Visited network is based on IMS, some IPTV services can be offered from both IPTV platforms from home or visited network.

14.3.1 IMS-Based IPTV Architecture and Functional Entities

The IP Multimedia Subsystem (IMS)-based IPTV is one of the most advanced NGN-based IPTV implementation. The IMS is common service control subsystem used by NGN architecture. IMS-based IPTV was standardized by several standardization organizations, for example, by ETSI TISPAN Workgroup [11,12]. ETSI TISPAN has been trying to standardize functions and interfaces among them. The basic schema provided herewith describes the used functions and interfaces (Figure 14.5).

First of all, network must be aware, if the user is allowed to roam. This work is done by IMS, where the UE must be registered. Later, the user needs to know which services are available for him or her. This is shown by the UE. The UE gets this information from the Service Discovery Function (SDF), which consults the list of all available services for the user with the User Profile Server Function (UPSF) and also where to find the services selection information (in Service Selection Function—SSF). Therefore, the SSF is asked, in time of some service is selected, for the corresponding service/content identifier (e.g., Universal Resource Information—URI). Then the UE should know how a certain service can be reached (URI).

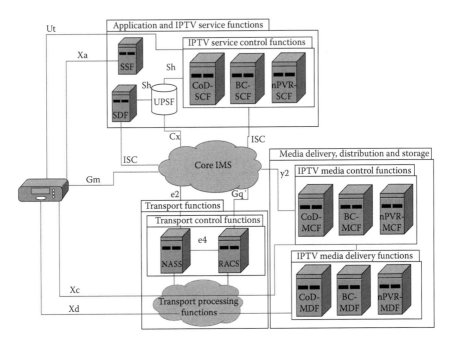

FIGURE 14.5
The schema of IMS-based IPTV.

When the UE contacts the application server (Service Control Function—SCF) via IMS core to get content directly, IMS core allocates some necessary resources. Then the SCF would request content from the media servers (Media Functions consist from Media Control Function MCF and Media Delivery function MDF). So SCF ask MCF to deliver certain content to certain user. MCF knows where the content is stored and arranges to start its delivery from the selected MDF.

All media functions are split on layer level into three groups SCF-s, MCF-s, and MDF-s.

The content is delivered from content sources (encoders for live content, disks for offline content). These functions are commonly called media delivery functions (MDF). Content travels via the network (Transport functions) and reaches the user sitting behind the UE. For correct transport over network (QoS), IMS core is involved to reserve resources in all network elements.

14.3.1.1 Service Discovery and Selection Functions (SDF and SSF)

The service list available for certain User behind certain UE is provided by Service Discovery Function (SDF) and Service Selection Function (SSF).

The main task of the SDF is to provide personalized service discovery and that of the SSF is to provide the service selection information, e.g., a list of available services that the UE can browse and select (e.g., Electronic Programme Guide, Content Catalogue).

14.3.1.2 Service Control Functions (SCF)

The SCF provides the service authorization during session initiation and session modification, which includes checking IPTV user's profiles in order to allow or deny access to the service by providing the credit limit and credit control.

SCF is also responsible for selection of relevant IPTV media functions for a given session.

14.3.1.3 User Profile Service Functions (UPSF)

The UPSF holds the IMS user profile and possibly IPTV specific profile data. It communicates with IPTV Service Control Functions and with the Core IMS. When multiple instances of a UPSF exist, the Core IMS and the IPTV Service Control Functions may use the services of a Subscription Locator Function (SLF) to fetch the address of the UPSF.

14.3.1.4 Media Control and Delivery Functions (MDF and MCF)

The IPTV Media Control Functions are responsible for controlling of media flow from MDF to clients. The IPTV Media Delivery Functions are responsible for the delivery of content to the client. MDF provides what is reported to MCF. MDF can also be used for transcoding for specific device requirements. MDF stores frequently used content (cache) and user specific content. MDF is exactly the element where n-PVR content is stored.

14.3.2 Scenario A—Visited Network Is Not IMS Based

Most of the networks nowadays do not support IMS functionality (metropolitan networks, Internet, private data network, etc.). If we want to provide IMS IPTV–based services (limited version of delivery like download or only for remote control of IPTV services) for the customers in these networks, it is necessary to "access" IMS functions over these networks. For this reason, we create tunnel over non-IMS network. Via this tunnel all IMS data will be transferred in a hidden form. UE then works as if it is directly connected to IMS network. The IMS network and UE feel that they are on one network, which is then controlled by IMS elements.

From technical side, because the Visited Network has no functionality of the IMS, the subscriber must use some remote data IP connection (e.g., VPN or secure remote access) to connect to his or her Home Network. Over this connection there can be transferred all media and signaling to the subscriber directly from his or her Home Network. Because such a connection should go over any IP network (also best effort) without resource reservation mechanisms, no QoS can be really ensured. This scenario may be used for IPTV services without guaranty, for example, remote control or download of n-PVR recordings.

In this scenario, there are two new components are added: VPN concentrator and VPN client. Because there can be many clients connected to home network, there are some strong capacity requirements on VPN concentrator. All data streams are packed into the tunnel. Tunnel is then transferred as a single TCP or UDP stream from the visited network point of view. In addition, advantages of multicast are lost, because all packets are split in VPN concentrator and transferred like a unicast stream.

If there are many issues with QoS, home platform should offer special type of access to content. It is useful to offer "download and watch" service. This service can be used only for offline content like n-PVR and CoD. This service downloads the content via the best effort service to the local storage and after all content is downloaded, it is ready for viewing.

In Figure 14.6, all the IPTV platforms are on the right side of tunnel UE is on the left side. Among this element is the "IMS gap," network which does not have functionality of IMS. To traverse this gap, there must be a VPN client on the UE side, this took all data from UE and packs them to forma transferable over "IMS gap" procedure On the platform side, there is the VPN concentrator, which unpacks data and sends to the correct elements via

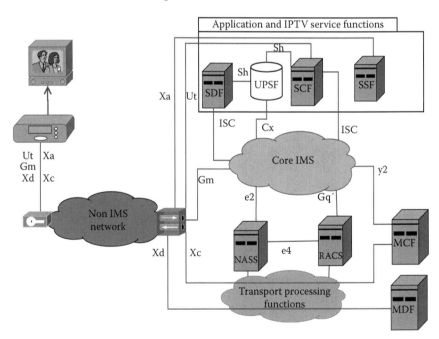

FIGURE 14.6
Scenario A—Visited network is not IMS based.

correct interface. When the data are "going back" from platform to the UE, VPN elements exchanges their roles. Of course, the communication is full duplex (exchanging of roles is just used for better explanation).

14.3.3 Scenario B—Visited Network Is Based on IMS, but Provider Have No IMS/Converged IPTV Platform

Additionally, to the previous scenario the following expects the IMS-based platform in Visited Network; however, all content and services are delivered only from Home Network. From the Visited Network, only the IMS functionalities for authentification and service initiation could be used.

The IPTV functionality is only offered by application servers on application level. If there is any IMS without IPTV application, it is necessary to offer this functionality from the home network. In this case, the Visited network serves only as a transport medium for data and signaling. But also the visited network is involved in communication on signaling level (IMS there can decline request for certain IPTV service).

This scenario is the simplest one with IMS involved. The subscriber will use all IPTV elements from his Home Network. The elements of visited network must provide just few functions:

- Check if certain subscriber is allowed roaming (via IMS roaming) to his visited network (Network-to-Network check).
- Collect and watch all data necessary for charging.
- Control limitation of its own network and apply limits to subscriber based on roaming policy agreed by Hosted and Visited Network operators.

The quality of the IPTV service for the end customer is the same as in home network, but no reuse of local media resources for the provider is possible (because in visited network there is no NGN-based IPTV platform).

The model of interconnection with the Visited Network within the Core IMS involved is used only for IMS type of roaming to access home IPTV service provider platform (Figure 14.7).

The UE can request the IPTV services from the Home Network when connected to the Visited Network. The Core IMS in the Visited Network or Home Network can request resources from the Resource and Admission Control Subsystem (RACS) in the Visited Network through the interface Gq'. The UE can be attached to the Visited Network, the Network Attachment Subsystem (NASS) in the Visited Network can assign the IP address for the UE and discover the address of P-CSCF (Proxy—Call Session Control function) in the Visited Network.

The Core IMS in the Visited Network can transfer the IPTV service request from the UE through the interconnection interface to the home Core IMS

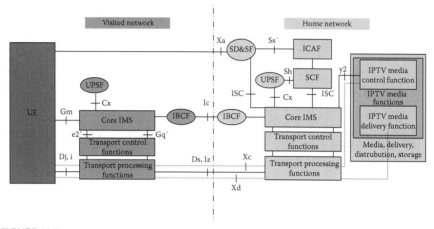

FIGURE 14.7
Scenario B—Visited network is based on IMS, but provider have no IMS/Converged IPTV platform.

through the connection of Intermediate Breakout Control Function (IBCF). The UE can connect to home SSF through the interface Xa. The UE can connect Service initiation, control as well as media delivery, over the existing Gm interfaces to the Home Network.

14.3.4 Scenario C/D—Visited Network Is Based on IMS, Some IPTV Services Can Be Offered from Both IPTV Platforms from Home or Visited Network

The model of interconnection with the Visited Network based on the Core IMS and IMS-based IPTV infrastructure is used for most advanced type of roaming to access the home IPTV service provider platform (Figure 14.8), where all services are provided only from Home Network (Scenario C) or from elements from both (Home and Visited) networks in Scenario D (using specialized MDFs for interconnect called I-MDF that performs content adaptation on the edge of both domains).

The UE can attach to the Visited Network to the NASS in the Visited Network, which can assign the IP address for the UE and discover the address of P-CSCF in the Visited Network. The Core IMS in the Visited Network or Home Network can request resources from the RACS in the Visited Network through the interface Gq'. The Core IMS in the Visited Network transfer the entire IPTV service request from the UE through the Core IMS in the Home Network through the connection of their own IBCF.

The UE is connected to the Home SDF and SSF (which can provide information about available content in this particular roaming case, because some of content could be restricted for roaming) to acquire service selection

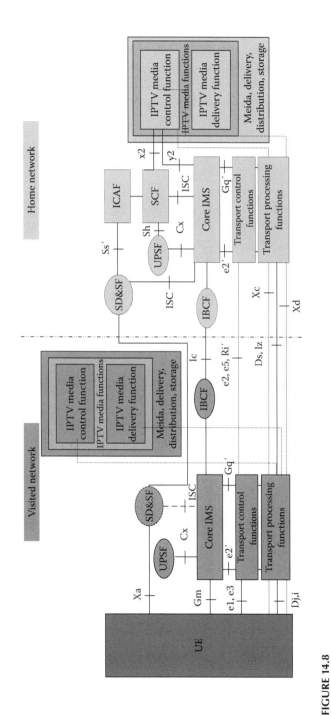

FIGURE 14.8
Scenario C/D—model of interconnection with the Visited Network with IMS-IPTV

information through the SSF interface Xa. The UE can be connected to the SCF in the Home IPTV network to configure parameters through the interface Ut. The Core IMS in both networks connect their own UPSF to get the user. But Home UPSF is responsible for service initiation authorization and user profile information required for personalization of IPTV services. The service initiation and control are provided over existing interfaces from Home Network (Gm to IMS and then others to Service Control Function (SCF), IPTV Converged Application Function (ICAF), or Media control Function (MCF)). The setup of media control and delivery channels is responsibility of the home SCF with MCF but media delivery itself could be provided from MDFs via Home or Visited Network through Xc (for control) or Xd (for delivery). The End User has to subscribe the roaming services at the Home Network before he/she moves to the Visited Network. The exact details and mechanism agreed between IPTV service providers have to be negotiated in advance together with interconnect agreements, policy rules, and most probably also service level agreements (SLA) to assure QoS on interconnect.

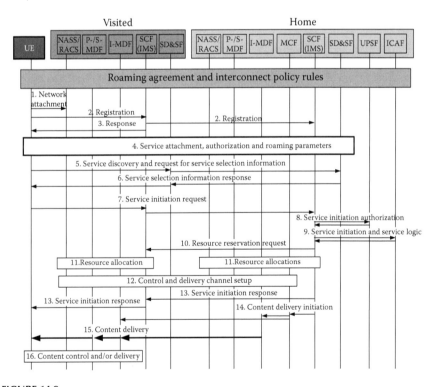

FIGURE 14.9
Procedures for interconnection with the Visited Network with IMS-based IPTV.

The following steps are proposed as the potential procedures [16] for NGN-IPTV interconnection with the Visited Network with IMS-IPTV (Figure 14.9):

1. Network Attachment: In this step the UE attaches to the network (with NASS, receiving IP configuration, P-CSCF address discovery of Visited Network, etc.).

2. UE performs IMS Registration in Visited Network. P-CSCF within Core IMS in Visited Network submits the registration request to the I-CSCF within the Core IMS in the Home Network.

3. Core IMS in the Visited Network return parameters (e.g., P-CSCF address of Home Network, SDF address) to the UE.

4. Home and Visited Network elements can required some exchanges of relevant information before allowing the user and his or her terminal attached to IPTV services and select or initiate any service.

5&6. UE perform service discovery and service selection contacting home SSF directly or via visited SDF and/or SSF (to compile a set of services and metadata, which can be provided to users).

7. User initiated request from UE for selected service is sent via visited IMS to home IMS. In the home IMS and SCF are the location where requests are processed (could be authorized by home UPSF—8 - and also apply service logic with interaction with IPTV apps—9).

10. After a successful authorization, the SCF initiates resource reservation in Home and/or Visited Network using IMS available mechanisms toward RACS

11. Later, the setup can also control and delivery channels to UE.

12. Finally, Home SCF sends response to UE.

13. The Home SCF initiates the content delivery through MCF via appropriate MDFs (14) to UE (15, 16).

14.3.5 Accessing Services

The SDF and SSF in the Visited Network should collect information from Visited Network about the services that the subscriber can use in roaming and also have access to relevant service information from Home Network. If this service is offered in Visited Network, the subscriber will get it directly from the Media distribution and Storage Subsystem of the Visited Network. If such service is not present in Visited Network, the UE will use it from the Media distribution and Storage Subsystem of Home network.

Since the UE or Visited Network can have some specific restrictions to format of content, the I-MDFs in Visited Network must be able to transcode any stream to the requested format and serve it to the UE. If there is no conjunction in the functions of the Visited Network, I-MDF, and UE possibilities, this media cannot be delivered to the customer from visited network

TABLE 14.1

Examples of Codec Profiles for IPTV

Profile	Video Codec	Audio Codec	Container	Bandwidth
Mobile	H.263	AAC	3GPP	256 kbps
WEB	H.264, baseline	mp4a	mp4	768 kbps
SD	H.264, main	MP2/AAC	MPEG2 TS	3.5 Mbps
HD	H.264, high	MP2/DD+	MPEG2 TS	12 Mbps

IPTV platform. Therefore, it is very useful for the IPTV providers in roaming relation to offer all the content in several bandwidth profiles (Mobile/WEB, SD, HD) with standardized common used codecs (3GPP, FLV, H.264, VC-1, MPEG2), preferably the same (to avoid transcoding). All UEs should support the agreed and standardized codecs used in the Home Networks and in the Visited Network during roaming. Table 14.1 is an example, but the worldwide discussion and agreement should be found and a global table be proposed for interconnection agreements. This will help all service providers to save costs for technology and will increase interconnection possibilities.

To save the resources in the provider's networks, it is useful to define a closed group of bandwidths and codes used for the streaming, for example as provided in the following.

If there is a service in the Visited Network that is not the same as that one, which was requested from the UE, it is possible to combine content/service provided partially from the local source and from the Home Network source. If the video stream with some common content (e.g., CNN TV) is present in the Visited Network, but the language is different as the requested one, the MDF can request only this elementary stream with missing language. This process requires global synchronization information, which has to be put in stream, (e.g., RTP timestamp).

14.3.6 Use Case of IPTV Roaming

We can imagine that a customer (Peter) from Slovakia has purchased IPTV service from the home operator (SK_TEL) [14,15]. SK_TEL has its network based on IMS. In summer, Peter decided to visit some holiday destination, for example Spain. In Spain, there is another operator ES_TEL, who has his network based on IMS and also has IPTV platform. Both operators have SLA in IPTV roaming. Peter takes his UE with him. On a raining day, he decides to see some TV content. He unpacked his UE and connects it to the hotel plug. The entire hotel is provided with ES_TEL lines or via wifi access. First of all Peter checks what is new in CNN, because CNN is popular worldwide; ES_TEL network also has such a channel. After UE requests for CNN channel, ES_TEL network started to offer this channel from local MDF.

Subsequently, Peter wants to know what is new in Slovakia; therefore, he switched to TA3 Slovak information channel. Because this channel is only available in SK_TEL network, ES_TEL network has to serve this channel from SK_TEL network. ES_TEL IMS IPTV contacts SK_TEL home network to serve TA3 channel. MCF of SK_TEL asks his MDF to send channel to ES_TEL MDF. ES_TEL MDF forwards that channel to UE. Peter can now see TA3 channel on his TV in Spain.

Peter realized that there is an interesting show on MTV. Therefore, he recorded the program. His UE has no HDD; therefore, recording must be done through network. Which network will be chosen? SK_TEL network is the holder of the user space in network; therefore, this recording will be done in home network. If Peter wants to show this content at his home, it will be streamed directly from Home network to his device connected in Home network.

14.4 Conclusions

This chapter provided an introduction on interconnection of NGN-based IPTV systems.

Section 14.1 gave a short overview of the complex worlds of content distribution, involving various business roles and a wide variety of business models. It introduced the concept of "interconnection," applied to content delivery and IPTV services. Two groups of business models are most prominent for this type of interconnection:

- CDN interconnection: parties collaborate to distribute content with higher geographical reach and better QoE.
- IPTV roaming: parties collaborate such that users of one network can use resources of another network.

Then, we elaborated some use on cases related to IPTV roaming, including linear/broadcast TV, trick play of live TV stored in network, time-shift TV service served by network, CoD service, PVR service, and UGC.

The main part of the chapter covered the various IPTV roaming scenarios. The simplest scenario involves just data connectivity, and all services being delivered by the home network. A more advanced scenario involves IMS interconnection to enable bandwidth and Quality of Service (QoS) guarantees. The most advanced scenarios have full IPTV service interconnection, where some content and services are obtained from the home network, whereas other content and services are obtained from the visited network.

Acknowledgments

This chapter has been partly supported by SenterNovem, an agency of the Dutch Ministry of Economic Affairs. Its writing has been carried out in the context TNO participation on the European EUREKA CELTIC RUBENS project.

This paper also presents some of the results from participation in various research projects at Slovak University of Technology in Bratislava such as NGNlab project [14], European CELTIC EUREKA project Netlab [15], Slovak National research projects: AV project 4/0019/07: Converged technologies for next generation networks (NGN), and Slovak National basic research projects VEGA 1/0720/09 and VEGA 1/4084/07.

The authors thank colleagues from ETSI TISPAN for critical discussion and suggestion in the area of IPTV interconnection.

Abbreviations

3GPP	Third generation partnership project
AAC	Advance Audio Coding
Bcast	Broadcast
CDN	Content Delivery Network
IMS-IPTV	IMS based IPTV
CoD	Content-on-Demand
CPE	Customer Premises Equipment
c-PVR	Client-based PVR
DRM	Digital Right Management
ETSI	European Telecommunications Standards Institute
FLV	Flash Video
GSM	Global System for Mobile Communications
HD	High Definition
ICAF	IPTV Converged Application Function
CSCF	Call Session Control Function
IMDF	Media Delivery Function
MCF	Media Control Function
IMS	IP Media Subsystem
IP	Internet Protocol
IPTV	Internet Protocol Television
ISCF	IPTV Service Control Function
NASS	Network Attachment Subsystem
NGN	Next Generation Networks
nPVR	Network PVR
OTT	Over the top
P2P	Peer to peer
PC	Personal Computer
P-CSCF	Proxy-call session control function
PVR	Personal Video Recorder
QoE	Quality of Experience
QoS	Quality of Service

RACS	Resource and Admission Control Subsystem
RTP	Real-time Transport Protocol
SCF	Service Control Functions
SD	Standard Definition
SDSF	Service Discovery & Service Selection Function
SIP	Session Initiation Protocol
SLA	Service Level Agreements
SS7	Signaling System No. 7
STB	set-top box
TCP	Transmission Control Protocol
TISPAN	Telecommunications and Internet converged Services and Protocols for Advanced Networks
TV	Television
UDP	User Datagram Protocol
UE	User Equipment
UGC	User Generated Content
UPSF	User Profile Server Function
VC-1	SMPTE 421M video codec
VPN	Virtual Private Network

References

1. Wikipedia 2009, Metcalfes's law, Wikipedia Foundation, Inc., San Francisco, CA, http://en.wikipedia.org/wiki/Metcalfe's_law (Accessed August 3, 2009).
2. S. Stephen, Researchers: Metcalfe's law overshoots the mark, 2005. http://news.zdnet.com/2100-1035_22-141783.html (Accessed August 3, 2009).
3. S. Van den Berghe, P. Nooren, S. Latré, B. Crabtree, M. Kind, E. Viruete, and C. Pons, QoE-driven Broadband Access, *NEM summit 2008*, Saint-Malo, France, September 28–30, 2009. http://www.nem-summit.eu/Presentations/Day2/SessionsA/A4/QoE.pdf
4. BBC iPLayer 2009. BBC. http://www.bbc.co.uk/iplayer/
5. Uitzending Gemist 2009, Omroep nl. http://www.uitzendinggemist.nl/
6. Eureka-Celtic project RUBENS, Re-thinking the use of broadband access for experience optimized networks and services, http://www.celtic-initiative.org/Projects/RUBENS/
7. ETSI TS 181 016, TISPAN; Service Layer Requirements to Integrate NGN Services and IPTV, ETSI, Sophia Antipolis, France, 2009.
8. Input to these discussions was a joint TNO-KPN-Slovak Telecom contribution, 19WTD057, WI2070 Cross-Domain MF-MF Reference Point, ETSI TISPAN 19W, Sophia Antipolis, France, November 3–5, 2008.
9. Draft ETSI TS 182 019, TISPAN; Content Delivery Network (CDN) architecture—Interconnection with TISPAN IPTV architectures, ETSI, Sophia Antipolis, France, September 2009.

10. Wikipedia 2009, Roaming, Wikimedia Foundation, Inc., San Francisco, CA. http://en.wikipedia.org/wiki/Roaming (Accessed Jun, 18 2009).
11. ETSI TS 182 027, TISPAN; IPTV Architecture; IPTV functions supported by the IMS subsystem, ETSI, Sophia Antipolis, France, 2009.
12. Draft ETSI TS 183 063, TISPAN; IMS based IPTV Stage 3 Specification, ETSI, Sophia Antipolis, France, September 2009.
13. Draft ETSI TR 182 030, TISPAN; NGN based IPTV mapping or interconnect between IPTV systems, ETSI, Sophia Antipolis, France, September 2009.
14. NGNlab—NGN laboratory, Slovak University of Technology, Bratislava, Slovak Republic, http://www.ngnlab.eu
15. Eureka-Celtic project NetLab: Use cases for interconnected testbeds and living Labs, http://www.celtic-initiative.org/Projects/NETLAB/default.asp
16. E. Mikóczy, R. Kadlic, and P. Podhradský, Concept for mobility and interconnection aspects in Converged NGN based IPTV architecture, *International Journal of Digital Multimedia Broadcasting*, Special issue: Convergence of Digital TV Systems and Services, Hindawi Publishing, New York, 2010, ISSN: 1687-7578, e-ISSN: 1687-7586.

15

End-to-End QoS and Policy-Based Resource Management in Converged NGN

Dong Sun and Ramesh Nagarajan

CONTENTS

15.1 Introduction

Several industrial standard organizations and forums have been taking the initiative on next generation networks (NGN) in recent years. For instance, the European Telecommunications Standards Institute (ETSI)—Telecoms & Internet converged Services & Protocols for Advanced Networks (TISPAN) focuses on an NGN for fixed access network, which has published Release 1 in 2005 [1]. Meanwhile, International Telecommunication Union's Telecommunication Standardization Sector (ITU-T) started the NGN Global Standards Initiative (NGN-GSI) and has published its first release [2] in 2006. On the other hand, a similar effort has been made in wireless network domain, UMTS (i.e., W-CDMA) and CDMA 2000 defined in 3GPP and 3GPP2 are categorized as third-generation mobile network technologies and are now evolving to fourth-generation mobile network—Evolved Packet System (i.e., Long-term Evolution/Evolved Packet Core (LTE/EPC)), which can be regarded as network generation mobile networks. The common notion of a variety of NGN is to transport all information and services (voice, data, video, and all sorts of multimedia applications) by utilizing packet network Internet Protocol (IP) technology, i.e., an "all-IP" network. Figure 15.1 illustrates a high-level overview of NGN architecture defined in ITU-T. The relevant components are described in Section 15.4.

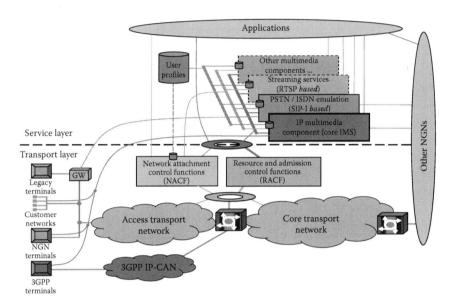

FIGURE 15.1
NGN overview.

15.2 QoS Requirements and Challenges in NGN

As the key differentiator against "best effort" legacy Internet that is also based on IP technology, NGN aims to provide consistent, manageable, and personalized service quality for the end users across different transport media, technologies, and operator's administrative domains, which is also the key enabler for ubiquitous services as the ultimate goal of fixed mobile convergence (FMC).

In order to ensure consistent user experience across fixed and wireless networks, the main challenge is to provide end-to-end Quality of Service (QoS) and to optimize transmission of all data, voice, and video applications to and among end users, no matter what their locations or devices are.

Generally speaking, the performance needs of applications are characterized by four key parameters: bandwidth, packet loss, delay, and jitter (i.e., delay variation), which determine the QoS.

Overall, the needs of the distinct applications are different with respect to QoS. For example, some popular data applications (such as e-mail and Web access) require low to medium bandwidth and are quite relaxed as far as the delay and jitter are concerned. On the other hand, Video-on-Demand (VoD) flows have relaxed requirements on delay, but they do need high bandwidth and cannot tolerate much packet loss or jitter. Voice over IP (VoIP), while tolerating some packet loss, needs much lower bandwidth than VoD, but it can tolerate neither long delay nor jitter. Delay becomes perceptible to the user somewhere between 150 and 400 ms, and some network operators specify maximum delay in the range of 250–300 ms. Jitter should stay under 50 ms.

In addition to the QoS, the networks often need to grant other resources (e.g., IP addresses and service-related port numbers) to the end users and the processes that execute on them. At the same time, the networks must protect themselves from unauthorized use of the resources and hide their topology. These diversified and already complex tasks are further complicated by the very structure of the NGN, which combines a variety of fixed and wireless access network types, including digital subscriber line (DSL), Cable, and UMTS, CDMA 2000, LTE, and WiMAX.

There are three main components of delivering QoS-oriented services. First, the network elements should be able to recognize service requirements and provide adequate resources for each service on demand for each subscriber; note that subscribers themselves may belong to different classes (e.g., consumers or business). Second, networks should be designed with QoS requirements in mind so that both links as well as nodes have sufficient resources; this will include reliability, security, and redundancy requirements. Finally, appropriate Service Level Agreements (SLAs) will have to be instituted with partner operators to provide end-to-end services.

In IP-based transport networks, there are many implementations of QoS; most of them use one of these two models: Integrated Services (IntServ) [3] and Differentiated Services (DiffServ) [4]. Due to the complexity and scalability issues of IntServ model, DiffServ is widely employed in the packet transport network especially in the aggregation and backbone networks. However, DiffServ can only provide a coarse-grained QoS control and is handicapped to guarantee the QoS at the micro-flow level. This inherent characteristics of DiffServ may not have a big impact on non-time-sensitive applications (e.g., e-mail, Web browsing, FTP) in the backbone network, but it does have significant impact on time-sensitive applications (e.g., voice, video, and gaming) especially in a resource contention access network, e.g., DSL, Cable, and Radio access network. It is widely acknowledged that connection admission control (CAC) is essential to assure the QoS to time-sensitive applications in resource contention access networks.

Furthermore, with the advent of numerous new applications and end users' devices, it is unrealistic to mandate each application can understand and support all of distinct QoS mechanisms and parameters in different network technologies, e.g., DSCP in MPLS, ToS in IP, 802.1p in Ethernet, DOCSIS (Data Over Cable Service Interface Specification) in Cable, and UMTS QoS in 3GPP GPRS.

This is more significant in the circumstance of FMC since the end users may roam from one type of access network, e.g., UMTS, to the other type of access network, e.g., WiFi/DSL, requiring service continuity with consistent quality. Therefore, it is sensible to make the end user's application transparent to network technology and accommodate the service attributes into different network technologies within each transport network domain.

Finally, one of the most fundamental features in NGN is to be able to provide customized service plan based on end users' preference, application, access network type, QoS condition, and all kinds of subscription and network policies. Obviously, only using the legacy QoS mechanisms at the transport network layer is incapable of providing this kind of capability.

In summary, the following capabilities are essential to the success of NGN and FMC:

1. Ensure consistent service quality in different networks.
2. Insulate the applications from underlying transport technologies.
3. Provide personalized and customized service quality based on end users' subscription and network policy.
4. Guarantee the QoS at application session and media flow level upon request.

To achieve these objectives, a policy-based QoS and resource management solution is the way forward. In the following sections, the architecture and mechanisms for various wireless and wireline access network technologies, including UMTS/CDMA 2000, LTE, WiMAX, as well as cable and broadband access DSL, will be reviewed. A typical use case in FMC environment—dynamic QoS control for Femtocell service—is illustrated.

15.3 QoS and Policy Control Solution in 3GPP

The evolution of 3GPP network architecture can be generally divided into three phases: (1) Phase 1—Rel 99/ Rel 4 items cover general aspects of wireless packet data service similar to Generic Packet GPRS; (2) Phase 2—Rel 5/ Rel 6/Rel 7 issues involve the IP Multimedia Subsystem (IMS) using a wideband radio access, i.e., High Speed Packet Access (HSPA); (3) Phase 3—since Rel 8 "all-IP" packet only architectures are defined for the LTE/EPC.

Owing to the advance of radio technology, the peak date rate increases rapidly. Rel 99 in theory provides 2 Mbps, but in practice only gives 384 kbps. HSPA in Rel 5 and Rel 6 raises the peak rates to 14 Mbps in downlink and 5.7 Mbps in uplink. HSPA+ evolution in Rel 7 pushes a maximum 28 Mbps in downlink and 11 Mbps in uplink. LTE in Rel 8 and beyond will add the peak rates beyond 100 Mbps in downlink and 50 Mbps in uplink using a 20 MHz bandwidth. Moreover, the LTE is evolved toward more simplified "flat" network architecture, which improves the efficiency for packet services significantly.

With the increase of radio access capacity, it enables the possibility of new applications, e.g., video and online gaming that requires higher bandwidth with stringent QoS requirement compared to regular data applications, e.g., Web browsing. As a consequence, the QoS in 3GPP for packet services is evolving with the evolution of radio access and core network technology. The initial QoS concept is mainly limited to GPRS/UMTS-specific transport technique; four QoS classes are defined in Rel 99: Conversational, Streaming, Interactive, and Background. The main distinguishing factor among these classes is the delay, jitter, and packet loss sensitivity. The Conversational class has the most delay and is jitter sensitive, while Background is the least sensitive to those. Conversational and Streaming classes are intended for real-time or near-real-time traffic. They both need preservation of time relation (variation) between information elements of the stream, but Conversational class has stricter delay requirements. For the Interactive and Background classes, end-to-end delay is not the major factor; instead, they both require preservation of payload content. Interactive class traffic follows a request-response pattern and defines priorities to differentiate between bearer qualities within the class itself; it does not provide explicit delay and jitter guarantees. The main characteristic of Background class is that the destination does not expect the data within a certain time. Table 15.1 depicts some example applications and important characteristics of each traffic class for UMTS network.

Rel 99 and Rel 4 only utilize in-band signaling, e.g., PDP Context to establish a bearer path from the end point (i.e., UE) to the Gateway (i.e., GGSN). It may provide basic QoS of bearer traffic, similar to ATM kind of connection-oriented network. The Gateways (i.e., SGSN/GGSN) may check UE's subscription obtained from HSS/HLR against the QoS request initiated by the UE. However, it imposes the limitation for new applications, e.g., IMS when the UE requires dynamic and customized QoS authorization for each individual application session. Furthermore, it remains a challenge to offer flow-based charging capability that is regarded as the main motivation for network providers to offer QoS differentiation.

The Policy Control Function (PCF) [5] is introduced in Rel 5 in conjunction with IMS Application Function (AF) together, which an open "southbound"

TABLE 15.1

Key Characteristics and Example Applications of Each Traffic Class in UMTS Network

Traffic Class	Conversational	Streaming	Interactive	Background
Typical characteristics	Symmetric bidirectional	Mostly unidirectional	Asymmetric traffic	Not time sensitive
	Stringent delay and jitter requirements	Time relation between info. elements very important	Sensitive to errors	Sensitive to errors
Sample applications	VoIP, videoconferencing	Audio/videostreaming	Web browsing	File transfer

interface Go is standardized for policy enforcement control toward the GGSN. In Rel 6, motivated by the layered control methodology between transport network and application service, the PCF is entirely separate from the IMS AF, which is essential to enable a universal intermediary of network layer for a variety of applications (e.g. IMS and non-IMS). A new, open "northbound" interface Gq is defined to support the QoS resource authorization from AF to PCF. Rel 7 makes further improvement on the consolidation of PCF with Charging Control Function (CGF) and names as Policy and Charging Rules Function (PCRF) since the information elements and procedures for policy control and charging control are very similar; two instances of PCRF home PCRF (H-PCRF) and visited PCRF (V-PCRF) are defined in support of roaming scenario; the Policy and Charging Enforcement Function is abstracted as a logical building block in the Gateway. To some extent, Rel 7 forges the baseline of current Policy and Charging Control (PCC) [6] architecture, which is the first inclusive standard in 3GPP on this subject (Figure 15.2). In Rel 8/Rel 9 (Figure 15.3), in order to support all-IP-based LTE architectures [7,8], in particular for Proxy Mobile IP (PMIP) variant [8], the PCRF is generalized for supporting non-GPRS like access technologies. One of key parameters/notion is the introduction of QoS Class Identifier (QCI): a generic scalar that is used as a reference to a specific packet forwarding behavior (e.g., packet loss rate, packet delay budget) to be provided to a media flow. Using QCI, the network node (i.e., PCEF) is responsible for translating the generic QCI class into transport-specific QoS class, e.g., UMTS QoS, DSCP, ToS or 802.1p, which provides the potential to realize a converged policy framework for both fixed and mobile access networks.

The main functional entities of PCC in Rel 8/9 consist of the following:

- Policy and Charging Rules Function (PCRF): provides QoS resource and charging control based on end users' subscription and network policy.

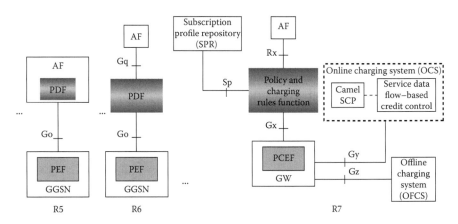

FIGURE 15.2
Evolution of policy control architecture in 3GPP.

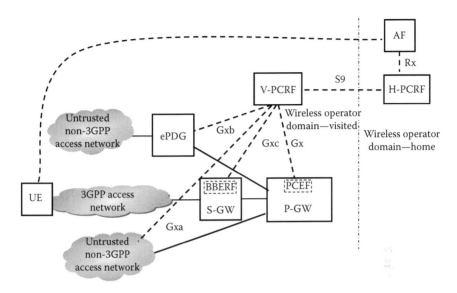

FIGURE 15.3
PCC architecture for LTE networks (PMIP, local breakout).

- Policy and Charging Enforcement Function (PCEF): provides application flow detection (e.g., deep packet inspection), data traffic handling, QoS handling, and media flow measurement as well as online and offline charging interactions.
- Subscription Profile Repository (SPR): contains all subscriber/subscription-related information for transport network level policy control, for example, subscriber's allowed services, pre-emption priority, QoS (e.g., Subscribed Guaranteed Bandwidth QoS), charging-related information (e.g., location information relevant for charging), and subscriber category.

In addition, Bearer Binding and Event Reporting Function (BBERF) as a subset of PCEF with limited capability of QoS control, is introduced for the PMIP (Proxy Mobile IP)-based LTE architecture. The main functions of BBERF are to provide QoS marking, policing, and special event reporting without charging control capability. In terms of bearer-binding functionality, one major change in LTE is that the network nodes (e.g., PDN Gateway and/or Serving Gateway) perform the bearer binding instead of the PCRF. Basically, the network nodes link an application flow into a particular transport bearer based on the QoS information as well as IP address and user identity, which simplifies and generalizes the PCRF functionality to separate from specific transport technology, and enables the possibility for a generic policy control framework for different types of access networks.

The basic notion of policy control function in 3GPP is as follows: when the UE starts a new bearer session, it sends a request through in band signaling, e.g., PDP Context toward the gateway; the gateway may verify the user subscription including maximum allowed bandwidth, QoS class with policy server (i.e., PCRF) before confirming the service request and allocating the resource; when the UE starts a new application session, e.g., SIP call, it sends a request through application signaling, e.g., SIP Invite to the AF, the AF requests the QoS authorization to the PCRF, the PCRF then performs the authorization and instructs resource allocation in the gateway.

15.4 Resource and Admission Control Architectures in ETSI TISPAN and ITU-T

As part of NGN Initiative, ETSI TISPAN and ITU-T have defined policy-based resource and admission control architectures, respectively, as Resource and Admission Control Subsystem (RACS) [9] and Resource and Admission Control Functions (RACF) [10]. In light of similarity of these two architectures, the ITU-T RACF model is described as the example.

As aforementioned, the NGN requires that application services and transport networks are insulated from each other, i.e., the application does not need to understand the specialty of transport network technology and the transport network does not need to know the attributes of individual application, which allows third parties to provide services independent of the underlying transport network. NGN has reflected this requirement in its overall architecture, schematically depicted in Figure 15.1, which we will consider here only insofar as the RACF is concerned. Note that the third-party applications access the NGN service stratum, where various service control functions, both IMS and non-IMS. The actual transport networks with routers, transmission facilities, and supporting servers belong to the transport stratum. RACF is positioned exactly on the border as the arbitrator of the service and transport strata.

Performance assurance (i.e., QoS) is achieved through the adequate use of resources within the transport stratum. There are two aspects of performance assurance: policy enforcement at the network borders and the verification that there are sufficient resources along the end-to-end path (and, if applicable, the reservation of these resources for the duration of a session).

Policy enforcement may involve a number of QoS and non-QoS-related functions, such as policing, filtering, QoS marking, metering, NAPT and media latching in support of remote NAT traversal. (Note that filtering and NAPT also contribute to network security.)

The RACF architecture, as illustrated in Figure 15.4, consists of three logical functional entities: policy decision functional entity (PD-FE), transport

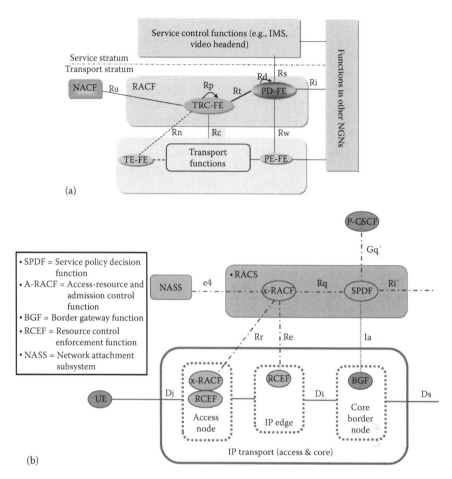

FIGURE 15.4
(a) ITU-T RACF and (b) TISPAN RACS architectures.

resource control functional entity (TRC-FE), and policy enforcement functional entity (PE-FE).

The PD-FE is the superset of the PDF in the 3G model and the SPDF in the TISPAN model. It is transport-independent. The northbound interface to the service control functions (SCF) i.e., application functions in 3GPP such as video headend, is an inter-provider interface. Thus, an application provider that does not own a network infrastructure can use this interface to request resources from the network provider. The PD-FE performs resource authorization and media reservations including gate control, QoS and priority handling, selection of the firewall working mode, and NAPT commands. In addition, the application may use this interface for requesting resource statistics and to address mapping information.

Before a user gains access to a service, the respective end-point device needs to be accepted by the network (i.e., it needs to be authenticated and authorized, at which point it will be provided with an IP address). To deal with the registration and initialization of user equipment as well as with the necessary security bootstrapping, the NGN architecture defines the network attachment control functions (NACF), which is an equivalent of the NASS in the TISPAN model. The PD-FE in the access network interacts with the NACF to obtain the transport subscription information, and binding information between the physical and IP addresses.

The TRC-FE monitors the status of resources and collects the relevant network information. Based on this information, it performs resource-based admission control. In addition, it controls transport-dependent policies. To this end, the TRC-FE verifies whether a request from a PD-FE matches the transport-specific policies. Multiple instantiations of the TRC-FE may exist in different sub-domains of the transport network. These instantiations communicate with one another via an open interface, and interact with the transport functions in order to collect the network topology and status information. In general, the role of the TRC-FE is to monitor network utilization and to use this knowledge for flow admission. To achieve that, the TRC-FE can use a combination of accounting, measurement-, and reservation-based methods. Similarly, A-RACF in TISPAN RACS also provides this kind of functionality as well as the policy enforcement control functions.

The last functional entity that remains is the policy enforcement functional entity (PE-FE), which enforces the network operator's policy on both a per-flow and per-subscriber basis. The PE-FE opens and closes the gates, limits the packet rate, and performs traffic classification, packet marking, NAPT, NAT traversal, and usage metering. It also maps the network-layer QoS information into that of the link layer. The PE-FE combines both Border Gateway Function (BGF) and Resource Control Enforcement Function (RCEF) in TISPAN RACS, which are more physical implementation oriented.

The RACF framework supports both push-and-pull models for the interactions between PD-FE and PE-FE. A typical example of the use of the pull model is the 3GPP mechanism, which we mentioned earlier in this chapter. In wireline applications, the push model dominates, since there are few examples where an endpoint directly requests resources through transport-layer signaling. Instead, resources are allocated based on an application request from the service layer.

15.5 QoS and Policy Control Solution in CableLabs

Multiple Service Operator (MSO) is the first one to apply the policy control notion to the real world. Major cable service providers in the United States including Comcast and Time Warner have been deploying fair usage cap and bandwidth limit based on CableLabs' standardized policy architecture.

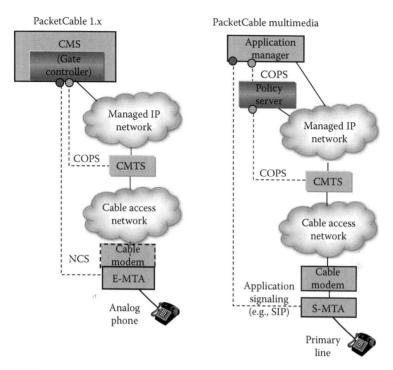

FIGURE 15.5
PacketCable 1.x and PCMM.

CableLabs is the earliest industry standard organization to define a policy-based QoS control architecture. There are two variants of policy control architecture in the PacketCable standard: PacketCable 1.x [11] and PacketCable Multimedia (PCMM) [12] (Figure 15.5).

PacketCable 1.x mainly provides QoS control for a dedicated VoIP service offered by cable modem using soft-switch technology, which bases specific mechanism in cable modem to initiate a QoS request through in-band DOCSIS signaling. The mechanism used in PacketCable 1.x is named as Pull model since it requires the end-user device to trigger the QoS process and pull the policy decision from policy server.

Envisioning the trend of multimedia applications, e.g., VoD PCMM provides more flexible mechanism to make the QoS control process less dependent on the end user's QoS capability. The main functional entities of PCMM consist of the following:

- Application manager (AM): provides application policy control to trigger the QoS request on behalf of end user application.
- Policy server (PS): provides network policy control upon receipt of the QoS request from AM and instructs CMTS to enforce the QoS rules.

- Cable modem termination system (CMTS): provides network policy enforcement functions to allocate network resources and handle user data traffic.

In the PCMM, AM can initiate QoS resource to the policy server on behalf of end user, and then policy server sends the decisions to the CMTS. This model is named as Push model that is viewed as main stream for advanced applications, e.g., video as well as IMS applications.

The following three types of end user devices are specified:

- Type 1, which represents existing "legacy" endpoints (e.g., PC applications, gaming consoles) that lack specific QoS awareness or signaling capabilities.
- Type 2, which is similar to a PacketCable 1.x telephony MTA in that it supports QoS signaling based on the PacketCable DQoS specification.
- Type 3, which directly requests QoS treatment from the access network without AM interaction. This client is aware of IETF standards–based RSVP and uses this protocol to request access network QoS resources directly from the CMTS.

Type 1 device communicates with an AM to request service, and does not (cannot) request QoS resources directly from the MSO access network, which can be supported by PCMM. Type 2 device can be supported by either PacketCable 1.x or PCMM. Type 3 device is not supported by either PacketCable architecture due to complexity and unavailability of this kind of device.

In addition, Broadband Forum, as the main stakeholder for fixed broadband transport networks including DSL, PON, and MPLS, has been intended to define a policy framework. It is still at the preliminary stage to consolidate use cases and high-level requirements and unclear what exactly its policy framework will look like at the moment.

15.6 Use Case of Dynamic QoS Control in Fixed Mobile Convergence: Femtocell QoS Control

Femtocells, which operate in the same RF (Radio Frequency) technology as regular "macro-cells" for cellular networks (e.g., GSM, CDMA, UMTS, and LTE), but utilize an indoor "mini-compact" base station with similar transmission and reception parameters and for channel demodulation as the WiFi access point, has received a great deal of attention in the recent years.

The rationale for femtocell technology by the mobile wireless operators is to provide a cost-effective solution to improve the indoor coverage implemented in residential and business service areas and offload the traffic from expensive macro cell sites. Now fixed network operators are showing more and more interest in femtocell technology as the evolution toward a FMC network infrastructure.

A femtocell (i.e., Home Node-B (Femtocell) or Home eNode-B (HeNB) in 3GPP term) will be connected to a wireless network via a wireline backhaul network that may belong to the same network provider (NP) as the wireless network or to a different NP. From a femtocell standpoint, the wireline backhaul network refers to the network that connects a femtocell to the SeGW in the wireless network. Figure 15.6 depicts femtocell reference architectures for UMTS [13] and LTE networks [14].

It is of essence to the success of femtocell service to provide competitive carrier grade of service as regular cellular service since the end user is under the same subscription and expects the same level of quality. Although a Femtocell network is able to provide QoS at the wireless network layer, there is no standard approach defined to support QoS in the wireline backhaul network. The default approach is that the IPSec tunneled traffic is transmitted as best-effort Internet-like service over backhaul networks.

However, this approach leads to a big "hole" in terms of providing end-to-end QoS on par with the macro cells of a wireless network since it cannot guarantee the "carrier grade" services due to the inherent characteristics of best effort class. For example, packet delay variation can lead to significant jitter, latency, or packet reordering issues, thereby impacting the QoS of conversational and streaming traffic classes. In turn, an overall delay in packet transmission will severely impact the end-to-end latency for voice and video services. Quality of the backhaul IP connection typically varies upon several factors, e.g., actual DSL bandwidth, overall network load and latency, and efficiency of routing in intermediate network nodes.

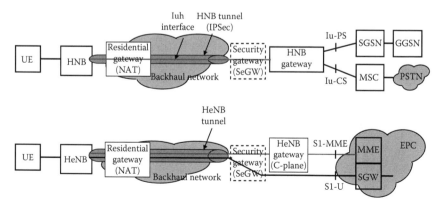

FIGURE 15.6
Femtocell reference architectures for UMTS and LTE.

Moreover, femtocell service flows must compete for resources with service requests from fixed access devices also deployed in that access network. According to HNB requirement in 3GPP [13], the total bandwidth from the femtocell toward the network to support four simultaneous voice calls (which includes signaling and overhead), shall not exceed 200 kbps. As a result, femtocell traffic may use up to 2/3 of total upstream capacity over a DSL link in certain cases, which, based on operators experience of DSL links, is typically ~300 kbps for upstream and 1 Mbps for downstream.

If the end user desires to access high bandwidth services using both mobile and wireline devices, regular broadband services such as HD streamed video, HQ videoconferencing, media sharing, and remote media access each can easily consume a significant portion of the available bandwidth. In this case, the resource limitations in the wireline access network will be compounded due to the fact that Femtocell traffic and regular BroadBand traffic will be competing for the same resources. When High Speed Packet Access (HSPA) and LTE are used in packet mode for data services, the effective data rate of downlink traffic can be ~8 Mbps or more, which will have additional impact on the downstream direction.

Therefore, it is important that resources are dynamically allocated in the wireline backhaul network before a new or modified flow is admitted in the wireless network. The initiation of the QoS request by the wireless network and conveying the request to the wireline backhaul network for dynamic resource admission/allocation is especially critical when the femtocell is connected to a DSL network that has resource constraints with respect to the available bandwidth.

The resolution is to utilize the interaction between wireless policy system and wireline policy system for enabling the wireless/femtocell network to initiate a QoS authorization request on the fly, which is viewed as one of the basic use cases for supporting end-to-end QoS in FMC circumstance.

Figure 15.7 depicts an example of end-to-end procedures for UMTS network. Although TISPAN RACS is used in this example, the same methodology is applied to ITU-T RACF or other policy system, and even PCMM for cable network.

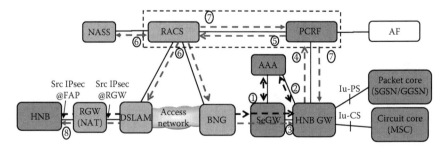

FIGURE 15.7
High level procedures of dynamic QoS control for Femtocell.

1. During the IPSec authentication, SeGW and AAA server derive the global routable Source IP address of IPSec (i.e., Src IP@Femtocell or Src IP@RGW) and stores in the AAA.

2. During femtocell registration, Femtocell GW and AAA server set up the association of Src IP address of IPSec and Femtocell ID and store in the Femtocell GW (or store in the AAA and retrieve by Femtocell GW when receiving the QoS request).

3. When UE needs to make a call, it sends service request to SGSN. SGSN sends a Radio Access Bearer Assignment Request to Femtocell, it triggers the QoS request (by sending a dedicated QoS Request message) toward Femtocell GW, including Femtocell ID, UE ID, BW, UTMS QoS class, and broadband connection ID if available.

4. Femtocell GW retrieves the Src IP address of IPSec based on Femtocell ID and forwards the QoS request (including UE ID, BW, UMTS QoS class, bearer ID etc) to PCRF in the same SP domain. (Note that the trigger of this QoS request is not necessarily from femtocell. It could be by Femtocell GW directly or by AF or the GGSN/P-GW dependent on the service requirement and network technology).

5. PCRF checks the interconnect SLA profile with backhaul operator, translates UMTS QoS attributes to generic QoS attributes based on SLA, discovers the backhaul SP and forwards the QoS request to the peer RACS.

6. RACS checks the resource availability and sends to the related nodes, e.g., digital subscriber line access multiplexer (DSLAM), Broadband Network Gateway (BNG) router to enforce the rules if appropriate. In case the Broadband connection ID is not provided in the request, the RACS can retrieve it from NASS using Src IP addr as the key.

7. RACS acknowledges the request and sends back confirmation to femtocell through PCRF/Femtocell GW path. Femtocell performs RRC procedures and sends back RAB assignment response to Core Network.

8. When the Femtocell receives the uplink Femtocell packets, it ensures the inner IP QoS marking is inline with authorized QoS class, and mapped to outer header (IPSec) based on agreed mapping rule. The backhaul (RGW, DSLAM, BNG) forwards the packet based on the ToS/DSCP in the outer header. In addition, the DSLAM/BNG may police the Femtocell traffic at the aggregate level to assure the max BW.

Besides femtocell, it is perceived some other use cases in the context of FMC are appealing, for example, application mobility, which an end user desires to use an application on their mobile device, and then wishes to change the device they are using to a fixed, home network–attached device. A multimedia call is handed over from the mobility macro network to a home network, but instead of remaining on the same device, the user chooses to transfer the multimedia call to a Set Top Box with a large screen display and resumes the call on that device. Bandwidth and QoS maintained for the large screen experience is to be important.

It is perceived that end-to-end QoS and policy-based resource management play a pivotal role in providing consistent user experience for the applications of FMC. The trend is to leverage existing wireline and wireless policy architectures defined in different SDOs with the interworking solution, and then evolve toward a single converged policy architecture (most likely based on 3GPP PCC) to achieve additional benefits in terms of cost-effectiveness, simplicity, and flexibility.

References

1. ETSI ES 282 001 V1.1.1, NGN Functional architecture release 1, 2005-08, European Telecommunications Standards Institute, Sophia Antipolis, France, 2005.
2. ITU-T Recommendation Y.2012, Functional requirements and architecture of the NGN release 1, 2006-09, International Telecommunication Union, Geneva, Switzerland, 2006.
3. R. Braden, D. Clark, and S. Shenker, Integrated services in the Internet architecture: An overview, RFC 1633, IETF, Fremont, CA, 1994.
4. S. Blake, D. Black, M. Carlson, E. Davies, Z. Wang, and W. Weiss, An architecture for differentiated services, RFC 2475, IETF, Fremont, CA, 1998.
5. 3GPP TS 29.207, 3rd Generation Partnership Project, Policy control over go interface, 3GPP, Sophia Antipolis, France, 2004.
6. 3GPP TS 23.203, 3rd Generation Partnership Project, Policy and charging control architecture, 3GPP, Sophia Antipolis, France, 2010.
7. 3GPP TS 23.401, 3rd Generation Partnership Project, General packet radio service (GPRS) enhancements for evolved universal terrestrial radio access network (E-UTRAN) access, 3GPP, Sophia Antipolis, France, 2010.
8. 3GPP TS 23.402, 3rd Generation Partnership Project, Architecture enhancements for non-3GPP accesses, 3GPP, Sophia Antipolis, France, 2010.
9. ETSI RES 282 003, Telecommunications and Internet converged services and protocols for advanced networking (TISPAN); resource and admission control sub-system (RACS); functional architecture, European Telecommunications Standards Institute, Sophia Antipolis, France, 2009.
10. ITU-T Recommendation Y.2111, International Telecommunication Union, Telecommunication standardization sector, resource and admission control functions in next generation networks, International Telecommunication Union, Geneva, Switzerland, 2009.
11. PacketCable 1.5 Dynamic Quality of Service, PKT-SP-DQOS-1.5-I02–050812, Cable Television Laboratories, Inc, Louisville, CO, August 12, 2005.

12. PacketCable Multimedia Specification, PKT-SP-MM-I03–051221, Cable Television Laboratories, Inc., Louisville, CO, December 21, 2005.

13. 3GPP TS 25.467 V8.1.0, UTRAN architecture for 3G Home NodeB: Stage 2, 3GPP, Sophia Antipolis, France, March 2009.

14. 3GPP TS 36.413 V8.6.0, Evolved universal terrestrial radio access network (E-UTRAN); S1 application protocol (S1AP), 3GPP, Sophia Antipolis, France, June 2009.

16

Presence User Modeling and Performance Study of Single and Multi-Throttling on Wireless Links

Victoria Beltran and Josep Paradells

CONTENTS

16.1 Introduction

Presence information is a well-known concept in the Internet world and is widely used by applications like Instant Messaging and Push to Talk. This kind of application allows a user to ascertain the willingness of other users on his or her buddy list to communicate. This is achieved by basic presence states like online, offline, busy, or absent. This basic understanding of presence is evolving toward a much more generic, flexible concept. Presence is used to characterize entities that may affect the management of communications with a user, which allows applications to take intelligent decisions about the start, continuation, and end of user communications. This information forms the basis for deploying a range of advanced applications that are innately associated with the mobility of users. Moreover, presence is seen as a key enabler of fixed mobile convergence (FMC) services in future next generation networks (NGNs). Global connectivity is the primary aim of NGNs: a unified network that is always on, to which users are connected in the most efficient or appropriate that is available. The basis for such a new generation of communication networks is the Internet Protocol (IP) multimedia subsystem (IMS) [9] introduced in the Universal Mobile Telecommunication System (UMTS) by Release 5/6. The IMS is leading mobile operators to rely exclusively on IP technology for the support and integration of advanced multimedia services. The service of presence is part of the IMS specification and plays an increasingly important role in existing and emerging multimedia services. The Session Initiation Protocol (SIP) is the signaling protocol chosen by the IMS and, hence, the extension of SIP for presence, the SIP for Instant Messaging and Presence Leveraging Extensions (SIMPLE) [13], has become the de facto protocol for managing presence.

SIMPLE [13] is a subscription-based framework where watchers send SUBSCRIBE messages to the Presence Servers (PSs) of the presentities in which they are interested. A presentity is an entity with associated presence information that announces its presence by sending PUBLISH messages to its PS. The PS is the logical entity in charge of binding presentities to watchers and of notifying watchers of proper presentities' presence changes by means of NOTIFY messages. Figure 16.1 outlines this operation. It can be seen that whenever a presentity changes his or her presence, a large number of messages must be exchanged.

As the number of subscriptions grows, the amount of traffic may overload the network servers that manage subscriptions. Moreover, when watchers or presentities are running on mobile wireless devices, they may have to deal with an excessive amount of traffic that uses up their processing and

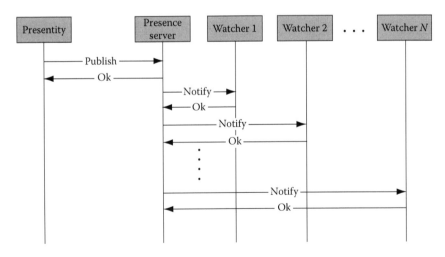

FIGURE 16.1
Presence publication in SIMPLE.

battery resources in addition to the wireless bandwidth. This situation worsens when mobile presence applications interact with Internet-designed applications which are not concerned with the volume of presence traffic. Facing this problem, several proposals for making the presence traffic overload lighter have been issued as IETF drafts or request for comments (RFCs). Throttling is one of such proposals.

Section 16.2 describes the throttling strategy in the SIMPLE framework and discuss some weaknesses in its operation. Section 16.3 explains how to model the behavior of presence users (i.e., the dynamics of their presence information). By way of example, we give the Markov chain for a presence user, which is used as model throughout our study. Section 16.4 shows the mathematical formula, based on Markovian probability distributions, for calculating the probability of presence changes that occur during a particular throttle interval. From this formula, we derive the total number and byte rate during each throttle interval. Although this mathematical model can be used to study the presence traffic due to notifications in presence subscriptions, we focus on the presence publications of presentities in Section 16.5. In addition to standard throttling, Section 16.6 studies the performance of throttling when multiple intervals are used to limit the publications of a particular presentity. Section 16.7 gives an overview of other works related to throttling strategies, and Section 16.8 draws conclusions on the major contributions made in this chapter.

16.2 Throttling in the Simple Framework

The SIP event framework (RFC 3265) [2] allows specific event packages to define their own throttle mechanism in order to limit the rate of notifications in presence subscriptions. This mechanism simply sets a minimum time between two presence notifications. This time is a passive interval during which no presence notifications are allowed. For example, the watcher-info event package (RFC 3857) [10] recommends that the server generate notifications at a rate no faster than once every, while the message-summary event package (RFC 3842) [11] suggests a maximum of one notification per second. In addition, RFC 3265 considers the possibility of allowing subscribers to determine the rate at which they receive notifications. However, to date, no event package specification has stated anything about such a possibility. Nevertheless, the draft [1] defines a mechanism that allows subscribers to limit the rate of notifications sent by their Resource List Servers (RLSs) [12]. An RLS allows watchers to subscribe to all their presentities by means of a single resource list subscription. This server significantly reduces the number of messages that are exchanged between watchers and the network. In [1], the subscriber sends the desired throttle interval in a "throttle" header of the

SUBSCRIBE messages sent to its RLS. During a throttle interval, the RLS may receive notifications from any of the presentities on the resource list. In such cases, the RLS aggregates the presence changes to the resource list's status. Once the throttle interval expires, it sends the whole aggregated presence in a single NOTIFY message.

Currently, throttling in the SIMPLE framework has two major drawbacks:

1. Throttling for presence subscriptions only. As there is no SIMPLE specification for the treatment of throttle mechanisms outside of a subscription presentities', it is not currently possible to apply throttling to publications. Presentities publish presence changes via asynchronous PUBLISH messages and do not need to create any subscriptions to maintain state variables. Although no standard document to date has described throttle mechanisms for presentities, such a mechanism would be very useful for saving presence traffic between client devices and PSs in cases in which the devices or the communication channels have limited resources.

2. A single throttle time per presence subscription. SIMPLE specifications describe throttling as a single time interval that is associated with a whole subscription. As a result, all presence information associated with such a subscription have the same throttle interval. However, watchers may consider that some presence attributes are more important than others. For example, a watcher may be very interested in his or her presentities' location, while their state or activities may be insignificant. When a watcher needs to obtain presence attributes with different levels of urgency, a single throttle interval does not provide the desired result. If we choose the interval that matches the most urgent attribute, insignificant attributes that change frequently will generate a great deal of unnecessary traffic. If we set the interval that matches the least important attribute, the urgent attributes will not be updated frequently enough, which may lead watcher application or the users to behave in an undesirable way. This fact is even more relevant in resource list subscriptions that have a wide range of information about all of the presentities included on a watcher's buddy list.

16.3 Modeling a Presence User

While a presence user is active, his or her presence information goes through a series of independent states. Each state reflects a different status of the presence information, and the sequence of states over time is determined by the changes that the user (or even other authorized agents) makes in his or her presence. This type of system can be easily represented by state diagrams: each possible presence status is a state in the diagram and the states are connected by direct arcs that represent the actions that alter the presence. Actions are sequential, that is, it is not possible that two actions occur at the same time.

If the presence information is made up of N attributes, where each attribute a_k can take r_k possible values, the representative state diagram will have $\Pi_{1 \le k \le N} r_k$ states. The number of states will be smaller if there are incompatibilities between some presence attributes because two or more values cannot coexist. If we assume that presence changes occur one after the other, the maximum number of arcs for each state in the diagram would be $\sum_{j=1}^{N} (r_j - 1)$.

In contrast, if we assume that a presentity is able to change several attributes at the same time, each state could be connected to all the others and, therefore, the maximum number of arcs in the diagram would be $N*(\Pi_{1 \le k \le N} r_k - 1)$. The presence diagram for a presentity contains its maximum number of arcs when there is no restriction on the possible presence changes. However, sometimes only a subset of the possible actions on the presence information is allowed. This graphical representation of the dynamics of a presentity closely matches the transition diagrams of Markov chains. Transforming a presence diagram into a Markov chain simply involves assigning one instantaneous transition rate per action. The transition rate measures how quickly a presence change takes place. However, we can only use Markov chains to model the behavior of a presence user if we assume that the future of the chain does not depend on its past, only on its present. In the proposed presence user below, we have chosen the states in a way that each state summarizes the chain's past and, hence, the user's future actions only depend on his or her current state. Therefore, the user below can be modeled by a Markov chain, and the probabilistic Markovian properties can be used to find out the probability of presence changes as time passes.

Presence users do not follow any particular statistical pattern that is valid for all of them. Therefore, there are no general models to describe the behavior of presence systems. We propose a reasonable model of presence users, which is applicable to different real-life environments. Although this model is applicable to different types of presence information, by way of example, we consider three presence attributes: audio, moving, and sphere. The first indicates whether the level of noise is acceptable (with the *OK* value) or inappropriate for incoming voice calls (with the *noisy* value). The second indicates whether the user is moving so fast that it is likely that he is using a mode of transport, such as a motorcycle or a car. The last attribute, sphere, is defined in the specification rich presence information data format (RPID) [14] for characterizing the current role of the presentity. We defined two roles, "logged onto the application," via the *online* value, and "logged off the application," via the *offline* value. We assume that the user is connected to his smartphone that has two sensors; one of them acts as a sound level meter

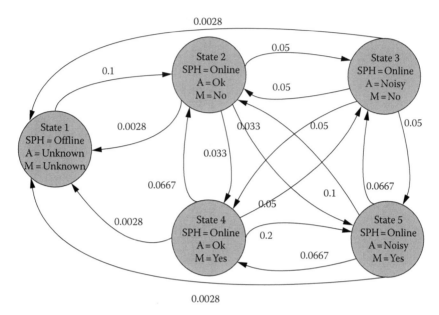

FIGURE 16.2
A Markov chain.

and the other as an accelerometer. These two sensors pick up information about the ambient acoustic conditions and movements, such as inclination, vibration, and shock that the user is experiencing. The presence application that is running on the user's device retrieves the information received by the sensor and modifies the presence attributes accordingly. We also assume that the application automatically changes the value of the sphere attribute when the user logs on/off. We modeled changes in the described presence information in the Markov transition diagram shown in Figure 16.2. This model shows information that is useful to improve user communications in addition to monitor people like the sick, elderly, and disabled.

This diagram shows the transition rates between states. Ideally, the transition rates of a Markov process should be chosen after a number of testbeds with reliable patterns of state changes over time have been examined. When presence changes are modeled, the ideal sources of testbeds are logs of real presence applications. To date, however, there are no real testbeds or statistics about the behavior of users of presence applications. An alternative way of estimating the value of transition rates is to approximate the average amount of time a presentity will spend in each presence state. Once we have these times, we can calculate the transition rates based on the Markovian property $T = 1/\lambda$, which states that the average sojourn time (T) in one state is the inverse of the rate of departures (λ) from that state. We considered a particular scenario that serves as reference for estimating the average times between state changes. In this scenario an employee is in charge of the technical maintenance of some kind of electrical machine (e.g., washing machines, refrigerators, etc.) supplied by his company. Therefore, throughout his working day, this employee visits various homes, shops, etc., in order to resolve technical problems in some machines. The employee is provided with a smart phone which runs a presence application by his or her supervisors. This application notify the of changes in the audio, moving and sphere attributes. For each presence change, we assume an average time that is reasonable for the described use case. The rate at which state changes occur is deduced using based on these average times, which enables us to characterize the Markov chain in Figure 16.2. For example, when the employee is traveling on his way to a new client, it is likely that the environment will quickly become noisy. Thus, we assume an average time of 5 min to change from state 4 to state 5 and, therefore, the corresponding transition rate is 0.2.

16.4 Probability of Notifying Presence Changes

The probability of presence changes during a particular throttle interval is estimated by means of a Markovian probability distribution. Provided that each state in the Markov chain reflects a different presence state, a presence change occurs when the Markov chain switches from one state to another. With throttling, a presence change occurs if the state at the beginning of the interval is different from the one at the end. Provided that *throttle* is a throttle interval that expires at instant t, and n is the number of states, the probability that a presence change has occurred during this interval is as follows:

$$\sum_{i=1}^{n} P_i(t - throttle) * \sum_{j(j!=i)}^{n} P_j(t) \tag{16.1}$$

Expression (16.1) is the probability of finding the chain in state i at the beginning of the throttle interval and in state j when this interval ends. This is therefore the probability of a presence change occurring at the end of a throttle interval.

Based on the Markov chain theory, the probability distribution can be expressed as the matrix equation $(\partial \tilde{P}(t))/\partial t = \tilde{P}(t) * Q$. The Expression (16.2) shows the transition rate matrix or infinitesimal generator $\tilde{Q}$, in which q_{ij} is the instantaneous transition rate of changing from state i to state j. The general solution to the above probability distribution is given by Expression (16.3), where λ_i is an eigenvalue of matrix $\tilde{Q}$, and $\tilde{z}_l$ is the corresponding eigenvector. The constant coefficients c_i are those that satisfy the initial state from which the chain starts.

$$\tilde{Q} = \begin{pmatrix} -\sum_{j\neq 1} q_{1j} & q_{12} & \cdots & q_{1n} \\ q_{21} & -\sum_{j\neq 2} q_{2j} & \cdots & q_{2n} \\ \vdots & \vdots & \ddots & \vdots \\ q_{n1} & q_{n2} & \cdots & -\sum_{j\neq n} q_{nj} \end{pmatrix} \tag{16.2}$$

$$\tilde{P}(t) = \sum_{i=1}^{n} c_i * e^{\lambda_i t} * \tilde{z}_i \tag{16.3}$$

Provided that the initial state of the Markov chain is summarized as the scalar vector $\tilde{I}_0$, the constants $\tilde{c}$ fit the expression $\tilde{I}_0 = \tilde{c}*\tilde{z}$. We assume that the initial state in the Markov chain is always the offline status (state 1 in Figure 16.2). We consider this to be a reasonable assumption because by default, the vast majority of presence applications assign an offline status to inactive users.

The probabilities in Expression (16.1) must be redefined for including the appropriate initial conditions. The first term, $P_i(t - throttle)$, gives us the probability of the chain reaching state i in $t - throttle$ time units. The chain starts in the offline state and hence it is the initial condition for the first term. The second term in (16.1), $P_j(t)$, is the probability of the chain reaching state j in *throttle* time units provided that the chain attains state i at $t - throttle$ time units. Markov chains have a memoryless property whereby the future probabilistic behavior of Markov chains only depends on the current state of the chain, regardless of how the chain reached that state. Therefore, the probability $P_j(t)$ should only be based on the fact that the chain was in state i at the beginning of the throttle interval. The probability is not affected by how and when the chain reached state i; what is important is what happens after the chain reaches state i. Thus, we set the condition that the chain initially starts in state i for the second term in (16.1). Assuming that the chain starts in state i, let us define the probability of the Markov chain reaching state j at instant throttle as $^i P_j(throttle)$. By applying this definition, we can rewrite (16.1) as the following expression to calculate the probability of a presence change when the interval *throttle* finishes at instant t:

$$\sum_{i}^{n} {}^1 P_i(t - throttle) * \sum_{j(j!=i)}^{n} {}^i P_j(throttle) \tag{16.4}$$

where state 1 is the initial state *offline* and the probabilities $^i P$ are based on the initial vectors $^i I_0$, which state that the chain is initially in state i. Thus, the terms in (16.4) are two different probability distributions and are therefore independent. Expression (16.4) disregards the possible intermediate states between states i and j. Instead, it only calculates the probability of the chain being in a different state from the one at the beginning of the throttle interval. Figure 16.3 shows the idea behind (16.4).

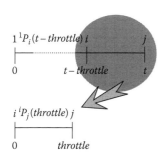

$$1\ {}^{1}P_i(t-throttle)\ i \qquad j$$

0          $t-throttle$ t

$i\ {}^{i}P_j(throttle)\ j$

0         $throttle$

FIGURE 16.3
Probability of presence change with the Markovian memoryless property.

Once we have the change probabilities, we can estimate the number of bytes sent each time a throttle interval expires. We need to know the size of the presence document that would be sent for each presence change. Let us define B_{ij} as the number of bytes in the presence document that is sent when the presence information changes from state i to state j. The byte rate during each throttle interval must be reduced in order to minimize the use of the communication channel. Therefore, our objective is to minimize Expression (16.5).

$$\frac{\left(\sum_i^n {}^{1}P_i(t-throttle) * \sum_{j(j!=1)}^n {}^{i}P_j(throttle) * B_{ij} \right)}{throttle} \tag{16.5}$$

16.5 Source Throttling

Although no standard document to date has described throttle mechanisms for presentities, such a mechanism would be very useful for saving presence traffic between client devices and PSs in cases in which the devices or the communication channels have limited resources. As this approach places throttling on client devices which are the sources of presence publications, we call it "source throttling." The performance of source throttling in location-based systems may be crucial. In such systems, presentities publish location information periodically. These publications are often so frequent that the large number of messages damages the network core and the batteries of client devices. By way of example, a GPS device can publish a new location every second, which involves 3600 PUBLISH messages per hour. Source throttling allows a PS to request a minimal interval between two consecutive location publications from the client device. This mechanism is mainly intended to reduce presence traffic but could also be applied to protect users' privacy. An example of the latter case is a user who allows his or her boss to see every location change, while other watchers are only allowed to see location updates after longer throttle intervals. The application of source throttling to location publications could be adapted to users' daily routines. During working hours, when a user stays indoors (i.e., in the office), the PS could request longer throttle intervals since the user is likely to stay in the building where he or she works. If the user finishes work or goes away, the PS could request a shorter throttle interval.

16.5.1 Analytical Results

We study the presence traffic exchanged between a presence user and his or her PS when the presence publications are regulated by throttling. The user application is forbidden to send any presence publications during the throttle interval. Every time the interval expires, the application checks whether any change occurred during the interval. In this case, the application adds all the changes to a single PUBLISH message that is sent to the PS. The changes made by the presence user are modeled by the Markov chain in Figure 16.2. The presence application user publishes partial-state documents [17] with only the presence attributes that have changed since the last publication. Let us consider presence documents from 350 to 590 bytes depending on the particular attribute that is being published. An example of a presence document when the noisy attribute changes is shown in Figure 16.4.

Let us assume a session time of 3h, which is sufficient for the chain to reach its stationary state. Expression 16.4 calculates the probability that the presence information has changed during a throttle interval. In addition, the formula in (16.5) gives us the byte rate during a throttle interval and its numerator gives the bytes sent during the interval.

Figure 16.5 shows the probability of presence changes that occur during the throttle interval for different time intervals. This means the probability of the presence application sending a PUBLISH message, which contains one or more presence changes, every time the throttle interval expires. Throttle intervals of 1, 5, 10, 15, and 30 min are considered and the case of no throttling is simulated with at an interval of 6s that is near to zero. Figure 16.6 shows the total number of bytes sent every time the throttle interval expires and Figure 16.7 gives us the byte rate during each throttle interval.

```xml
<?xml version="1.0" encoding="UTF-8"?>
    <p:pidf-diff
          xmlns="urn:ietf:params:xml:ns:pidf"
          xmlns:p="urn:ietf:params:xml:ns:pidf-diff"
          xmlns:r="urn:ietf:params:xml:ns:pidf:rpid"
          xmlns:d="urn:ietf:params:xml:ns:pidf:data-model"
          entity="pres:someone@example.com"
          version="568">

    <p:replace sel="presence/person/place-is/audio/text()">
        <r:noisy/>
    </p:replace>

</p:pidf-diff>
```

FIGURE 16.4
Example of partial state presence document

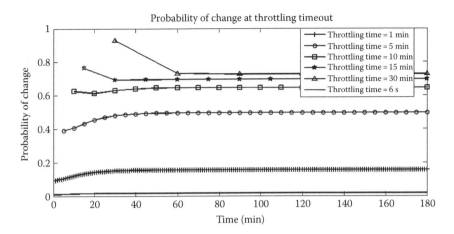

FIGURE 16.5
Probability of change at every throttle timeout.

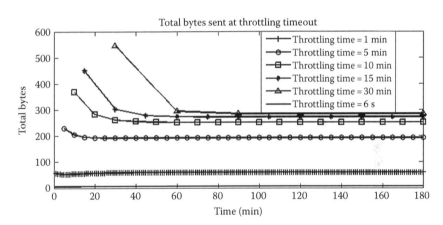

FIGURE 16.6
Total bytes sent at every throttle timeout.

The byte rate per minute is the most relevant variable to study because it is a clear signal of the level of presence traffic in the network. The byte rate is given by two determining factors in (16.4): the probability that changes in the presence information are published (the numerator) and the throttle time during which publications are not allowed (the denominator). In order to reduce the byte rate, the publication probability, or numerator, should be decreased and the throttle time, or denominator, should be increased. However, these two elements are correlated a way in that the longer the throttle time, the higher the probability of changes occurring during this time. The maximum value for a probability is one unit and, thus, the increase in the throttle time is the most relevant factor for reducing the byte rate.

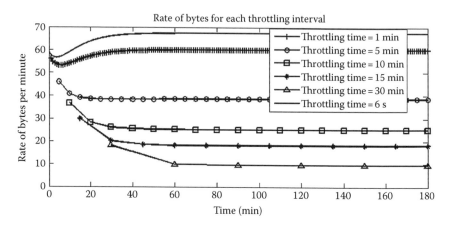

FIGURE 16.7
Byte rate for every throttle interval.

The byte rate for a throttle interval of 1 min is very close to the case of no throttling (simulated at an interval of 6 s). Therefore, the reduction in traffic using such short intervals does not compensate for the processing cost to implement throttling in both the PS and the client application. It is preferable not to apply a throttling mechanism with an interval equal to or less than 1 min. The longer the throttle interval applied to presence sources, the lower the byte rate at which they send the presence traffic. However, the delay perceived by the watchers that are subscribed to the PS is greater. Specifically, we improve the rate of bytes by around 41%, 61%, 71%, and 85% with throttle intervals of 5, 10, 15, and 30 min, respectively. A trade-off must be found between the maximum delay that watchers are prepared to accept and the bytes saved in the communication channel between presence sources and their PSs. In order to ascertain a suitable time interval, it can be assumed that the average delay for a throttle interval is half of this interval. In the case of a throttle time of 30 min, there is an average delay of 15 min, which may prove to be excessive for many watchers. However, a throttle time of 15 min results in a reduction in traffic similar to that experienced over 30 min, but it has a more acceptable average delay of 7 min. The watchers' requirements with regard to delays in presence notifications and network traffic congestion may change continuously. Thus, the PS and the client presence application should be sufficiently intelligent to properly change the throttling time.

16.6 Source Multi-Throttling

There is currently no specification for the application of several throttle intervals to the same entity. As a result, all presence information associated with an entity have the same throttle interval. However, this operation may not be flexible enough in many scenarios. A presentity's presence information is useful when other users (i.e., his or her watchers) know this information and use it to establish some kind of communication with him or her. Consequently, the state and interests of the presentity's watchers is an indicator of the importance or urgency of the presence publications. Let us assume that a user uses a symmetrical buddy list, which means that the user is both a watcher and a presentity of his or her buddies. The PS could use the buddies' presence and conversation history to deduce how much they are interested in the user's presence. Depending on the buddies' state and interest in the user's presence, the PS could request shorter or longer throttle intervals for different pieces of presence information. For example, if all of the user's buddies are offline, the PS could set a very long interval for all the information. If none of the online buddies have participated in a conversation with the user for a long time, the PS also could request a long interval, because it is unlikely that these buddies will be very interested in the user's presence. However, if the online buddies regularly participate in communications with the user, or have some kind of close relationship with the user (e.g., they are friends or relatives), the PS could request shorter throttle intervals for the presence attributes that are of interest to them. The choice of the interval length could also depend on the number of online buddies: the larger the number of online buddies, the greater the probability that one of the buddies will be interested in the user's presence.

The maximum delay in presence updates that is desirable for the watchers should also be considered in choosing a throttle interval. A presentity's presence information could be divided into parts associated with different throttle intervals. Each group would be a subset of the whole presence information for which the watchers have requested the same level of urgency. All of the presence attributes in a particular group would therefore have the same throttle interval. The presentity's watchers could have different requirements for the same presence attributes. Thus, the PS would request throttle times from the presentity according to a trade-off between the needs of all of his or her watchers.

16.6.1 Probability of Publication

Formula (16.4) calculates the probability of one or more presence changes occurring during a throttle interval. This expression works on the assumption that any presence change triggers a presence publication or notification when the throttle time expires. However, when we apply multi-throttling, only the changes in the piece of presence information that is associated with the expired throttle interval trigger a presence publication. Once a throttle interval expires, the changes in the presence attributes associated to that interval are aggregated into a single publication, which could include the changes in the rest of presence information. This is a common practice for reducing the overhead of multiple messages and we call it nonforced multi-throttling. In contrast, when a minimum throttle interval for a piece of presence information is obligatory, the publication of this information can only be sent when this interval expires. We call this strategy forced multi-throttling.

Figure 16.8 presents an algorithm for calculating the probability of presence changes every time a throttle interval expires in the case of nonforced multi-throttling. Here, we assume that all presence attributes have an associated throttle interval. *Not_times* is the set of times at which throttle intervals expire. *Attr_not* is the set of presence attributes whose throttle interval has expired at a particular time t_k. *TRANS* contains all the state transitions that are possible in the Markov chain. A transition *tr* is identified by a source and an end state: $tr = (e_o, e_d)$, $e_o, e_d \in STATES$, where *STATES* is the set of all the states in the Markov chain. TR_i is the set of state transitions $tr = (e_1, e_2) \in TRANS$ that change the value of the presence attribute a_i. The expression ${}^{e_1}P_{e_2}(t)$ gives us the probability of the chain making a transition from state e_1 to e_2 in a time t. The PNS (Probability of last Notified State) table contains the probability of each state being the last one that was sent, and therefore the state that is known by the presence user's watchers. The PSC (Probability of State Change) table contains the probability of each state being published.

```
∀ tₖ∈ Not_times
    Attr_not={aᵢ: the interval associated with aᵢ expires at tₖ}
    TR_not={tr∈ TANS:  tr∈ TRᵢ  ∀  aᵢ ∈ Attr_not}
    ∀  eₐ ∈ STATES
            PNS(eₐ,  tₖ) =PNS(eₐ,  tₖ₋₁)*eₐ Pₑₐ(tₖ-tₖ₋₁)
            PSC(eₐ,  tₖ) =0
            ∀  eₒ ∈ STATES/eₒ!=eₐ  ∧  ∃  tr=(eₒ,  eₐ) ∈ TRANS
                If (tr=(eₒ,  eₐ) ∉ TR_not)
                        PNS(eₒ,  tₖ) =PNS(eₒ,  tₖ)  +  PNS(eₒ,  tₖ₋₁)*eₒ Pₑₐ(tₖ-tₖ₋₁)
                Else
                        PSC(eₐ,  tₖ) = PSC(eₐ,  tₖ) +PNS(eₒ,  tₖ₋₁)*eₒ Pₑₐ(tₖ-tₖ₋₁)
                        PNS(eₐ,  tₖ) =PNS(eₐ,  tₖ) +PNS(eₒ,  tₖ₋₁)*eₒ Pₑₐ(tₖ-tₖ₋₁)
                endIf
            end∀
    end∀
                Pnot(tₖ) =  ∑        PSC(e,tₖ)
                          ∀e∈STATES
    end∀
```

FIGURE 16.8
Algorithm for calculating the probability of a presence change with Multi-throttling.

Whenever the timer associated to a throttle interval expires at time t_k, the algorithm fills in the PNS and PSC tables for each state. For state e, the entry t_k in the PNS is calculated as the sum of three probabilities:

1. *No transition probability*: The probability that the Markov chain was in state e at time t_{k-1} and is still in the same state at time t_k.

2. *Notifying transition probability*: The probability that the chain was in another state e', other than e, at time t_{k-1} and the chain has moved to state e at time t_k, provided that the transition (e', e) changes the value of some presence information with a throttle timer expiring at the associated time t_k.

3. *Non-notifying transition probability*: The probability that the chain was in state e at time t_{k-1} and the chain has moved to a state e', other than e, at time t_k, provided that the transition (e, e') does not change the value of any piece of presence information with a throttle timer expiring at the associated time t_k.

Regarding the PSC table, for state e, the entry t_k is calculated in the same way as the *Notifying transition probability* above.

In case of forced multi-throttling, the probability of change for each piece of presence information with a different associated throttle interval would be given by the Expression (16.4). However, the values of states i and j would be limited to these states provided that the transition (i, j) changes that piece of presence information.

16.6.2 Analytical Results

As in the Section 16.5, let us we assume a presence user modeled by the Markov chain described in Figure 16.2 and partial-state documents from 350 to 490 bytes. Such a user uses a presence application that sends PUBLISH messages to notify the PS of the changes in the user's presence. The presence information is made up of three presence attributes: sphere, audio, and moving. Therefore, the user's presence application could handle a maximum of three throttle intervals. We propose that there are four levels of urgency on presence information, each of which has a throttle interval according to its urgency:

- *Insignificant*: The insignificant presence information is not relevant to the watchers. It is assumed that the watchers' applications use rarely this information. A throttle interval of 30 min is associated with this kind of information.

- *Normal*: The normal presence information does not make any particular requirement on urgency. We use a throttle interval of 10 min because an average delay of 5 min seems reasonable for the majority of presence applications.
- *Important*: The important presence information is quite relevant for the watchers. We associate a throttle interval of 5 min with this information that involves an average delay of 2.5 min.
- *Urgent*: The urgent presence information is crucial for the proper performance of the watchers' presence applications. We do not associate any throttle interval with this kind of information. The presence entity is simply requested to directly publish this information.

The assignation of a level of urgency to a piece of presence information is not based only on the watchers' requirements. Other criteria could be taken into account, such as traffic congestion and privacy, for determining the throttle intervals. The efficiency of multi-throttling depends on the level of urgency set for the presence attributes that change most frequently. In the best-case scenario, the most dynamic attributes are not important for watchers, and they therefore have longer associated throttle times. In the worst-case scenario, these attributes are the most urgent and are notified at short throttle intervals. Figure 16.9 shows the probability of change for each presence attribute every 10 min. This graph helps us to ascertain the frequency of change for each presence attribute. The sphere attribute is far less dynamic than the others. The audio attribute changes most frequently, but is closely followed by the moving attribute.

Let us consider different use cases in which the presence attributes are associated with different levels of urgency. Although these use cases could be useful for many real applications, in order to make them easier to understand,

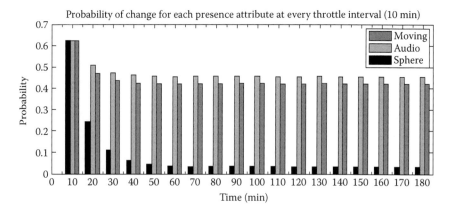

FIGURE 16.9
Probability of change in each presence attribute with forced multi-throttling.

we explain them through the scenario described in Section 16.3. In the first use case, the audio and moving attributes are insignificant for the employee's supervisors, as they are only interested in the sphere attribute. The supervisors are not interested in the employee's acoustic or moving conditions, they only want to know whether he or she is online or offline so that they are aware of when he or she is working and available to communicate. In the second use case, the tasks that supervisors assign to employees change rapidly and they must be assigned as fast as possible. Thus, the supervisors must know all the time which employees are the most available. Therefore, the employees' acoustic and moving conditions are important because this information determines their capacity to maintain high-quality voice calls. By applying the urgency levels that were discussed at the beginning of this section, the first scenario assigns normal importance to the sphere attribute and insignificant importance to the audio and moving attributes. In contrast, the second scenario assigns normal importance to the sphere attribute and urgent importance to the audio and moving attributes. Given that the audio and moving attributes are the most dynamic ones, in multi-throttling, the first scenario is the best-case scenario for reducing presence traffic and the second is the worst-case scenario.

The worst-case scenario would not apply throttling to the audio and moving attributes, that is, each change in these attributes would be notified directly (without waiting for the expiration of any intervals). The lack of throttling for these attributes means that all the sphere changes are also notified directly, because all the state transitions associated with this attribute are associated with both the audio and moving attributes (see Figure 16.2). Therefore, the statistics for the worst-case scenario are equivalent to the "no-throttling case" ones that are shown in Figures 16.5 through 16.7.

In the best-case scenario, the watchers (i.e., the supervisors) are not interested in the audio and moving attributes, but they attach normal importance to the sphere attribute. If an employee's presence application is only able to apply a single throttle interval, it has to decide between optimizing presence traffic and satisfying the watchers' urgency requirements. In order to satisfy the urgency requirements, the presence application would publish all the presence attributes every 10 min. This involves sending a great number of unnecessary updates about the presence attributes that change frequently (audio and moving), although these are not of interest to the watchers. In contrast, if the presence application chooses to reduce the byte rate, it could associate a longer throttle interval (such as 20 or 30 min) to the presence information. However, this choice could lead watchers to make an incorrect or inadequate decision as some obsolete information remains active.

If the employee's presence application implements multi-throttling, we can achieve a trade-off between the optimization of presence traffic and the satisfaction of watchers. In this case, the audio and moving attributes are controlled by throttle intervals of 30 min because these are insignificant

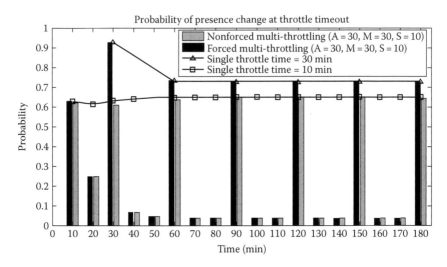

FIGURE 16.10
Probability of presence change with multi-throttling in the proposed best-case scenario.

for the watchers. The sphere attribute has an associated interval of 10 min because its importance is normal. In the graph in Figure 16.10, the bars are the probability of sending a PUBLISH message every time a throttle timer expires when nonforced and forced multi-throttling intervals are used. We are mainly interested in nonforced multi-throttling because it is applicable to the vast majority of presence applications. In Figure 16.10, the lines represent the probability of sending a PUBLISH message when only a single throttle interval is applied to the overall presence information (i.e., the three presence attributes). The square-marked line shows the case of throttling with intervals of 10 min and the triangle-marked line represents intervals of 30 min. The former shows the interval that satisfies the needs of the most urgent information (the sphere attribute), and the latter satisfies the needs of the less urgent information (the audio and moving attributes) and optimizes the byte rate.

Figure 16.10 shows that the probability of change in nonforced multi-throttling is much lower (about 92%) than the two cases of single throttling in most expiration times that are not a multiple of 30. This is due to the fact that when single throttling is applied, the probability of change at each expiration time is mainly affected by changes in the audio and moving attributes. Therefore, when only the sphere attribute is checked, that is, every 10 min, the probability is drastically reduced. When the audio and moving attributes are checked, the probability of publishing a change increases significantly. However, this probability is always lower than in single throttling, because every time the sphere changes are published, the rest of the attributes are also published. In contrast, forced multi-throttling does not take advantage of the publication of a presence attribute to publish all the attributes that

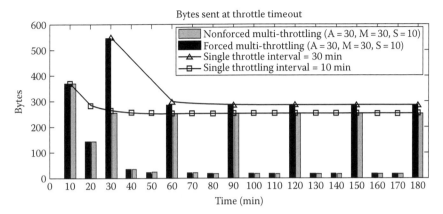

FIGURE 16.11
Bytes sent at every timeout with multi-throttling in the proposed best-case scenario.

have changed and, therefore, in this case, the probability of change every 30 min is equal to that in the case of a single throttle interval of 30 min.

Figure 16.11 shows the bytes sent every time a throttle interval expires. This graph has the same pattern as the probabilities shown in Figure 16.10, because the bytes after a throttle time are calculated by simply multiplying the probability of a particular change by the bytes that would be sent due to that change. The number of bytes at every expiration time is always lower than the two cases of throttling.

Figure 16.12 shows the rate of bytes per minute at each throttle timeout. For each point on abscise X (time), the corresponding one on abscise Y (bytes per minute) is calculated as the total number of bytes sent until the time in X divided by this time. As time passes, the rate for nonforced multi-throttling is almost equal to the case of single throttling with an interval of

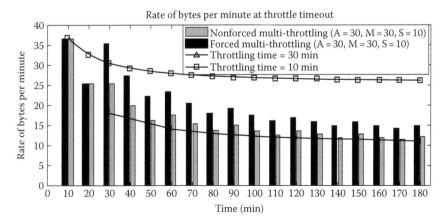

FIGURE 16.12
Rate of bytes at every timeout with multi-throttling in the proposed best-case scenario.

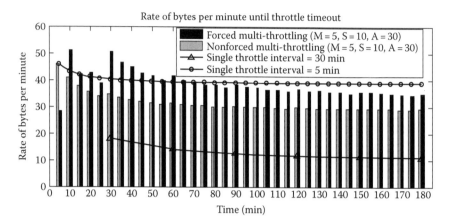

FIGURE 16.13
Rate of bytes at every timeout with multi-throttling in a medium-case scenario.

30 min, which is approximately 10 bytes per minute. This is the minimum rate among the rates considered because it has the longest associated throttle interval. Thus, a similar rate to the best-case scenario is achieved while the required accuracy for the sphere attribute (which is checked every 10 min) is maintained. In addition, the rate of traffic is reduced by around 54% if the shortest interval is associated with all of the presence attributes. This is due to the fact that a great amount of unnecessary publications of audio and moving attributes are saved.

Let us now consider another use case in which the supervisors are strongly interested in knowing whether the employee is using some means of transport, because this indicates if the employee is going to the next technical problem. This information is given by the moving attribute. However, the supervisors do not care about the level of ambient noise to communicate with the employee, so that the audio attribute is not important. Based on these assumptions, the following levels of urgency and, hence, throttle intervals can be assumed: important for the moving attribute (5 min), insignificant for the audio attribute (30 min), and normal for the sphere attribute (10 min). Figure 16.13 shows the byte rate per minute at every throttle timeout. The required accuracy for the moving attribute is achieved. In addition, around 27% of the traffic that is generated when this accuracy is achieved by applying a single interval of 5 min is saved.

16.7 Related Work

The studies [3,4] explain the reasons for which presence attributes should be treated differently depending on their importance. The authors of [3] use decay functions to describe how the accuracy of presence attributes decreases over time in order to prevent watchers from retaining obsolete values. This work explains the need for different decay functions depending on the presence attributes, since each attribute has a different nature and, hence, is published at a different rate. For example, a user normally publishes changes in basic personal information (such as state or activity) much more frequently than changes in the properties of his or her device. If the decay function of a presence attribute drops below its threshold value, this means that the probability that this attribute has changed is too high and is no longer reliable. If this happens, the PS must recalculate the presence information that could be affected by a change in the attribute. If the presence attribute is binary (i.e., only two values are possible), its new value can be calculated automatically. Otherwise, the PS must retrieve the exact value of the attribute in a way that is not specified in this work. This strategy prevents PSs from maintaining obsolete values for presence attributes whose publications have somehow been lost in the network or when long throttle intervals have been applied. The underlying idea here is the same as in throttling: some presence attributes change value faster than others. Moreover, decay functions and throttle intervals could be applied together. We could use the decay function of a presence attribute as the basis for determining its throttle interval that the PS would request from the presentity. If the presence source does not implement multi-throttling, we could configure a single throttle interval for all presence notifications and use a pull model to retrieve the most urgent attributes from the presence source (provided that the presence source implements the pull model) when their decay functions reach the threshold values. This strategy is more advisable when only a few attributes change quickly.

The author of [6] examines the efficiency of the throttle mechanism in [1] at reducing presence traffic between the user and his or her resource list server (RLS). Furthermore, they propose an extension, called extended event throttling, which allows users to set a throttle interval that the RLS should request from the presentities in the user's resource list. In the opinion of the authors of the throttle mechanism described in [1], when a watcher specifies a throttle interval in his or her RLS, this server requests the same throttle value from the watcher's presentities. The authors state that the RLS is not able to quickly update the presentities' presence when the watcher sets a very long throttle interval (e.g., when the presence application is in background). In this case, if the watcher requests a shorter throttle interval again, the RLS will not have the most up-to-date presence information at its disposal and the watcher will perceive delays while that information is being recovered. The authors define a new tag in the SUBSCRIBE messages sent by the watcher to indicate the rate at which the RLS should request notifications from the presentities. In our opinion, the grounds for this extension are dubious, because [1] does not require the RLS to request notifications from presentities at the same rate specified by the watcher. In addition, we believe that the choice of the throttle time requested from presentities should be left to network administrators rather than to watcher applications.

The study described in [4] deals with the regulation of presence publications in IMS networks. The authors explain that the presence attributes associated with a user can be requested by watchers with different urgency and accuracy needs. Generally, watcher applications are not interested in users' entire presence, because they have different rate, urgency, accuracy, etc., restrictions on presence items. In addition, each presence source has its own reliability and accuracy. The authors stress the need to allow watchers to only subscribe to the presence subsets of interest to them in order to avoid unnecessary presence traffic. RFC 4660 [7] defines the operations that a subscriber has to perform in order to configure filtering rules that express the presence attributes that the subscriber is requesting and indicate when this information should be delivered (i.e., trigger conditions). However, this specification does not provide any way of describing requirements of urgency, cost, rate, or accuracy. The authors also point out the lack of strategies for configuring these types of requirements in presence sources. This work proposes a hierarchical structure of servers, called context mediators, which are specialized in managing particular kinds of presence. Each context mediator only communicates with the presence sources that publish the kind of information that it manages. These servers require advanced intelligence in order to perform presence aggregation and balance the source publications with the needs of watchers. The authors proposed a draft [8] that expired without any further progress. This draft defines a new event package that allows presentities to be up to date with their PSs' presence requirements. According to [8], presentities should subscribe to their PSs in order to be notified of changes in requirements on urgency, rate, cost, etc., and thus be able to publish presence changes according to these requirements. We believe that one major reason why this draft was not accepted is that the benefits of this new subscription do not make up for the overload it causes. It seems that the regulation of presence publications has not yet been sufficiently covered. Apart from [4], we found no research papers that deal with this problem. However, [5] is a patent that presents this problematic but does not provide any implementation details. The authors highlight the need to regulate the publication of particular presence attributes by applying threshold conditions in order to reduce the traffic load. Threshold conditions are applied to sets of presence information, so that a presence source may publish presence attributes at different rates. A remarkable example of threshold condition is a minimum throttle interval that must expire before the associated presence information is sent. Other examples are a minimum change of a parameter such as the distance moved by a presentity and load conditions on the network or on the PS.

In [15], a mathematical model of the behavior of presence users during the day and night is described. The authors use two discrete-time Markov chains—one for day hours and the other for night hours—to model the times that users log in/out and, hence, are online/offline. The behavior of a presence user during the day is not a homogenous process because the transition rates vary over time. Therefore, the authors need to model the day and the night hours separately. In contrast, we study the behavior of a presence user during three working hours, which is reasonably assumed to be a homogenous process. The authors only consider the simplest presence state: online and offline. They model this state through a discrete Markov chain and, therefore, the transitions between states can only take place at times expressed as integer numbers. However, presence changes can occur at any instant and, hence, continuous-time Markov chains are more suitable. This work is based on the stationary probability distribution of the proposed Markov model. Therefore, the authors assume that the limiting values of the state probabilities are independent of the initial probability vector. We believe that the suitability of this assumption is dubiously suitable for the majority of presence users. The authors draw a number of conclusions about user behavior during daytime and nighttime hours, but these conclusions are based on particular times between state transitions. The authors set these times without justifying the values chosen. These authors have published another related work [16] about the IMS PS's performance. In this work, they calculate the ratio at which several types of messages are received or sent by the PS. This ratio is given by simple lineal equations in which all the parameters are constant except for the login rate. The authors also study the throttling concept, which they call vacation, for limiting the rate of NOTIFY messages. They use a Poisson process to model the arrival of presence publications and, hence, the generation of notifications to watchers. Using this model, they analyze how the NOTIFY message performs and what times for vacation are the most suitable in order to avoid a queue reaching its maximum length. Although this work gives us an idea of the complex nature of traffic analysis in PSs, the authors make many unrealistic assumptions. They assume that presence users publish presence changes at constant rates, that each presence change is notified to all the watchers (privacy filtering is not considered) and that any presence change results in notifying watchers (presence aggregation reduces the number of necessary notifications).

16.8 Conclusion

Forthcoming NGNs will deliver a wide range of seamless services as a result of the convergence of fixed and mobile technologies. Presence information is considered as a key enabler for user personalization and adaptation, which are indispensable features of any next-generation service. Presence applications generate a great deal of signaling traffic for updating user presence, which can have harmful effects on mobile devices. This traffic can consume the battery and processing resources in these devices and waste the wireless bandwidth. Throttling is one of the strategies for optimizing presence traffic that has recently been defined in the SIMPLE presence framework. This strategy aims to save presence notifications in a presence subscription by setting a minimum interval between two consecutive notifications. In our opinion, there are two major weaknesses in the implementation of throttling in SIMPLE. First, throttling is not applied to presentities' publications. However, regulating presence publications would be highly useful for saving traffic on wireless links. Second, all the presence information of a resource (e.g., a resource list) is notified at the same rate. However, the application of several throttle intervals for a single presentity could be useful for matching the watchers' urgency requirements on particular pieces of presence information.

In this study, we present throttling as both a mechanism for reducing presence traffic and for allowing watchers to set their preferences about the rate at which they are notified. The watchers' preferences could be inferred from policies other than the optimization of traffic, such as application-specific policies, privacy policies, or the optimization of processing resources on mobile devices. In order to study the efficiency of throttling strategies, we build a mathematical model for a user's presence changes based on Markov chains. The behavior of presence users differs a great deal because the nature of presence applications is very diverse. Even the changes in the presence of a particular application's users do not follow any particular patterns, because the operations that the users can perform and that affect their presence information (such as modifying personal state, privacy filters, calendar, or policies) are highly subjective and depend on temporary circumstances. Thus, there are no known models for presence users, which makes the study of presence systems more difficult. In view of this problem, we assume that the behavior of presence users is a stochastic process that follows the Markovian properties. Although we believe that this assumption is reasonable for many presence users, we are aware that it may be inappropriate for others. Nevertheless, we were only able to approximate a mathematical model for the behavior of presence users by adopting this assumption, and we are aware that our model is only valid as long this assumption is true.

We give the mathematical formula that calculates the probability of presence changes that occur over a particular interval of time, provided that the presence user is modeled by a continuous-time Markov process. We derive the total number and rate of bytes during the session time from this formula when the presence user uses a throttle interval for limiting his or her publications. In addition, we give general ideas about how to model presence users with Markov chains, and show the Markov chain for a presence user, which we use by way of example as a model for all the calculations. We focused the analytical study on the presence traffic exchanged between a presence user and his or her PS. This study is especially interesting for cases in which the overconsumption of bandwidth and batteries is critical, as when a user is connected to a wireless mobile device. Using the continuous-time Markov chain described and its associated transient probability distribution, we calculate the probability of the presence information that has changed at the end of the throttle intervals during the session time. We also calculate the number and rate of bytes that are sent after each throttle interval. We studied throttling using a single time interval, that is, single throttling, and using several throttle intervals, that is, multi-throttling.

- *Single throttling*: Without any mathematical model, a number of reasonable assumptions about presence users could be inferred. It would be reasonable to assume that the number of presence changes is higher during the first minutes of a session, because the presence user is likely to modify his or her status at the beginning of the session and will subsequently be more inactive. Another reasonable assumption is not to apply a throttling mechanism with a very short interval because it would be equivalent to the no throttling case. The last assumption that we find reasonable is that the longer the throttle interval applied to presence sources, the lower the byte rate at which they send presence traffic. However, the delay perceived by the watchers is greater. The results of the analytical study that we carried out confirm the three above-mentioned assumptions and demonstrate the goodness of the mathematical model proposed. We studied throttle intervals of 5, 10, 15, and 30 min, which reduce the traffic of presence publications by 41%, 61%, 71%, and 85%, respectively. We believe that it is necessary to find a trade-off between the maximum delay that watchers are willing to accept and the bytes saved in the communication channel between presence sources and their PSs. If the average delay for a throttle interval is considered to be half of such an interval, in the use case studied, we observed that a 15 min interval achieves the best balance between delay and rate of traffic. In general, the PS and the presence source should be sufficiently intelligent to change the throttle interval value at runtime, because the watchers' urgency requirements and the network overload may change continuously.

- *Multi-throttling*: We give the algorithm that calculates the byte rate generated by a presence user when his or her publications are regulated by different throttle intervals at the same time. We applied the multi-throttling algorithm to different scenarios with different associations between presence attributes and throttle interval values. Multi-throttling makes it possible to set short intervals for the most urgent information, while the rest of the information is regulated by longer intervals. We thus save many publications due to changes in information that is not so urgent or important for the PS or watchers. The efficiency of multi-throttling depends on the level of urgency that is set for the presence attributes that change most frequently. In the best-case scenario, the most dynamic attributes are not important for watchers, and therefore they have longer associated throttle times. In the worst-case scenario, these attributes are the most urgent and are notified at short throttle intervals. In general, combining several throttle intervals according to the required urgency in the presence attributes achieves a traffic rate that is close to the rate of single throttling with longer intervals.

Acknowledgment

This work has been supported by the Spanish Government through CICYT project TIC2006-04504 and the FPU AP2006-02846.

References

1. A. Niemi, K. Kiss, and S. Loreto. 2009. Session initiation protocol (SIP) event notification extension for notification rate control. Internet Engineering Task Force Draft (May), IETF, Fremont, CA, http://www.ietf.org/id/draft-ietf-sipcore-event-rate-control-00.txt
2. A. Roach. June 2002. Session initiation protocol (SIP)—specific event notification. Internet Engineering Task Force RFC 3265, IETF, Fremont, CA.
3. O. Bergmann, J. Ott, and D. Kutscher. 2005. A script-based approach to distributed presence aggregation. *Proceedings on International Conference on Wireless Networks, Communications and Mobile Computing*, Maui, HI, June 2005, pp. 1168–1174.
4. J. Brok, B. Kumar, E. Meeuwissen, and H. J. Batteram. March 2006. Enabling new services by exploiting presence and context information in IMS. *Bell Labs Tech. J.*, 10(4): 83–100.
5. J. Christoffersson and D. Henriksson. November 2006. Selective throttling presence updates. U.S. Patent 20,070,150,605, United States Patent and Trademark Office, Washington, DC.
6. F. Wegscheider. 2006. Minimizing unnecessary notification traffic in the IMS presence system. *Proceedings on the First International Symposium on Wireless Pervasive Computing*, Phuket, Thailand, January 2006.
7. H. Khartabil, E. Leppanen, M. Lonnfors, and J. Costa-Requena. September 2006. Functional description of event notification filtering. Internet Engineering Task Force RFC 4660, IETF, Fremont, CA.
8. J. Brok. September 2006. Regulating publication of event state information. Internet Engineering Task Force Draft, IETF, Fremont, CA, work expired.
9. The Third Generation Partnership Project. March 2009. TS 23.228 IP multimedia subsystem (IMS), stage 2, release 8, 3GPP, Sophia Antipolis, France.
10. J. Rosenberg. August 2004. A watcher information event template- package for the session initiation protocol (SIP). Internet Engineering Task Force RFC 3857, IETF, Fremont, CA.

11. R. May. August 2004. A message summary and message waiting indication event package for the session initiation protocol (SIP). Internet Engineering Task Force RFC 3842, IETF, Fremont, CA.
12. A. Roach, B. Campbell, and J. Rosenberg. August 2006. A session initiation protocol (SIP) event notification extension for resource lists. Internet Engineering Task Force RFC 4662, IETF, Fremont, CA.
13. The IETF Simple working group. http://www.ietf.org/html.charters/simple-charter.html
14. H. Schulzrinne, V. Gurbani, P. Kyzivat, and J. Rosenberg. July 2006. RPID: Rich presence extensions to the presence information data format (PIDF). Internet Engineering Task Force RFC 4480, IETF, Fremont, CA.
15. Z. Cao, C. Chi, R. Hao, and Y. Xiao. 2008. User behaviour modeling and traffic analysis of IMS presence servers. *Proceedings on the IEEE GLOBECOM*, New Orleans, LA.
16. C. Chi, R. Hao, D. Wang, and Z. Cao. 2008. IMS presence server: Traffic analysis and performance modelling. *Proceedings on the IEEE International Conference on Network Protocols 2008*, Orlando, FL, October 2008.
17. A. Niemi, M. Lonnfors, and E. Leppanen. September 2008. Publication of partial presence information. Internet Engineering Task Force RFC 5264, IETF, Fremont, CA.

Index

Printed and bound by CPI Group (UK) Ltd, Croydon, CR0 4YY

21/10/2024

01777090-0010